New Horizons in Mobile and Wireless Communications

Volume 3

Reconfigurability

For a listing of recent titles in the
Artech House Universal Personal Communications Series
turn to the back of this book.

New Horizons in Mobile and Wireless Communications

Volume 3

Reconfigurability

Ramjee Prasad
Albena Mihovska

ARTECH HOUSE

BOSTON | LONDON
artechhouse.com

Library of Congress Cataloging-in-Publication Data
A catalog record for this book is available from the U. S. Library of Congress.

British Library Cataloguing in Publication Data
A catalogue record for this book is available from the British Library.

ISBN-13: 978-1-60783-971-2

Cover design by Igor Valdman

Artech House, Inc.
685 Canton Street
Norwood MA 02062

10 9 8 7 6 5 4 3 2 1

Contents

Preface

ज्ञानं ज्ञेयं परिज्ञाता त्रिविधा कर्मचोदना ।
करणं कर्म कर्तेति त्रिविधः कर्मसंग्रहः ॥१८॥

jñānaṁ jñeyaṁ parijñātā
tri-vidhā karma-codanā
karaṇaṁ karma karteti
tri-vidhaḥ karma-saṅgrahaḥ

Knowledge, the object of knowledge, and the knower
are the three factors that motivate action;
the senses, the work, and the doer
are the three constituents of action.

The Bhagavad Gita (18.18)

European Research Framework Programs are a public policy instrument to strengthen European competitiveness through cooperation. Although they have a fixed timeframe, determined research themes, and specific expected impact, the achievements in research and development (R&D) made by the funded projects pave the way for a research continuum.

The Information Society Technologies (IST) research program was launched in 1999 as a successor to the Advanced Communications Technologies and Services (ACTS) research framework. Within this program, two consecutive frameworks, namely FP5 and FP6, were focused on advancements in the state of the art in the area of mobile and personal communications and systems, including satellite-based systems and services (FP5) and mobile and wireless systems beyond 3G and broadband for all (FP6).

Under FP6, the European Union has been funding collaborative R&D activities in the field of telecommunications, with a financial allocation of over 370 million. The objective of these activities is to make significant progress towards advanced communication technologies, systems, and services.

FP6 IST R&D was a primary initiative that launched large integrated projects (IPs) alongside the smaller specific targeted research projects (STREPs), specific support actions (SSAs), and networks of excellence (NoEs).

The enormous research effort concentrated in the various R&D project activities required a special supporting initiative that spanned the whole domain of projects, promoted structure, and disseminated the research effort and results. This was the main idea behind the European FP6 IST project SIDEMIRROR. This book is the first of a series of books from the project effort supported by the European Union under FP6. It collects and integrates the final research results of the numerous projects in the above-mentioned R&D European initiative, with the objective of creating a permanent record of their achievements in four books.

In particular, this book is about advancements in reconfigurable systems. At the end of FP5, the foundations for allowing the radio network, including terminals and base stations, to adaptively/automatically adjust to traffic and user requirements were laid. At the end of FP6, standardization activities in the area of reconfigurability had gained momentum, as it was recognized that technologies, such as software-defined radio (SDR), cognitive radio (CR), and opportunistic radio (OR), are key in solving the challenges for future communication systems in terms of spectrum availability and management, multimode terminals, ubiquitous service provision, and ambient intelligence.

The FP6 project work encompassed areas such as how to reduce the transceiver complexity and cost by exploiting the advantages offered by multiple-input multiple-output schemes, policies for context and service management in a reconfigurable environment, the design and exploitation of opportunistic radio for dynamic spectrum allocation, hardware design and implementation of reconfigurable platforms, achieving an autonomic communication framework, and so forth. Additionally, work has been performed for the design and measurements of reconfigurable air interfaces, including studies of the propagation environments and use of other supporting technologies, such as localization and positioning, for the support of reconfigurable systems.

Particular emphasis was placed on the integration of reconfigurability in the overall picture of future-generation wireless broadband systems and networks, from cellular to broadband fixed radio access and broadband wireless local area networks, for both interactive and distributive services.

A key aspect of the work undertaken within the FP6 R&D initiatives *Mobile and Wireless Systems Beyond 3G* and *Broadband for All* was the intention to validate and demonstrate the proposed concepts and solutions.

The material collected in this book was edited to provide useful reading material to senior and junior engineers, undergraduate and postgraduate students, and anyone else interested in the development of current and future mobile communications.

We hope that all readers will experience the benefits and power of this knowledge.

Acknowledgments

First of all, the editors would like to acknowledge Ms. Dua Idris from CTIF at Aalborg University for her big effort toward the completion of this book. Dua was involved in the FP6 IST project SIDEMIRROR as a research engineer. Her consistent hard work and passionate efforts toward improving the quality of the content are very commendable.

Further, this book would not have been possible without the strong support of the European Union project officers from the Directorate General Information Society (DG INFSO) of the European Commission in Brussels, namely, Dr. Jorge M. Pereira, Dr. Manuel Monteiro, and Dr. Francisco Guirao Moya, who guided the project work and ensured strong cooperation with the rest of the FP6 IST projects in the area. Their effort was an essential prerequisite for the successful editing of the available research results. The material in this book was collected and structured with the support of the European Commission within the framework of the IST supporting project SIDEMIRROR, in an effort that spanned a period of five years. It originated from the technical reports and documentation of the projects involved in the FP6 IST R&D initiatives *Mobile and Wireless Systems Beyond 3G* and *Broadband for All*.

The project SIDEMIRROR consisted of two member organizations, Aalborg University and Artech House Publishers. The editors would like to acknowledge the support of their colleagues from Artech House toward the completion of this manuscript.

Finally, the editors would like to thank the administrative and IT staff from Aalborg University for providing the required administrative and IT project support.

Ramjee Prasad
Albena Mihovska
Aalborg, Denmark
2009

Introduction

Reconfigurability aims to provide a seamless experience to its users, as well as much higher flexibility, scalability, configurability, and interoperability, as an improvement to communication systems. The challenge is how to transform embedded flexibility into end-to-end reconfigurability and how to find the right balance between integrated and distributed approaches.

Reconfigurable systems can provide common platforms and associated execution environment for multiple air interfaces, protocols, and applications, which would yield to a scalable and reconfigurable infrastructure that optimizes resource and spectrum usage and increases network and equipment capability and versatility by software modifications. It would be possible to provide the users with special services via customized solutions that are open, flexible, and programmable at all layers.

Research activities on reconfigurable technologies were sustained by a collaborative research within the European-funded research and development (R&D) 6th Framework Programme (FP6) [1], a large-scale R&D initiative in the area of Information Society Technologies (IST) that was partially funded by major industries worldwide.

This book describes the results and contributions from the European research effort within the frames of FP6 in the areas of reconfigurability and cognitive radio as a means to achieve efficient, advanced, and flexible harmonization of legacy and new standards, end-user service provision, multistandard platforms, radiocentric operations, and networkcentric operations. The material is based on achieved results and the advancement of the state-of-the art by the collaborative effort of these and other FP6 IST projects:

- End-to-End Reconfigurability (E2R) [2];
- Flexible Network and Gateways Architecture for Enhanced Access Network Services and Applications (FLEXINET) [3];
- Self Configurable Air Interface (SURFACE) [4];
- Opportunistic Radio Communications in Unlicensed Environments (ORACLE) [5];
- Advanced Resource Management Solutions for Future All IP Heterogeneous Mobile Radio Environments (AROMA) [6].

Chapter 1 introduces the thematic topic and puts the achieved results in the perspective of the worldwide effort of standardization and regulatory bodies such as the International Telecommunication Union (ITU), the European Telecommunications Standards Institute (ETSI), the Open Mobile Alliance (OMA), the 3rd Generation Partnership Project (3GPP), as well as different fora, among others, the

Software-Defined Radio (SDR) Forum [7] and the Wireless World Research Forum (WWRF) [8]. Chapter 1 is organized as follows. Section 1.1 describes the role of SDR as a basis for reconfigurability. Some of the major developments worldwide are presented, including the applications. Section 1.2 describes the role and benefits of cognitive radio technologies as an enabling technology for reconfigurable communication systems. Section 1.3 describes the state of the art in reconfigurable devices and required technologies for the design of such. Section 1.4 talks about the importance of security as the backdrop for introducing reconfigurability to communications. The challenges and some ways to overcome them have been defined. Section 1.5 describes the process of radio and spectrum management and its evolution. Section 1.6 gives a preview of the book.

1.1 From Software-Defined Radio to End-to-End Reconfigurability

The worldwide growth in wireless communication technologies and equipments has fueled the existence of multiple radio access technology (RAT) standards. The various RATs have been optimized individually by trading data rate, range, and mobility to superlatively suit target applications. Some of the major applications are cellular technology standards, such as the Global System for Mobile communications (GSM), the Enhanced Data rates for GSM Evolution (EDGE), and Wideband Code-Division Multiple Access (W-CDMA) for long-range transfer, and wireless local area network (WLAN) standards (e.g., 802.11a,b,g) for home and office area networks. The Wireless Personal Area Network (WPAN) standards (e.g., IEEE 802.15) are being developed for short-range connectivity.

The accelerated rate of technology innovation necessitates that wireless systems manufacturers and service providers frequently upgrade their systems to incorporate the latest innovations. This frequent redesigning of the network and terminal equipments is very expensive and inconvenient. The situation has led to the idea of a reconfigurable system, which can provide services to the users in the heterogeneous environment.

On the other hand, to better adapt to the continuously increasing demands of the end user, the telecommunication systems are migrating towards the Beyond Third Generation (B3G) era, characterized by the convergence of mobile communications systems and IP networks. Reconfigurability is a key concept facilitating such convergence.

The concept of reconfigurability can exploit the full benefits of the valuable diversity of the radio eco-space, which is composed of a wide range of systems such as cellular, broadband radio access networks, WLANs, and broadcast and multicast.

1.1.1 Heterogeneity of RAT Standards

Many of the widely used communication standards differ from each other with respect to used carrier frequencies, required channel bandwidth, and implemented modulation schemes. For example, the GSM is carried over 850, 900, 1800, and 1900 MHz worldwide.

In a scenario of disparate RATs and newly emerging technologies, reconfigurability support for resource and spectrum management, flexible spectrum management, dynamic spectrum allocation, cognitive radio, and resource management strategies are crucial technologies to ensure end-user service delivery.

Wireless communication standards require the use of frequency bands spread over a wide range, such as W-CDMA, WLAN, GPS, DTV, and so on. As there are multiple mobile communications standards in use worldwide, and each system has its own set of specifications in terms of frequency, modulation and demodulation schemes, and communication protocol, the users must buy a dedicated terminal for each system.

This situation has propelled the need for multiband, multistandard devices that can be used for "anywhere anytime" connectivity. The multistandard devices offer only a short-term solution, as their cost and form factor continues to rise with the increase in the number of covered standards.

SDR, an evolving technology that enables the creation of multistandard and multimode terminals, is one of the best ways to achieve seamless mobile communications, and, has generated enormous interest in the field of telecommunications. To realize a multiband RF front end, for example, SDR would be needed to achieve global roaming. The multiband RF front-end circuit is expected to cover every communication standard—even new standards. The reconfigurable circuit design is an important concept in realizing this multiband RF front end. Hence, SDR is believed to be the ultimate solution; it uses a single radio to cover any communication channel in a wide frequency spectrum, with any modulation and bandwidth.

Although initially SDR was developed exclusively for military applications, gradually its scope has broadened to include commercially oriented perspectives, such that the standards for designing SDRs (e.g., software communications architecture), as well as the representative open source implementations, reflect industry-standard object-oriented software practices. SDR helps to shorten the production time, which is very critical for the industry and also increases the product lifetime due to product upgradeability [9].

Figure 1.1 shows the industry's evolutionary recognition of ways to approach and implement the concepts of software radio reconfigurability.

Some examples of the benefits that are expected from the development in SDR technologies are categorized as follows [10]:

- *Subscribers* will gain easier international roaming, improved and more flexible services, increased personalization, and choice.
- *Mobile network operators* gain the potential to rapidly develop and introduce new, personalized and customized services; a tool for increased customer retention; and added-value services.
- *Handset and base station manufacturers* gain the promise of new scale economies, increased production flexibility, and improved product evolution.

To further explain SDR, it is compared here with the normal radio concept. The basic elements of the standard radio are shown in Figure 1.2. In a SDR implementation, the components shown in Figure 1.2 will be realized by hardware and software technologies that enable reconfigurable system architectures for wireless networks

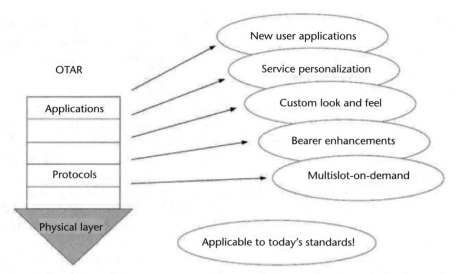

Figure 1.1 Reconfigurability can offer a range of benefits at differing levels of the protocol stack, some of which are already applicable to today's digital standard [10], OTAR (Over-the-Air) download.

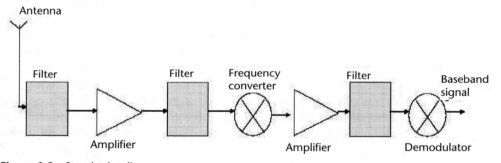

Figure 1.2 Standard radio.

and user terminals [7]. An SDR implementation would be like the one shown in Figure 1.3.

The conventional analog front-end circuitry behind the antennas in a standard radio is replaced by the digital circuitries in the form of a programmable chipset to perform the channelization function so that the powerful digital signal processors (DSPs) can be used to further process the signal.

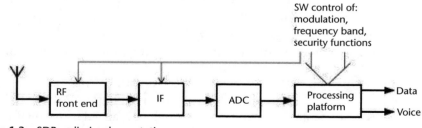

Figure 1.3 SDR radio implementation.

In an ideal SDR, the digitization will be performed right after the antenna, so that even the channelization can be performed by the DSP; this will consequently provide a platform to implement a pure software radio where every feature involved in the radio is real software programmable. An ideal SDR architecture is shown in Figure 1.4.

SDR makes it possible to reconstruct the analog RF circuits by controlling bias voltages and variable passive devices, which can be used as a RF front end for SDR [9].

SDR technology involves the implementation of some of a radio system's functional modules in the software—for example, modulation and demodulation, signal generation, and so forth—which is extremely helpful in building reconfigurable software radio systems. A reconfigurable radio system facilitates the dynamic selection of parameters for the functional modules implemented in software. The ability to reconfigure the system extends its utility compared to a radio system, which is completely hardware based and has fixed parameters for each of the fuctional modules, resulting in only limited utility. One important feature of this technology is that increasingly, programmable hardware modules are being used to build radio system software based on an open architecture.

The SDR terminal comprises programmable devices and employs a multiband transceiver, the components of which are capable of broadband performance, to cover the frequency bands of various systems. To realize an SDR mobile terminal, however, it is important to minimize both power consumption and terminal size. The most promising feature would be a signal processing unit that offers high-speed data processing while minimizing the power consumption.

Bagheri, R., et al. [9] explain that the direct conversion architecture is much more promising for designing an SDR transceiver than a super heterodyne architecture would be, as the latter has IF channel filters and image rejection filters that cannot be programmed to change their frequency band. The Sigma-Delta type ADC has been shown to be a good choice for SDR application [11].

Multiband wireless technologies have also been investigated. Recently, Si CMOS (complementary metal-oxide-semiconductor) technologies have provided high-density integration, high-frequency performance, and low fabrication cost. These topics are further detailed in Section 1.3.

SDR represents a disruptive trend that will create new business models [7]. For example, as operators look to add services to increase the average revenue per user,

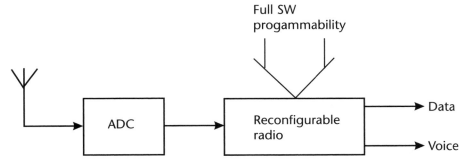

Figure 1.4   An ideal SDR implementation.

wireless infrastructure vendors will be able to capture some of this value through software upgrades even after the infrastructure has been installed.

1.1.2 SDR as an Enabling Technology

SDR can act as a key enabling technology for a variety of other reconfigurable radio equipment commonly discussed in the advanced wireless market. While SDR is not required to implement any of these radio types, SDR technologies can provide these types of radio with the flexibility necessary for them to achieve their full potential, the benefits of which can help to reduce cost and increase system efficiencies. Figure 1.5 shows the relationship between different reconfigurable technologies based on SDR.

- *Adaptive radio* is a radio in which communication systems have a means of monitoring their own performance and modifying their operating parameters to improve this performance. The use of SDR technologies in an adaptive radio system enables greater degrees of freedom in adaptation, and thus higher levels of performance and better quality of service in a communications link.
- *Cognitive radio* is a radio in which communication systems are aware of their internal state and environment, such as location and utilization on RF frequency spectrum at that location. They can make decisions about their radio operating behavior by mapping that information against predefined objectives. Cognitive radio is further defined by many to utilize SDR, adaptive radio, and other technologies to automatically adjust its behavior or operations to achieve desired objectives [12, 13]. The utilization of these elements is critical in allowing end users to make optimal use of available frequency spectrum and wireless networks with a common set of radio hardware. As noted earlier, this will reduce cost to the end user, while allowing the user to communicate with whomever he or she needs, whenever they need to, and in whatever manner is appropriate.

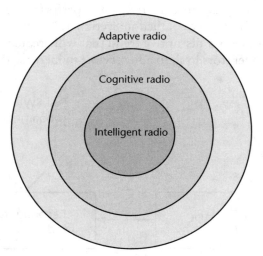

Figure 1.5   The relationship between different SDR-based wireless reconfigurable technologies [7].

- *Intelligent radio* is a cognitive radio that is capable of machine learning, which allows the cognitive radio to improve the ways it adapts to changes in performance and environment to better serve the needs of the end user.

These types of radio—adaptive radio, cognitive radio, and intelligent radio—do not necessarily define a single piece of equipment, but may instead incorporate components that are spread across an entire network.

The FP6 IST project E2R [2] investigated reconfigurability from an end-to-end perspective in relation to a set of chosen relevant scenarios and use cases. In this context, a novel system architecture was defined that is adapted to autonomous operation of system entities, comprising a corresponding hardware/software implementation architecture, resource management algorithms in combination with a functional architecture (FA), and proof-of-concept solutions.

E2R contributed to work of many standardization bodies, including ETSI, 3GPP, IEEE, OMA, the Object Management Group (OMG), and so forth. Particular effort was given to the IEEE P1900.4 standardization, in which the project actively contributed with the definition of a *cognitive radio* vision: in this context, network and terminal reconfiguration management entities were defined, as well as a *cognitive pilot channel* (CPC). Section 1.2 explains the concept of cognitive radio and its role for end-to-end reconfigurability.

1.2   Cognitive Radio

Interest in cognitive radio (CR) has been motivated by the interest of standards groups and regulatory bodies around the world toward new ways of using, allowing access to, or allocating spectrum.

Combined with both the introduction of SDR and the realization that machine learning can be applied to radios, CR has emerged as an adaptive, multidimensionally aware, autonomous radio system that learns from its experiences to reason, plan, and decide future actions to meet user needs [7].

Although the term CR—coined by Joseph Mitola III in 1999 [14]—has evolved over time and now has several specific meanings in a variety of contexts, it is most commonly perceived as "the next step up" for SDRs emerging today [12–14]. Within the wireless industry, consensus is growing that many of the key attributes of cognitive radio—its promise to deliver a radio that is aware of its radio frequency (RF) environment, can adapt to this environment, and consequently can adjust its operating parameters—are best (if not only) enabled through the SDR technology. This concept is shown in Figure 1.6.

The SDR Forum [7], which has several initiatives underway to support the continued development of cognitive radio, is in the process of drafting a definition that summarizes the basic concept of cognitive radio. Other perspectives within the industry describe a cognitive radio as a device that can accomplish the following:

- Autonomously exploit locally unused spectrum to provide new paths to spectrum access;

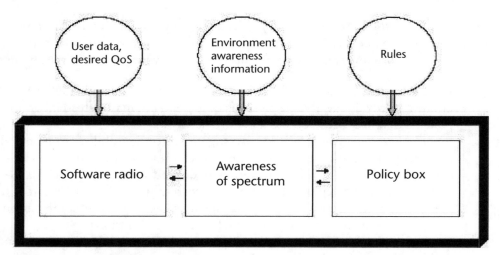

Figure 1.6 Concept of cognitive radio.

- Roam across borders and self-adjust to stay in compliance with local regulations;
- Negotiate with several service providers to connect a user at the lowest cost;
- Adapt themselves and their emissions without user intervention;
- Understand and follow the actions and choices taken by their users and, over time, learn to become more responsive and to anticipate the users' needs.

Presently, there is a growing research community investigating the means of taking advantage of the processing resources in SDR platforms to develop the ideal cognitive radio (iCR), which is the ultimate vision of CR technology. SDR, as mentioned, forms a critical component of CR. The CR detects and uses the unoccupied frequency bands to increase the spectrum usage efficiency.

The radios in general can be broadly classified as follows:

- *Hardware (or fixed) radios:* The radio operators set the parameters.
- *Software (or adaptive) radios:* The parameters can be adjusted to accommodate and anticipate channels and environments.
- *Cognitive radios:* These can sense their environment and learn how to adapt.

Figure 1.7 shows the relationship between software radio and cognitive radio. Cognitive radio could use software radio technologies that allow users to switch communication systems by changing the software on the multipurpose and reconfigurable RF and digital signal processing units [12].

1.2.1 Basics of Cognitive Radio

The CR technology encompasses many facets of intelligent behavior, such as context awareness, adaptation of action due to stimulus and prior information, reasoning (including inferring information not explicitly stated), learning, natural

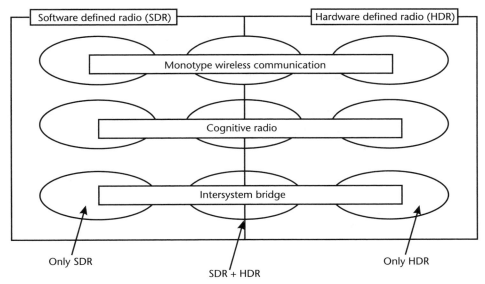

Figure 1.7   Relationship between software radio and cognitive radio [13].

language processing, and planning. To date, researchers have preferred to focus on one or a few of these facets of intelligence [12].

CR employs model-based reasoning to achieve a specified level of competence in radio-related domains. The concept was first introduced by Joseph Mitola, a scientist of the Defense Advance Research Products Agency (DARPA) [14]. Figure 1.8 shows the various important actions a CR needs to perform.

The fundamental concept of the cognitive cycle is to sense the neighborhood environment, adapt to dynamics and variations, and learn from past experiences to change future behavior and act accordingly. The CR-enabled device, therefore, is able to maintain quality of service (QoS) and at the same time reduce the interference to neighboring users.

Some of the technologies to be implemented in a CR include:

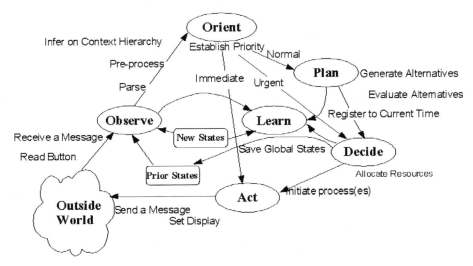

Figure 1.8   The cognitive cycle [13].

- A location sensor (e.g., GPS or Galileo);
- The capability to monitor its spectral environment (e.g., by using a real-time broadband fast-Fourier transform [FFT]);
- Learning and reasoning algorithms to track its location or the spectral environment's development;
- Use of game theory or similar opportunistic techniques to make decisions in a competitive environment, as it may have to compromise its own demands with the demands of other users.

The important requirements of a CR can be classified as follows [15]:

- Awareness;
- Intelligence;
- Learning;
- Adaptivity;
- Action;
- Real-time operation.

The challenges that arise from use of CR are complexity, restriction of the computational power, the limitations of the existing standards, and the associated cost.

1.2.2 Regulatory Landscape

Cognitive radio offers the possibility of dynamic spectrum access that can overcome the rigid table of allocations derived from historical allocation and assignment methods [7]. Advanced cognitive radio systems are aware of their spectral environment and can make decisions about radio operating behavior based on that awareness and the software policy controls embedded in the radio [16].

A snapshot of the regulatory and licensing landscape for cognitive radio is given by the SDR Forum as follows:

- From an international perspective, preregulatory activity has begun in all three ITU areas (Americas, Europe/Africa, and Asia), most notably in Australia, Korea, Sweden, the United Kingdom, and the United States.
- The U.S. Federal Communications Commission (FCC) [17] has been one of the more proactive regulator bodies in its support of cognitive radio via its spectrum policy task force and cognitive radio Notice of Proposed Rulemaking (NPRM). Rulemaking for the 3650–3700 MHz refers to contention-based operation that can be interpreted as benefiting from cognitive technologies.
- The partitioning and disaggregation clauses in recent commercial spectrum license rulemaking proceedings also indicate the FCC's desire for more flexible spectrum access, and the SDR/CR rules apply to device certification. Many of the bands have secondary-use regulations that could be exploited by cognitive radio solutions with little to no additional rulemaking.

Agile, adaptive radio technology could provide impressive economical and technical efficiencies. Cognitive or adaptive radios will allow the disparate frequency bands used by public safety to be "stitched" together. Although the use of Wireless Interoperability for Microwave Access (WiMAX) and Wi-Fi technologies has also been considered for public safety, the limitations come from their use in unlicensed bands [16]. Tunable radios and devices that can sense available transmit and receive frequencies can provide stunning improvements in interoperability among systems in a region or in a country.

Base stations (BSs) and edge devices that are either software-defined to licensed frequencies or adaptive to licensed frequencies will greatly decrease the cost per device and the cost per radio system. Their primary user, as well as their secondary users, could benefit from the economies of scale of being able to use the same, or only slightly adapted, equipment developed for the commercial and consumer sectors. An interoperability problem might occur if first responders at an accident scene are unable to use their radio networks to communicate with each other. If those BSs and devices can dynamically retune across all frequency bands (including public safety ones), they will solve interoperability problems between first responder networks within a region, between regions, and even across the country. The current problems of adding frequencies in an already congested band could be relieved if frequencies could be added from a different, noncontiguous band, and the electronic devices in the system could adjust. CR offers this possibility.

Finally, as system-controller devices use more sophisticated yet standardized user prioritization, or QoS schema, one imagines that spectrum usage could grow to be far more efficient.

Going forward, regulations that could speed CR development and deployment include *dynamic spectrum access*, *interference metrics*, and *authorization for experimental licenses* to prove-out the technology before adopting new rules.

1.2.3   State of the Art of CR

A variety of technical advances have been made in both the *cognitive* and *radio* sides of CR. On the radio side, the continued development of SDRs that exploit the processing flexibility of DSPs running on programmable chip technologies is creating new opportunities for high-performance, extremely flexible radios. On the cognitive side, cognitive processing and computing opportunities are being developed by a wide range of commercial, defense, and research organizations, such as the DARPA Information Processing Technology Office (IPTO).

Many of the capabilities of a CR are achievable with today's technology. In fact, numerous organizations around the world are currently developing what can best be described as *rudimentary cognitive radios*. Technical challenges still need to be addressed, however, for the development of an iCR. These challenges include the need to develop:

- Efficient, agile radio frequency (RF) front ends and verification, validation, and authentication (VV&A) of operational characteristics;
- Wideband linear, adaptive filtering and amplification;

• Advanced cognitive processing for applied learning, reasoning, and knowledge representation algorithms for specific wireless domain solutions.

It is possible to demonstrate the behavior of CR with legacy systems. This is shown in Figure 1.9 for the effects of CR on neighboring legacy users, including the interactions between groups of CRs [35].

The FP6 IST project E2R [2] developed and implemented a platform of a mobile terminal (MT) demonstrator. Based on a reconfigurable RF board and a baseband (BB) board, it prototypes air interfaces for SDR and CR. A dynamically reconfigurable *system on chip* (SoC) was designed for performance purpose at the baseband side. Advanced functions, such as flexible low density parity codes (LDPC) or complex multiple-input-multiple-output (MIMO) decoding can be programmed on this platform.

1.3 State-of-the-Art Devices

CMOS is the chosen technology for the RF front end (RF-FE) as well as for the wireless cellular business.

Portions of the analog transceiver functionality are slowly being moved to the digital domain to fully exploit the advantages of using deep submicron CMOS for RF-FE. Digital correction functionality will be used to cope with the impairments of the RF-FE. One of the driving factors to implement digital functions locally in the

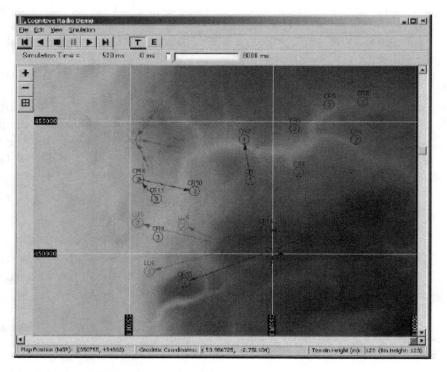

Figure 1.9   Behavior of CR demonstrated with legacy systems.

RF-FE is the ever-increasing demand for functionality at, simultaneously, lower power consumption and higher integration level.

1.3.1 RF-BB Interface

Currently, analog interfaces between the RF and digital baseband integrated circuits (ICs) are being used widely. But the analog interfaces are slowly being changed to a digital interface to achieve a SDR-compliant RF transceiver. Efforts required to standardize the definition of the interface for GSM/GPRS systems can be found in Doyle [18], which helps define the electrical characteristics and functional properties.

The objective of the DigRF (Digital RF) standard is to ensure compatibility at the physical level, and it does not prescribe anything within either the RF or the digital baseband IC. For example, the serial control interface between baseband and RF IC is assumed to be register-based, but nothing is specified about the address allocation or data length, and the interface definition allows great flexibility in this. Similarly, the form of the receive sample interface does not dictate the implementation of the RF IC receive chain, since it has extensive configuration options. Figure 1.10 shows the basic functionality of the DigRF interface.

The DigRF interface covers the following interfaces:

- A *control interface* is based on a standard 3-wire interface (baseband IC is master), which has a long history in RFICs (e.g., for programming phase-locked loops, or PLLs).
- A *data interface* has to be shared between transmit and receive data. There is only one data line foreseen.

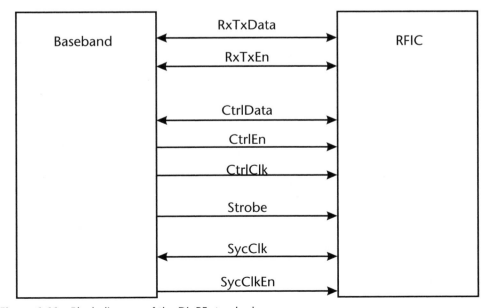

Figure 1.10 Block diagram of the DigRF standard.

This concept is not compliant with the SDR concept, as it is currently restricted to only time-multiplex systems. It is to be noted that currently, the DigRF implementation is not reconfigurable and does not support a wider range of different bandwidths.

However, for the definition of a SDR-compliant digital interface implementation, some of the concepts of the DigRF could be used and possibly the SDR variant could be made downward-compatible to the DigRF.

Two of the main challenges in this case would be the electromagnetic compatibility of the interface with the other RFIC functionality and the stringent requirements on the power consumption.

1.3.2 Converters

1.3.2.1 Analog-to-Digital Converters (ADCs)

The analog-to-digital converter (ADC) is a key block in the scope of SDR, considering the shift of the analog functionality (hard to reconfigure) to the digital domain (easy reconfiguration). Unfortunately, its performance increases the lag behind the enhancements of digital signal processing circuitry. Figure 1.11 shows the improvement in ADC resolution in terms of effective number of bits, which is only 4 dB per decade.

The multibit DeltaSigma($\Delta\Sigma$)-ADC is the most promising converter technology; it delivers high resolution at tolerable power consumption levels and may be employed in either zero-IF or low-IF based receivers. Also, it outperforms the conventional 3-dB SNR increase of conventional Nyquist converters, due to oversampling. The $\Delta\Sigma$-ADCs may be differentiated as continuous time and discrete time, based on the implementation of the loop filters of the modulators.

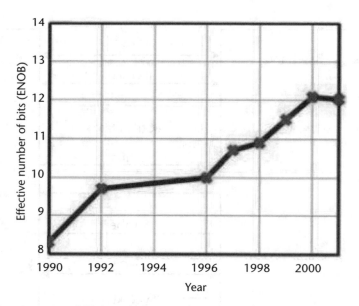

Figure 1.11   Improvement of ADC resolution sample rates > 10 MSPS.

Table 1.1, 1.2, and 1.3 summarize three state-of-the-art (SOTA) multistandard ΔΣ-ADCs. Some ΔΣ-ADCs are reconfigurable to support multistandard operations.

1.3.2.2 Digital-to-Analog Converters (DACs)

In most wireless systems, the requirements such as the dynamic range are less stringent in the transmit path than in the receive path. Due to this situation, the work performed in DAC-related research cannot be compared to that done for ADC, since DAC specifications have lower requirements. Though it is technologically feasible to realize DACs with suitable dynamic range and sampling rates for low-IF and direct modulation transmitters, the architecture of choice for the transmitter may influence the relative importance of the DACs.

For multimode capable transmitters, the polar modulation–based transmitters seem to be a challenging alternative as the architecture of choice.

Table 1.1 A ΔΣ-ADC for Zero-IF Receivers Described in ttpcom.com [19]

	GSM	UMTS
Signal Bandwidth	200 KHz	2 MHz
Dynamic Range	79 dB	50 dB
0Peak SNDR	75 dB	49 dB
Power Dissipation	2.4 mW	2.9 mW
Process	0.13 μm CMOS	

Table 1.2 Multistandard Continuous Time ΔΣ-ADC for Zero-IF and Low-IF Receivers Described in Gomez et al. [20]

	UMTS	CDMA2000	GSM	EDGE
Receiver Architecture	Zero-IF	Zero-IF	Zero-IF	Zero-IF
Sampling Rate	153.6 MHz	76.8 MHz	26 MHz	26 MHz
Signal Bandwidth	3.84 MHz	1.228 MHz	200 kHz	271 kHz
Oversampling Ratio	40	64	65	48
Dynamic Range	74 dB	83 dB	92 dB	90 dB
Power Dissipation	14.1 mW	13.1 mW	9.1 mW	9.1 mW
Process	0.18 μm CMOS			

Table 1.3 ΔΣ-ADC for Low-IF Receivers Described in Veldhoven [21]

	UMTS	GSM
IF	138.24 MHz	78 MHz
Sampling Rate	184.32 MHz	104 MHz
Oversampling Ratio	24	192
Dynamic Range	54 dB	86 dB
SNDR	52 dB	72 dB
Power Dissipation	13.5 mW	11.5 mW
Process	0.25 μm CMOS	

1.3.3   Digital Front End (DFE)

The available SOTA RF transceivers only modestly employ digital signal processing capabilities locally implemented on the transceiver, which are mainly focused on control interfaces (e.g., for PLL programming via 3-wire-interface types) and calibration functionality (e.g., for analog filter tuning). These DFE concepts should be extended to signal processing tasks and to tackle RF impairments. This functionality may include digital frequency correction, digital gain control, digital IQ phase and amplitude matching, and so on. One important requirement is that of signal decimation/interpolation and the respective filtering task in the DFE, due to excessive data rates at the converters.

1.3.4   Analog Front End (AFE)

1.3.4.1   Voltage-Controlled Oscillators (VCOs)

The two main performance measures for voltage-controlled oscillators (VCOs) are:

- Covered frequency range;
- Spectral purity.

The anticipated frequency range of SDR is in the low-GHz range. SOTA VCOs are amenable to monolithic integration into deep submicron CMOS processes; one example is presented in Burger et al. [22]. To extend the frequency range of the VCOs, the tuning capabilities must be integrated. In Konstanznig et al. [23], an example has been presented that shows digital tuning capability with subsequent analog fine-tuning.

1.3.4.2   Fractional-N–Based Phase-Locked Loops (PLL)

In recent years, fractional-N PLLs and modulators based on the fractional-N PLL principle have attracted a lot of attention.

This situation happens to be in line with the trend towards RF-CMOS, since large parts of the fractional-N architecture consist of digital circuitry. Furthermore, generating arbitrary phase-modulated signals through an extension of the fractional-N PLL by using two-point modulation is an attractive approach that is particularly beneficial in cases in which polar-based architecture is employed.

A fractional-N modulator using two-point-modulation has the advantage that most of the blocks are implemented in the digital domain, and thus are not susceptible to component variations. The digital implementation also allows the possibility for reconfiguration of the PLL [24].

1.3.5   Equipment Management for Reconfigurable Radio

Technical solutions in the field of *opportunistic radio* (OR) were provided by the FP6 IST project ORACLE [5]. Frequency-agile terminal architectures and RF front ends were studied and advanced Sigma-Delta DAC/ADC concepts were evaluated. The performance gains of ORACLE technology in different domains, such as

WLAN or cellular mobile communications, were investigated by means of extensive simulation campaigns and novel analytical capacity evaluation methods.

An ORACLE demonstrator was provided to show the capacity gains of the achieved spectrum agility, adaptive modulation techniques, and interference avoidance.

The FP6 IST project E2R focused on the secure, reliable, and seamless configuration and reconfiguration of equipment (terminal, base station, access point, and gateway). The terminal-centric development of the local (to the equipment) reconfiguration management and the development of extensions for execution environments to provide the basic mechanisms supporting the needed reconfiguration capabilities for dynamic adaptation and secure/reliable operation were the initial focal points that led to deeper integration and exploitation of the reconfigurable and flexible protocol stack.

The framework designed for equipment management within the frames of the FP6 IST projects of the area allows coordinated and efficient management as well as control of the equipment reconfiguration across all the layers. For example, within E2R work, to realize a reconfiguration that requires modification of various equipment components, the various configuration management and control functionalities were grouped and assigned to different submodules that cooperate to achieve coordination.

The different aspects of the required equipment management architecture are as follows:

- *Configuration management and control architecture:* This architecture is responsible for taking all the decisions for reconfiguration, which are based on triggers from and negotiations with other functional entities in the equipment and the network. The two major functional entities of the configuration management architecture are:
  - *Configuration Management Module (CMM):* This manages the reconfiguration processes according to specified semantics, protocols, and the configuration data model (which maybe stored in a distributed configuration database system). It must interact with other equipments (of different types, located in the terminals or network domains) through its external interface.
  - *Configuration Control Module (CCM):* This is responsible for the control and supervision of the reconfiguration execution, which is done using specific commands/triggers and functions of a given layer or a given execution environment.
- *The execution environment:* This environment provides a generic framework offering the basic mechanisms required for dynamic, reliable, and secure change of equipment operation. For applying the reconfiguration actions to the equipment, it provides a consistent interface to the CMM and CCM modules.
- *The security and reliability architecture:* This architecture provides a number of security related tasks to ensure only authorized access to the reconfiguration management system. Also, it safeguards the access to the functions of the reconfiguration management part within the terminal and helps to prevent attempts of fraudulent access from the outside world.

- *The protocol stack reconfiguration:* This is required to adapt the communication channel of the equipment to different radio access technologies.

The description of the functionality of the reconfigurable equipment with respect to the equipment reconfiguration scenarios addresses mainly the following areas:

- Multimode operation and standard switching, including more monitoring and discovery, negotiation and selection, and TCP parameters;
- Reconfigurable device management, including enhanced security/reliability features;
- Execution environment optimization;
- Modem reconfiguration.

The interfaces for the reconfigurable equipment should distinguish, on one hand, the external interfaces to reconfiguration entities on the network side and, on the other hand, the internal interfaces between different equipment local reconfiguration modules.

1.4 Security Threats

A key value proposition of reconfigurability is that it enables the users, manufacturers, and network operators to modify radios after they have been manufactured. The reconfigurability characteristic makes easy retrofits and upgrades possible. It also is the foundation for next-generation wireless communication, in which radios can sense their environment and autonomously change their behavior to optimize user preferences or network efficiency.

Though the possibility of reconfiguration of communication equipment has many advantages, it also introduces new security threats. It allows the equipment using a "good software" to be reconfigured in undesirable ways or to have malicious software installed. The threat arises when devices are reconfigured in a way that the changes made to their configuration contradict the interests and the expectations of the end users, network operators and service providers, equipment manufacturers, and regulatory authorities.

Maliciously or poorly designed radio software can render a radio inoperable or degrade its performance. It can cause a radio to transmit in ways that interfere with other radios or that violate spectrum rights. It can also compromise the security of data and applications that the radio software can access.

Some of the security threats that are specific to the reconfiguration are summarized as follows:

- Download and execution of malicious software;
- Invalidation of conformance requirements;
- Undesired reconfigurations;
- Illegitimate access to private information;

- No or insufficient protection of intellectual property;
- Easier attacks against wireless systems.

The above-mentioned factors demonstrate the need for investigation and under-standing of security issues involved with reconfiguration.

These factors can be used to derive the reconfiguration security objectives and to select or develop required and suitable security features. However, the overall objective would be to ensure reliable operation despite the ongoing, flexible down-load of software and reconfiguration information that could lead easily to extremely severe security problems.

Reconfigurability, therefore, demands a universal approach, because once radios are reconfigurable, they can be reconfigured to support applications across existing market boundaries. For example, a reconfigurable mobile device might be reconfigured to support public safety radio communications or vice versa. In this scenario, it is not appropriate to have different security models support each other, because either one might be circumvented by reconfiguring to the other. In the near term, most radios do not have the frequency agility to support multiband reconfigu-ration, but these transactions are likely in the future. Another problem arises with neighboring frequencies.

An example of applying a restrictive security measure would be to use only a single, trusted reconfiguration manager that has the capability to perform restricted types of reconfiguration and hence cannot introduce any security problems. But the real challenge is to develop security concepts for more open and decentralized approaches for reconfiguration, approaches that would ensure reliable and correct operation and meet the expectations of the users and operators while respecting regulatory boundaries.

For the equipment management and control architecture to include the existing and trusted protocols within a coherent framework that is open to any security protocol and satisfies the security requirements, a negotiation mechanism must be designed to select the available algorithms between entities. This would enable the reconfiguration architecture to properly operate over various kinds of networks and terminals, each being provided with native security mechanisms.

1.5   Evolution of Radio Resource and Spectrum Management

The emerging concept of SDR has triggered the design of new algorithms to manage and exploit radio resources in a better way. Their key characteristics are flexibility and reconfigurability.

These algorithms can be broadly classified into three main categories:

- *Dynamic Network Planning and Management (DNPM):* Resource manage-ment efforts are focused on the planning process of the radio networks. The objective is to introduce reconfigurability into the operational phase of the network so that the network capacity can be adjusted even after the network deployment. For example, there is an analytical evaluation of a case where the operator owns a GSM/GPRS and a universal mobile telecommunications system–frequency division duplex (UMTS-FDD) license, and it is shown that if joint resource radio management (JRRM) and joint admission control algorithm are applied between these two systems, the number of the connected systems and the load factor both increase, which indicates a better spectral efficiency. The study shows the behavior of the network while loading changes and during the reconfiguration scenario; it also explores traffic splitting and the application of heuristic approaches to optimize the planning process. In addition, it presents an optimization process for a network using reconfigurable terminals for UMTS, WLAN, and WiMAX.

- *Advanced Spectrum Management (ASM):* This concentrates on the spectrum allocation problem and involves concepts such as spectrum pooling, spectrum sharing, and spectrum auctioning. It is important to consider the technical and economic aspects of spectrum management. On the technical side, spectrum sharing has been shown to improve capacity, if managed well. On the economic side, auctioning approaches are being explored to optimize the spectrum allocation process. For example, spectrum auctioning could be done among operators or it could be auctioned directly to the end users.
- *Advanced Radio Resource Management (ARRM):* It considers ways to optimize traffic over several RATs. Fuzzy neural algorithms could be used to optimize the traffic over several RATs. Load-based RAT selection provides better optimization results than coverage-based selection. The load-balancing algorithms could also be adapted by considering the radio modeling aspects.

The above mechanisms were studied, as well as some others (e.g., cognitive pilot channel [CPC], flexible spectrum management, dynamic spectrum allocation [DSA], functional architecture [FA], and inherent solutions for optimized exploitation of spectrum and radio resources), by the FP6 IST projects active in the area, and the achieved results showed considerable potential to increase spectrum and radio efficiency and overall spectrum productivity.

Building on the concept of CR, cognitive networks and the concept of the CPC, the FP6 IST project E2R developed and implemented solutions for enhanced spectrum and resource efficiency in heterogeneous wireless systems [25]. The CPC has been presented to and discussed with different regulation and standardization organizations. There is a strong potential to standardize the CPC and the FA (IEEE SCC41, ETSI). For example, the project E2R II [1] set up and developed the standardization group IEEE P1900.4/SCC41, and activities have commenced in setting up the ETSI standardization path. E2R II also worked out a specific questionnaire to ECC regulators on the evolution of spectrum management.

The FP6 IST project AROMA [6] devised and assessed a set of specific strategies and algorithms as part of the ARRM for optimization of the capacity over specific air interface in the context of all-IP networks, as well as an efficient development of common radio resource management (CRRM) strategies in the case of heterogeneous networks. The optimization of the RRM/CRRM algorithms is one of the most important keys to successfully handle highly loaded networks, which are envisaged to dominate the wireless arena in the incoming years. The increase of the network capacity provided by these algorithms allows the network operators to optimize their networks in a cost-efficient manner and becomes a keystone of Europe's leadership in the mobile area. In addition to that, AROMA has also performed a technoeconomic analysis in terms of the capital expenditures (CAPEX) and operating expenses (OPEX) of the technology trends and steps related to the evolution of mobile system architectures, including the evaluation of the potential economic advantages of using specific RRM/CRRM algorithms and solutions already identified and only assessed from the technical point of view.

Radio resource management for self-configurable air interfaces was studied within the scope of the project SURFACE [4]. SURFACE addressed network resource allocation for QoS achievement as well as optimal modulation and coding in the presence of realistic channel modeling in the uplink (UL). To that end, the concept of stability region was introduced and proposed and a throughput optimal scheduler was evaluated.

The project FIREWORKS [26] investigated ARRM algorithms for relaying, including cooperative communication, ad-hoc association, and disassociation of nodes. ARRM strategies that extend the advance antenna system (AAS), adaptive modulation and coding (AMC), and PUSC/FUSC modes of the IEEE 802.16 standard were developed.

The FP6 project PULSERS (I and II) [27] succeeded to provide technical and scientific support to the CEPT ECC TG3 and the Radio Spectrum Committee (RS-COM), enabling the first release of a European Regulation for the application of ultra-wideband (UWB) devices. The introduction of new radio services enabled by the innovative UWB short-range radio technology ensures a more efficient use of the radio spectrum by applying the *spectrum underlay principle,* meaning that UWB radio devices are allowed to operate in frequency bands where other radio services already operate. This double use is possible due to the inherent technical features and the ultralow transmit power of UWB devices. The release of the first ECC recommendation on that topic (ECC/Dec/ (06)04 26 March 2006) and the amendment released in July 2007 enabling the use of UWB inside road and rail vehicles was a very big success for the EC policy and for European national administrations.

Spectrum and resource management continue to be key themes. The technical scope of research encompasses flexible spectrum and radio resource management, network (self-) management, cognitive radio, CPC, opportunistic spectrum use, interference mitigation techniques, and ultraefficient MAC design. Consolidating results and possibly creating momentum on some major challenges from technical, business and regulatory perspectives is very important to overcome the challenges and obtain easily deployable solutions.

1.6   Preview of the Book

The main focus of this book is the reconfigurability of radios and the related enabling technologies which have been investigated within FP6.

Contributions to this book have been made by IST projects undertaken within the Communications and Networks Technologies domain under the strategic objectives *Mobile and Wireless Systems and Platforms Beyond 3G* and also partially under *Broadband for All.* The book is organized as described in this section.

Chapter 2 discusses the technologies that will support the concept of reconfigurability. The enabling technologies, such as multicarrier transmission techniques (e.g., orthogonal frequency-division multiplexing [OFDM], and multicarrier code division multiple access [MC-CDMA], MIMO, UWB radio technology, and smart antennas, are discussed in this chapter. Related IST projects and contributions to the FP6 program were made by the IST FP6 projects Broadcasting and Multicasting over Enhanced UMTS Mobile Broadband Networks (B-BONE) [27], Flexible Relay Wireless OFDM-Based Networks (FIREWORKS) [28], Multiple-Access Space-Time Coding Testbed (MASCOT) [29], Multi-Element Multihop Backhaul Reconfigurable Antenna Network (MEMBRANE) [30], ORACLE, Pervasive Ultra-Wideband Low Spectral Energy Radio Systems (PULSERS), Reconfigurable Systems for Mobile Local Communication and Positioning (RESOLUTION) [32], and Self Configurable Air Interface (SURFACE) [4] . Chapter 2 is built upon the results of some of these projects.

Reconfigurability forms one of the key enablers for the autonomic communications concept. Chapter 3 goes into detail about autonomic communications. The goal of reconfigurability is to limit complexity for networks and users in a heterogeneous context. Achieving this goal requires minimizing human interaction with the equipment and the network. Autonomic computing emerges as a new paradigm for managing increasingly complex tasks at the business, system, and device level without human intervention. The IST FP6 projects contributing to advancements in the state of the art in this area were the projects E2R [2] and ORACLE [5], which have investigated this concept in the scope of advancing resource and spectrum management technologies.

Chapter 4 focuses on system capabilities. The topics discussed in this chapter are policy management; cognitive service provision and discovery; emergency services; context interpretation; self-configuring protocols; mass upgrade of mobile terminals; handover; formation of network compartments and base-station reconfiguration; traffic load prediction and balancing; network resource management; RAT discovery; and selection. IST FP6 project E2R [2] is the main contributor to this chapter.

Chapter 5 is focused on protocols, architectures, and traffic modeling for reconfigurable or adaptive networks. Reconfigurability supports the B3G concept by providing technologies that enable equipment to dynamically select and adapt to the most appropriate RAT, according to conditions in the specific service area region conditions and local time zone. The goal is to highlight aspects of the management and control architecture for end-to-end reconfigurable equipments. The related IST projects are FLEXINET [33] and Secure, Internet-able, Mobile Platforms Leading Citizens Towards Simplicity (SIMPLICITY) [34].

Reconfigurable radio equipments will be a key part of future end-to-end reconfigurable systems. Chapter 6 talks about reconfigurable radio equipment management, both from the network and equipment perspectives. Reconfiguration management covers all the means inside the equipment, enabling it to contribute to make appropriate decisions concerning the RAT reconfiguration. Reconfiguration control covers all the means inside an equipment that apply those decisions, taking advantage of the set of reconfigurable elements composing the flexible equipment. Reconfigurable elements correspond to the actually reconfigured modules of the flexible equipment, which perform the useful RAT, protocol, or user application processing. The IST project E2R had achieved significant advances in this area and further research is still needed.

Chapter 7 has as its main theme the efficient utilization of spectrum and radio resources. The introduction of flexible and reconfigurable systems requires a paradigm change in how radio resources are handled. Spectrum sensing and sensing, cooperation protocols for sensing, algorithms for spectrum and cooperative sensing, spectrum policies, and economic considerations are covered in this chapter. Major contributions to this chapter are from the IST FP6 project ORACLE [5].

With reconfiguration, previously fixed properties of communication equipment can be changed by software download. The improved flexibility poses the threat of downloading incorrect or malicious software. A secure reconfiguration is required to ensure correct, reliable operation that respects the interests of the end users, operators, and regulatory bodies. Chapter 8 discusses the reconfiguration threats and security objectives associated with the alteration of system due to software modifications.

Chapter 9 focuses on the prototyping and requirements of the developed platform. Equipment and network prototyping and their requirements are discussed in this chapter. The other related aspect of adaptive applications and its requirements is also described.

Chapter 10 concludes the book and gives guidelines for future research work in the area of reconfigurability. At the time of writing of this book, the 7th Framework Programme (FP7) had commenced its activities, with some of the earlier projects continuing research and development under the new umbrella. The guidelines provided are based on the outlines for future research provided as public documents by the European community.

References

[1]   European Union 6th Framework Programme (FP6) for Information Society Technologies (IST), at http://cordis.europa.eu/ist/ct/index.html.

[2]   IST project End-to End Reconfigurability (E2R), at http://cordis.europa.eu/ist/ct/proclu/p/mob-wireless.htm.

[3]   IST project FLEXINET, at http://cordis.europa.eu/ist/ct/proclu/p/mob-wireless.htm.

[4]   IST project SURFACE, at http://www.ist-surface.org/.

[5]   IST project ORACLE, at http://www.ist-oracle.org/.

[6]   IST project AROMA, at http://www.aroma-ist.upc.edu/.

[7]   SDR Forum, at http://www.sdrforum.org/.

[8]   WWRF, at http://www.wireless-world-research.org/.

[9]   Bagheri, R., et al., "Software-Defined Radio Receiver: Dream to Reality," *IEEE Communications Magazine,* August 2006.

[10] Tuttlebee, W., "Software-Defined Radio: Facets of a Developing Technology," *IEEE Personal Communications*, April 1999.

[11] Araki, K., et al., "Implementation and Performance of a Multiband Transceiver for Software Defined Radio," *IEEE* 2004.

[12] Ginsberg, A., et al., "Toward a Cognitive Radio Architecture: Integrating Knowledge representation with software defined radio technologies," *Military Communications Conference (MILCOM),* October 2006.

[13] Harada, H., "Software Defined Radio Prototype Toward Cognitive Radio Communications Systems," *First IEEE 2005 International Symposium on New Frontiers in Dynamic Spectrum Access Networks (DySPAN)*, November 2005.

[14] Mitola III, J., "Cognitive Radio—An Integrated Agent Architecture for Software Defined Radio," *KTH Stockholm*, 2000.

[15] Prompijit, A., "Cognitive Radio", *Keio University*, 4th July 2006.

[16] Jesuale, N., and B. Eydt, "A Policy Proposal to Enable Cognitive Radio for Public Safety and Industry in the Land Mobile Radio Bands," in NetCity Publications 2007, at http://www.netcityengineering.com.

[17] The Federal Communications Commission (FCC) at http://www.fcc.gov.

[18] Doyle, L., "Complexity Continued—A Very Basic Introduction to Cognitive Radio," *Trinity College*.

[19] http://portal.acm.org/citation.cfm?id=1465424.

[20] Gomez, G., et al., "A 1.5V 2.4/2.9mW 79dB/50dB DR $\Delta\Sigma$ Modulator for GSM/WCDMA in a 0.13 μm Digital Process.", *ISSCC*, February 2002, pages 306–307.

[21] Veldhoven, V., "A Triple-Mode Continuous-Time $\Delta\Sigma$ Modulator with Switched Capacitor feedback DAC for a GSM-EDGE/CDMA2000/UMTS Receiver," *IEEE Journal of Solid-State Circuits*, Vol. 38, No. 12, Dec 2003.

[22] Burger, T., et al., "A 13.5-mW 185-Msample/s $\Delta\Sigma$ Modulator for UMTS/GSM Dual-Standard IF-Reception," *IEEE Journal of Solid-State Circuits*, Vol. 36, No. 12, Dec 2001.

[23] Konstanznig, G., et al., "Design of 1.5V, 1.1 mA Fully Integrated LC-tuned Voltage Controlled Oscillator in the 4 GHz-band using a 0.12 μm CMOS-Process," *APMC* 2002.

[24] Konstanznig, G., et al., "A 10 mW, 4 GHz CMOS Phase-Locked Loop with Dual-Mode Tuning technique and Partly-Integrated Loop Filter," *RFIC* 2003

[25] FP6 Network and Communication Technologies Project and Cluster Results, Report, February 2008, at http://cordis.europa.eu/fp7/ict/future-networks/home_en.html.

[26] IST project Flexible Relay Wireless OFDM-based Networks (FIREWORKS) at http://www.fireworks.intranet.gr.

[27] FP6 IST project PULSERS at http://cordis.europa.eu/ist/ct/proclu/p/mob-wireless.htm.

[28] http://b-bone.ptinovacao.pt/

[29] http://fireworks.intranet.gr/

[30] http://www.ist-mascot.org/

[31] http://www.imperial.ac.uk/membrane/

[32] http://www2.ife.ee.ethz.ch/RESOLUTION/index.html

[33] http://www.everest-ist.upc.es/

[34] http://www.ist-simplicity.org/

[35] http://www.sportviews.org/

[36] http://www.ist-phoenix.org/

CHAPTER 2

Enabling Technologies

Software-defined radios (SDRs) provide flexibility and multimode, multifunctional capabilities to the mobile communications networks through reconfigurability. SDR enables reconfiguration of both the device air interface parameters and the entire mobile network, which is a key enabler to new personalized applications for the users. The key factor enabling SDR is the convergence of digital radio and downloadable software technologies.

This chapter covers the concepts of SDR and cognitive radio (CR) as the building concepts for reconfigurable systems. Further, it describes the advances made by the IST FP6 projects that were active in the area related to the relevant physical-layer (PHY-layer) related technologies enabling reconfiguration, such as multicarrier (MC) transmission techniques, MIMO, ultra-wideband radio technology, adaptive antenna systems, advanced local positioning radar and software tools and architectures. The major contributions were made by the projects E2R [1], E2R II [1], B-Bone [2], MEMBRANE [3], PULSERS [4], MASCOT [5], RESOLUTION [6], and SURFACE [7]. The activities of these projects have been continued in the follow-up research and development program funded by the European Union (EU) under the umbrella of the 7th Framework Programme (FP7). Further advancements in the area of reconfigurability are targeted by the FP7 projects E3 [8], ARAGORN [9], and some others [10].

2.1 SDR and Cognition as Building Concepts of Reconfigurable Systems

SDR and cognitive radio (CR) are two related concepts that enable reconfigurability from different viewpoints [11]. The first, proposed by Mitola, suggests that cognition on a wireless device/network can be realized by "integration of model-based reasoning with software radio," which would be "trainable in a broad sense, instead of just programmable." The second considers CR and SDR as enabling technologies to dynamic spectrum access techniques.

The term CR is often associated with the SDR concept, although they are two distinctly separate concepts, as explained in Chapter 1 of this book. Neither concept relies on the other; however, an SDR radio transceiver will generally perform most signal processing functions in software (which have been implemented until now by hardware, for obvious performance reasons), and cognitive radios can exploit this flexibility to provide additional benefits. SDR, thus, is considered as the practical realization of the idea of software radio that implements some radio functions on a general purpose or programmable hardware. The IST project E2R developed a complete software reconfigurable framework [12] and defines it as a radio communication system that uses software for the modulation and demodulation of radio signals. An SDR performs significant amounts of signal processing in a general-purpose processor or in a reconfigurable piece of digital electronics. The goal of this design is to produce a radio that can receive and transmit in any new form of radio access technology just by running new software.

From a communication point of view, SDR provides a flexible radio that can support multiple air interface standards operating on the same or different bands. A number of research projects from military, industrial, and public sectors have been developing various architectures and technologies to exploit the flexibility and adaptability provided by SDRs.

In general terms, SDR provides reconfigurability, in which multiple modes (or radio access technologies) are supported on a single hardware platform, and portability, in which new radio access technologies can be deployed to multiple platforms. A lot of the research effort has been put into defining a common set of interfaces for management and deployment of software components and representing different functions associated with radio/protocol standards in a generic manner.

SDR defines transmitter and receiver waveforms in software. Software implementation provides a number of advantages over a hardware solution, especially when developing adaptive and intelligent algorithms on communication systems. Usually, the modular design provides a greater separation of functionality, which allows reprogramming and reconfiguration of functions. The associated generic interfaces provide access to integral details of the algorithms and a greater control of functionalities. Along with the generic representations or models of standards, the above features provide greater flexibility for algorithm developers. When considering radio algorithms, the degree of adaptation required could vary from the tuning of one or two parameters of an existing software component to a complete reconfiguration by reprogramming with a different piece of software. The main application areas for SDR systems are in multimode devices, where multiple bands and standards are supported by reprogramming the device. Reconfigurable components also enable application of dynamic optimization techniques, for example, to improve performance or reduce power consumption. Software controls also create more opportunities for equipment manufacturers, system integrators, and operators to develop new solutions and services. At present, SDR-based solutions are popular in the fields of dynamic spectrum access and cognitive radio.

It should be noted that a cognitive radio does not necessary need SDR as an implementation platform.

One of the key missing ingredients in a pure SDR system is that it is designed to be very flexible and a PHY-layer adaptive system. SDR research, however—even if it is focused on providing flexibility—does not answer the new problem of how to handle the complexity that is generated by such flexibility. In principle, SDR technology allows to build highly adaptive radios that could flexibly change according to the user needs and environmental conditions; still, in a certain sense a "classical" SDR is a dumb device. It is capable of changes, but current research has not yet provided answers on how to put such intelligence inside the SDR so that it could automatically learn to adapt.

Currently there is no commonly agreed-upon definition for CRs, and the term means just any flexible air-interface technology.

A lot of research effort is toward enabling SDR and CR. In principle, cognitive radio research can benefit from any (spectrum) flexible transceiver hardware, and SDR prototyping environments are particularly suitable for this purpose.

The most popular platform available at the time of writing of this book is the so-called GNURadio [13], which includes a flexible open-source software package and also a small number of compatible hardware boards. The most used hardware board in this context is the USRP GNURadio experimental board, which has a motherboard that can take up to 4 Tx/Rx daughterboards, which operate in different frequencies, mainly between 1–2800 GHz. GNURadio is particularly popular in the academic community due to its low cost and readily available software. However, it is not without problems; for example, its bandwidth and RF-parts linearity are limited.

Another enabler for CR and SDR is MIMO-capable hardware. These so-called WARP-boards [14] provide extremely powerful platform for PHY-layer experimentation, but the price is high development complexity, and the currently available open-source code base is very limited.

Standardization activities on CR-related techniques have also gathered momentum recently. Different standardization bodies are tackling SDR-, CR-, and DSA-related techniques. In Europe, ETSI formed a new technical committee for reconfigurable radio systems (RRS) in March 2008 [15]. SDRF [16] and OMG [17] also have been playing their part in the reconfigurable area, which is closely associated with CR concepts. In addition, there are extensions to existing 802.x standards (e.g. .11x, .16h) to cover CR- and SDR-type concepts.

2.2   Design of an Optimal Transmit and Receive Architecture for a Reconfigurable Air Interface

Self-configurability of the radio interface can be achieved as a function of the channel status [7]. In the single-user case, the system can benefit from any partial or full information about the channel at the transmit side. This can be achieved by a unified framework encompassing most known multiplexing or coding schemes, such as OFDM or code-division multiple access (CDMA). This step is useful to derive a flexible scheme that is able to switch, if useful, from one system to the other, depending on the channel and the level of interference. A general space-time coding scheme, named Trace-Orthogonal Design (TOD), is valid for any number of transmit/receive antennas and guarantees lossless information transmission for any kind of channel.

The design of radio resource allocation (RRA) and scheduling (PS) strategies able to match the instantaneous channel conditions is prominent in the achievement of high spectral efficiencies (throughput). Imperfections in the knowledge of the network state limit the ultimate performance and constrain the PHY-layer techniques to be applied.

A multiuser transmit and receive strategy can be designed as a three-step procedure: first, the fundamental transmission limits of the uplink (UL) and the downlink (DL) are evaluated; then, optimal and efficient PHY-layer transmission techniques that operate close to these limits are proposed; and finally, the design of the PS and

RRA under a loose cross-layer approach and the transmission techniques proposed in the previous step is performed.

2.2.1   Limits of the Single-User MIMO Channels

The increasing demand for higher data rates in wireless systems, where the common resources of power and bandwidth are particularly limited, has ignited much interest in investigating other forms of increasing spectral efficiency. In particular, a key ingredient in the design of more spectrally efficient systems is the spatial dimension. Although the use of the spatial dimension in wireless systems is hardly new (cell sectorization and base stations arrays are common examples), it is really the simultaneous deployment of multiantenna elements at both sides of the communication links that can unleash levels of performance by opening up multiple spatial dimensions [18].

In addition to the channel conditions and the degree of channel information available at the transmitter and/or the receiver, the capacity of a given MIMO system depends also heavily on other additional factors, which can be referred to as signaling constraints.

The general ergodic MIMO channel is a very interesting setting, as it provides the maximum achievable rate since no additional constraints are imposed. The limiting rate is provided by the Shannon (ergodic) capacity, obtained as the maximum mutual information averaged over all channel states.

The signal model for a flat fading MIMO channel with n_T transmit and n_R receive antennas (dimensions) is given by:

$$y = Hs + w \tag{2.1}$$

where $s \in Cn_T \times 1$ is the transmitted signal vector, $H \in Cn_R \times n_T$ is the channel matrix, $y \in Cn_R \times 1$ is the received vector, and $w \in Cn_T \times 1$ is a zero-mean complex Gaussian vector (which can include other interference Gaussian signals), normalized so that its covariance matrix is the identity matrix. The normalization of any nonsingular covariance matrix R_w to fit the model of (2.1) is as straightforward as multiplying the received vector y by $R_w^{-1/2}$ to yield the effective channel $R_w - 1/2$ H and a white noise vector.

The channel matrix in the signal model of (2.1) gathers the transmission paths between each transmit antenna and each receive antenna. For instance, [H]ij represents the channel gain that experiences the signal transmitted from the ith transmit antenna to the jth receive antenna. Although some results have been derived for a general fading distribution, these channel matrix coefficients are usually modeled as zero-mean circularly symmetric complex Gaussian distributed random variables. The Gaussian distribution results from the application of the central limit theorem when there are a large number of scatters in the channel that contributes to the signal at the receiver [19].

Under the Gaussian assumption, the channel matrix HH† is Wishart distributed.

The signal model of (2.1) represents a single channel use or transmission. A real communication, however, consists of multiple transmissions. To represent this time-varying nature of the communication process, it suffices to index the signals in (2.1) with a discrete-time index:

$$(y(n) = H(n)s(n)+w(n)) \tag{2.2}$$

For the more general case of having a set of K parallel MIMO channels (e.g., separated in time or frequency) with n_T transmit and n_R receive dimensions, the signal model is:

$$y_k = H_k s_k + w_k \tag{2.3}$$

Nevertheless, this can be written as in (2.1) by defining the block-diagonal matrix H =diag $\{Hk\}$ and stacking the vectors as $s = [s_1{}^T \cdots s_k{}^T]^T$, $y = [y_1{}^T \cdots y_k{}^T]^T$, and $w = [w_1{}^T \cdots w_k{}^T]^T$.

In real propagation environments, the fades connecting pairs of transmit and receive antenna elements are not independent, due, for example, to insufficient spacing between antenna elements. In the information theory context, a convenient approach is to construct a scatterer model that can provide a reasonable description of the scattering environments for the wireless application of interest. The advantage of using abstract models is that with a simple and intuitive model that makes the problem analytically tractable, the essential characteristics of the channel can be clearly enlightened. For an overview of the numerous scattering models, see, for example, Ertel et al. [20]. The most common scattering model [21, 22] assumes that antenna correlation at the transmitter side and at the receiver side is caused by independent phenomena and, thus, correlation can be separated. A complete review of MIMO channel modeling can be found in [23].

A crucial factor affecting the performance of a MIMO system is availability of channel state information (CSI) at the receiver (CSI-R) and/or at the transmitter (CSI-T), that is, the knowledge of the fading gains in each one of the transmissions paths or the values taken on by the entries of the channel matrix H.

The most commonly studied situation is that of perfect CSI-R, both with perfect CSI-T and with no CSI-T at all. In this context, the MIMO channel capacity was derived in the pioneering work of Foschini [24] and Telatar [25], showing the linear (with the minimum of the number of transmit and receive antennas) capacity increase of MIMO channels characterized by i.i.d. complex Gaussian elements. It must be noted that the perfect CSI model is inaccurate in a real situation, and this motivates the situation of perfect CSI-R with partial CSI-T, in which the channel state can be accurately tracked at the receiver but the transmitter has an inaccurate channel knowledge based on the channel information fed back from the receiver.

In practice, the channel estimate available at the transmitter does not coincide with the true channel, for the following reasons: (1) any channel estimate, in the presence of noise, is inevitably affected by estimation errors; (2) in the case of FDD systems, the feedback channel is itself affected by transmission errors, and the channel estimate needs to be properly quantized before transmission; and (3) in the case of a time-varying channel, the channel behavior to be used to optimize the transmission must be predicted from its past estimates, and this operation introduces an inevitable prediction error. This error is more or less important depending on the ratio between the time interval T_0, over which the transmission will be optimized, and the channel coherence time T_c. If $T_0 < T_c$, one may assume that the transmit channel is approximately equal to the receive channel (beside estimation errors). But if $T_0 > T_c$, one needs to predict the channel evolution from its past. In this case, the channel information at the transmitter can be modeled as a set of random variables whose parameters depend on the true channel. We refer to this situation as a perfect CSIR with statistical partial CSI-T model. The statistical channel model depends on the time scale of interest. In the short term, the channel coefficients may have a non-zero mean and one set of correlations reflecting the geometry of the particular propagation environment. Over a long term, the channel coefficients may be described as zero-mean and uncorrelated due to averaging over several propagation environments. For this reason, uncorrelated, zero-mean channel coefficients are a common assumption for the channel distribution in the absence of distribution feedback or when it is not possible to adapt to the short-term channel statistics.

If the transmitter receives frequent updates, it can adapt itself to these time-varying short-term channel statistics to increase the capacity.

The channel coefficients are typically assumed to be jointly Gaussian, so that the channel distribution is specified by the channel mean and covariance matrices. In addition, almost all research has focused on three special cases:

- Zero-Mean Spatially White (ZMSW):

$$E\{H\} = 0, H = H_w$$

where H_w has zero-mean, unit-variance, complex circularly symmetric Gaussian entries. This model typically captures the long-term average distribution of the channel coefficients averaged over multiple propagation environments. Thus, it is appropriate in the absence of any feedback information.

Mean-feedback model:

$$H = \sqrt{\frac{K}{K+1}} H_{spec} + \sqrt{\frac{1}{K+1}} H_w \tag{2.4}$$

where the mean component matrix H_{spec} and the coefficient K are assumed to be known and H_w has zero-mean, unit-variance, complex circularly symmetric Gaussian entries.

The mean-feedback model can be used when an approximate (or outdated) channel estimate is available and the estimation errors can be regarded uncorrelated among antennas. A typical example is given by the Rice channel model, where the transmitter is assumed to know the specular component. In such a case, the mean component is the rank-one matrix:

$$H_{spec} = a_R a \frac{T}{T} \tag{2.5}$$

where a_T and a_R are the specular array response vectors at the transmitter and receiver, respectively. In the multiple-input-single-output (MISO) case, (2.5) reduces to the n_T-size row vector $a_T T$, whereas in the single-input-multiple-output (SIMO) case, (2.5) reduces to the n_R-size column vector a_R.

Covariance-feedback model:

$$E\{H\} = 0, \; H = \Sigma_R^{1/2} H_w \Sigma_T^{1/2}$$

where Σ_T is the transmit correlation matrix, Σ_R is the receive correlation matrix, and H_w has zero-mean, unit-variance, complex circularly symmetric Gaussian entries.

The matrices Σ_R and Σ_T are assumed to be known at the transmitter side. As a particular case of this model, there is the zero-mean spatially white (ZMSW) model where $\Sigma_R = \Sigma_T = I$.

Under the covariance-feedback model, the channel is assumed to vary too rapidly to track its mean, so the mean is set to zero and the information regarding the relative geometry of the propagation paths is captured by a non-white covariance matrix.

The standard constraints for a channel model can be summarized as follows:

- *Average power constraint* is applied to each of the transmitting antennas or averaged over all transmitting antennas.
  - *Short term power constraint*: the average is over a single transmitted code block.
  - *Long term power constraint*: the average is over many transmitted codewords.
- *Peak-power or amplitude constraints* are common practices in information-theoretic analyses, as they reflect the limitations of practical communication systems more accurately [26].
- *Bandwidth* plays a major role in the set of constraints applied in information-theoretic analysis. The bandwidth constraints can be given in terms of a distribution defining the percentage of time that a certain bandwidth is allocated to the system.
- *Delay constraints* pose a limitation on any practical system. The delay constraints give rise to new information-theoretic expressions, such as the capacity versus outage and delay-limited capacity.

Single-user MIMO channel capacity depends heavily on the statistical properties and antenna element correlations of the channel. Antenna correlation varies as a function of the scattering environment, the distance between transmitter and receiver, the antenna configurations, and the Doppler spread. The impact of channel correlation on capacity depends on what is known about the channel at the transmitter and receiver. We may assume perfect CSI, partial CSI or no CSI at all both at the transmitter and receiver.

The most relevant MIMO channel capacity results with perfect CSIR and several degrees of channel knowledge at the transmitter under uncorrelated fading between antennas were summarized in [27].

2.2.2 Precoding Schemes with Perfect and Partial CSI-T

In point-to-point MIMO communication systems with CSI at both sides of the link, theoretically, when perfect CSI is simultaneously available at the transmitter and at the receiver, the optimal transmission is given by a Gaussian signaling with a waterfilling power profile over the channel eigenmodes [25, 28]. From a more practical standpoint, however, the ideal Gaussian codes are substituted with practical constellations (such as QAM constellations) and coding schemes. To simplify the study of such a system, it is customary to divide it into an uncoded part, which transmits symbols drawn from some constellations, and a coded part, which builds upon the uncoded system. Although the ultimate system performance depends on the combination of both parts, it is convenient to consider the uncoded and coded parts independently to simplify the analysis and design.

The design of linear MIMO transceivers has been extensively treated in the literature based on different system performance measures. The classical approach refers to the minimization of the sum of the mean square error (MSE) of all subchannels or, equivalently, the trace of the MSE matrix [29, 30]. Some other results consider the minimization of the weighted trace of the MSE [31], the minimization of the determinant of the MSE matrix [32], and the maximization of the signal-to-noise ratio (SNR) [30]. Recently, a unifying framework for the systematic design of linear MIMO transceivers was proposed in Palomar et al. [33]. More exactly, the optimum linear precoder and linear equalizer for a wide range of different design criteria are derived, assuming that the number of transmitted symbols per channel use and the corresponding constellations have been previously chosen (i.e., fixed rate transmission). Results related to this were achieved and reported by the FP6 IST project SURFACE [7]. There, keeping the fixed rate transmission assumption, the multimode precoding scheme [34] was considered, which adapts not only the linear precoder but also the number of transmitted symbols to the instantaneous channel realization. Independently on the system quality measure, multimode precoding always outperforms classical linear precoding, since the multimode precoding scheme selects the best linear precoder with a fixed number of symbols per transmission (modes) over the set of all supported modes. An analytical analysis was presented in the SURFACE Deliverable [27] for the average bit error rate (BER) performance of both precoding schemes.

It can be assumed that the linear precoder is designed to minimize some system performance measure (e.g., the BER averaged over the active substreams) subject to a transmit sum-power or peak-power constraint. Note that, as opposed to fixing the transmit power and optimizing some measure of quality, it is also possible to minimize the transmit power subject to some minimum global quality of the system or, even better, subject to an independent quality for each of the subchannels [35]. These results can be straightforwardly obtained from the optimal linear precoder structures given in [27, 36].

The random time-varying nature of wireless channels, however, makes it difficult to obtain perfect CSI at the transmitter. In some cases (e.g., TDD mode), the transmit channel can be inferred from the receive channel through reciprocity. For a full-duplex MIMO communication system, such as the one shown in Figure 2.1, the reciprocity principle states that the channel from point A to point B is the same as the

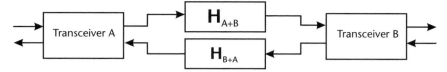

Figure 2.1   CSI-T acquisition using reciprocity.

channel from point B to point A, since both channels are measured at the same time, the same frequency, and the same antenna locations.

Although in a real full-duplex system the reverse and forward link cannot use identical spectral, temporal, and spatial resources, the perfect channel knowledge assumption is still accurate whenever the difference in any of these dimensions is relatively small compared to the channel variation. More often (e.g., frequency division duplex, or FDD, mode), the CSI at the transmitter is obtained using feedback from the receiver, as shown in Figure 2.2.

The feedback is not limited by reciprocity requirements, but it implies an additional delay on the CSI obtained by the transmitter and an additional loss of information due to the quantization of the CSI. To minimize the additional transmission overhead caused by the feedback, the channel estimated at the receiver must be somehow quantized before being communicated to the transmitter through the (usually low rate) feedback channel. Hence, in this case the assumption of perfect CSI at the transmitter is unrealistic. The IST project SURFACE presented some joint transmit-receive strategies [37] that can be used as a benchmark or as design guidelines for the limited feedback case. Figure 2.3 shows the results for different SNR values regarding the computation of a function C_{net} obtained via a Monte Carlo simulation that averaged more 200 independent channel realizations and considering $\gamma = 104$. These results show that the effectiveness of feedback strongly depends on the SNR; in particular, when the SNR is high enough, its contribution to increase the transmission rate is negligible, so it could be more favorable to avoid it.

The simulation considered a 2×2 MIMO system with $\gamma = 104$, and SNR=10 dB. The dashed lines denote the rate achieved with perfect CSI-T and without CSI-T, respectively. The upper solid line denotes the achievable rate with partial CSI-T. It increases as the number of bits increases, since it does not take into account the cost of feedback. Finally, the lower solid line is the net rate and, since it considers the cost of feedback, it exhibits different behavior. In fact, as the number of bits rises, the net rate rapidly increases until it reaches its maximum, and then, with a gentle slope, it decreases monotonically. As a consequence, the increase in feedback rate due to the rising of the number of quantization bits beyond a certain threshold is not

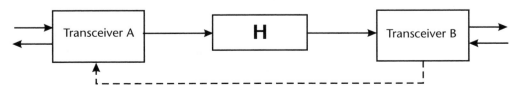

Figure 2.2   CSI-T acquisition using feedback.

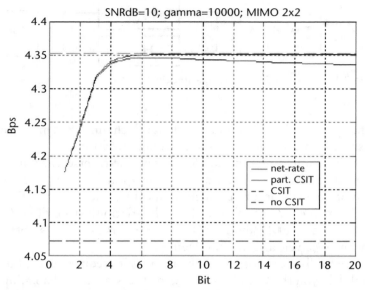

Figure 2.3   Cnet versus the number of quantization bits for a 2×2 MIMO system with SNR=10 dB and $\gamma = 104$.

worthwhile, because the gain in transmission rate that it entails is lower than the additional feedback needed.

With reference to Figure 2.3, it is noteworthy that by using only six bits, it is possible to achieve almost all the ergodic capacity. Even limiting the number of bits to three would yield a significant gain with respect to the case without CSI-T at all. Moreover, the gain is appreciable even using one bit.

Similar results hold for MIMO systems with more antennas. These are available in [38]. As an example, in Figures 2.4, 2.5, and 2.6, a 4 × 4 MIMO system has been considered, with $\gamma = 104$ and SNR=10 dB, 20 dB, and 30 dB, respectively. The comments made before still hold in this case.

It should be noted that because of the increased number of parameters fed back, the slope of the Cnet curve has changed. Moreover, it is confirmed also in this case that for high SNR the effectiveness of feedback tends to be less and less pronounced.

In a similar way, one can approach the problem of optimizing the single-user MIMO system when perfect CSI is available at the receiver and the transmitter is informed using a rate-limited feedback channel (see Figure 2.7).

The straightforward solution to this problem is to obtain an accurate representation of the channel matrix using vector quantization (VQ) techniques. However, the channel state parameters grow as the product of the number of transmit and receive antennas, and must be updated every time the channel changes. Thus, the classical VQ approach results in a prohibitive amount of feedback information. Even more important is the fact that we are not interested in approximating the channel at the transmitter, but the optimum transmission strategy and, hence, classical VQ (based on minimizing the MSE) cannot be optimal.

In summary, the design and implementation of a limited feedback linear MIMO system requires solving two main problems: (1) selection of the optimal precoder B

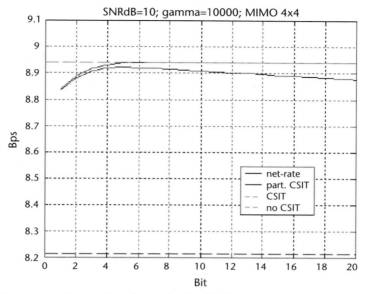

Figure 2.4 Cnet versus the number of quantization bits for a 4×4 MIMO system with SNR=10dB and $\gamma = 104$.

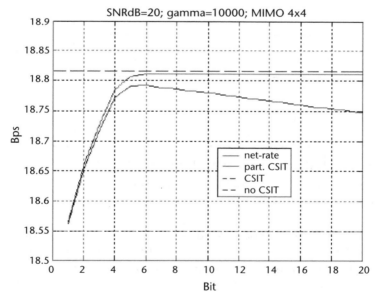

Figure 2.5 Cnet versus the number of quantization bits for a 4×4 MIMO system with SNR=20 dB and $\gamma = 104$.

from the codebook B for a given channel realization; and (2) design of the optimal codebook B. Because the design of the feedback information is tightly coupled with the transmission scheme, both problems should be addressed for different transmission strategies.

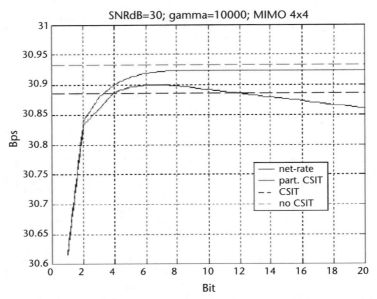

Figure 2.6 Cnet versus the number of quantization bits for a 4×4 MIMO system with SNR=30 dB and $\gamma = 104$.

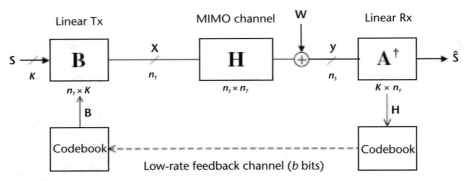

Figure 2.7 General limited-feedback MIMO system.

In summary, using some kind of channel information at the transmitter is useful to boost performance. The problem is different for the single and multiuser case, especially, in relation to the chosen optimization strategies.

2.2.3 General Framework for Air Interface Reconfigurability

A general framework for the formal description of several classes of transmission schemes for frequency selective channels can be used as a unified formalism for the representation of such different schemes [39]. It is helpful in understanding how a reconfigurable air interface works, at the physical layer. On one hand, it is useful in singling out the parameters that distinguish one transmission scheme from another, (i.e., the parameters whose alteration allows an air interface to adapt its transmission scheme to possibly different channel conditions, interference from other users, traffic, and so on). On the other hand, it can help to clarify what these schemes have in common. A unified description of different schemes is also useful from a practical implementation point of view, in the architectural development of an air interface that should be capable of changing its transmission scheme using the same hardware.

To outline the design strategies related to the spectral properties of the channel, single and MIMO antenna systems were considered by the IST FP6 project SURFACE. In particular, the design issues related to the spatial channel structure that arise in multiantenna (MIMO) systems were considered. It should be noted that all the issues treated in this respect can be extended to MIMO systems, as they refer to frequency selectivity, which is present also in the MIMO channels.

For a unified framework for a point-to-point communication over a multipath wireless fading channel, block transmission strategies allow the equalization of the discrete-time channel without the necessity of infinite impulse response (IIR) filters [7]. Further, systems that, thanks to the use of suitable precoding matrices, exhibit an equivalent channel matrix that is Toeplitz-circulant can enable the use of fast Fourier transform (FFT) algorithms for the necessary diagonalization operations at the receiver.

2.2.4   Design of Multiuser Optimal TX and RX

It is well known that when channel state information (CSI) is accessible simultaneously at the transmitter (CSI-T) and at the receiver (CSI-R), the MIMO system can be adapted to each channel realization to maximize the spectral efficiency and/or reliability of the communication. The optimal transmission from an information-theoretical point-of-view was reported in the IST SURFACE Deliverable 4.2 (June 2007) [40]. From a more practical standpoint, in [40] the ideal Gaussian codes were substituted with practical constellations (such as QAM constellations) and coding schemes. To simplify the study of such a system, it is customary to divide it into an uncoded part, which transmits symbols drawn from some constellations, and a coded part that builds upon the uncoded system. Although the ultimate system performance depends on the combination of both parts, it is convenient to consider the uncoded and coded parts independently to simplify the analysis and design. A MIMO multiple-access channel with K users is shown in Figure 2.8.

A joint optimization of a system equipped with a linear multiantenna transmitter and a linear multiantenna receiver under perfect CSI-T and perfect CSIR, in contrast to the single-user case, would require focus on the uncoded part of the system. The analysis of the structure of the capacity region in all the cases where perfect CSI-T is available has led to the proposal of optimal and efficient sum-capacity maximization algorithms based on water-filling solutions for the multicarrier case. In the single-carrier setting, a linear transceiver framework was introduced in [40]. These maximization procedures rely on the CSI at the user terminals to minimize multiple-access interference in order to provide the BS with maximum throughput. When perfect knowledge of the channel is not available at the transmit side, space-time codes can be used and the extend trace-orthogonal designs from [39] can be extended to a multiple-access setting. One way to carry out such a coding strat-

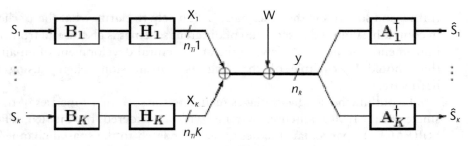

Figure 2.8 Scheme of a MIMO medium access control (MAC) communication system with linear transceiver.

egy to avoid information losses is a design that guarantees the achievement of the instantaneous capacity region regardless of the statistics of the channels.

2.3   Candidate PHY-Layer Techniques for Reconfigurable Air Interfaces

2.3.1   Downlink Scenario

PHY-layer techniques that are good candidates to be implemented in a reconfigurable air interface are directly related to achieving the two sources of capacity increases in a MIMO downlink channel: the multiplexing gain and the multiuser gain. The multiplexing gain denotes the ability of the system to transmit simultaneously to multiple users over the same bandwidth (orthogonal schemes like CDMA or transmission division multiple access [TDMA] are generally highly suboptimal). On the other hand, the multiuser gain arises as the result of smart user selection in downlink channels serving a large number of users. By selecting users who are experiencing the best channel conditions, the system throughput benefits significantly (multiuser diversity). The analysis of this latter gain was reported in [38]. Apart from the complexity issue, the main (and primal) bottleneck for the realization of the aforementioned capacity gains is the availability of CSI-T, (i.e., channel knowledge at the BS). The quantity and richness of the CSI-T not only constrain the set of suitable transmission techniques that can be applied, but also affect the theoretical throughput limits, reducing the achievable performance of the system.

Two different availabilities of CSI-T at the BS can be distinguished. Partial CSI-T can indicate access to some knowledge of a parameter (or parameters) related to the channel response of each specific BS-user terminal link. This knowledge may be perfect (analog, noiseless, and delay-less feedback of the parameter(s)) or imperfect (quantized, noisy, and possibly delayed feedback of the parameter(s)). The impact of imperfections of partial CSI-T on the achievable performance was reported and studied in the IST SURFACE Deliverable 4.2 (June 2007) [40], which also included a focus on a review of existing techniques able to work with perfect partial CSI-T, for which performance analyses and fundamental tradeoffs were given.

Partial CSI-T can be divided in two classes: channel quality indicators (CQIs), which describe the link quality using a non-negative real scalar; and channel direction indicators (CDirIs), which are nothing but a quantized version of the direction of the channel vector of the link. Note that other parameters exist that can reflect approximately the link performance but do not fall within this description. For example, knowledge of the second-order statistics of the channel might be considered. However, these parameters describe the link state in the long-term horizon, providing the system with statistical CSI-T, and are not relevant to quantify the gains arising from short-term adaptation to the cell conditions. For the latter, two different situations of partial CSI-T can be considered: CQI feedback only, and CQI plus CDirI feedback. In the first situation, the BS is only aware of the quality of each link, but it does not have any knowledge on the directions of the users. Therefore, in this case it cannot apply user multiplexing techniques based on spatial separation (but other channel-unaware multiplexing techniques can still be performed, e.g., V-BLAST) and, basically, the CQI information is used for user selection. In contrast, in the second situation, the BS has access to both (quantized) channel characteristics (direction and magnitude), and this fact has a significant impact on the throughput, since now it is able to reduce multiuser interference by, for instance, pointing to the users through narrow beams.

Perfect CSI-T denotes the complete nonquantized knowledge of each channel response. Although quite optimistic, this situation is worth considering for benchmarking the more practical scenarios. So far, two assumptions have been made. First, it was assumed that the user terminals are able to obtain accurate estimates that track the possible variations of either the channel response or the CQI/CDirI (i.e., inaccuracy effects). Hence, CSI-R is available. Second, it is assumed that there is a feedback link from each of the user terminals to the base station. This feature, already implemented in existing wireless systems, is further being enhanced in evolved standards. For this reason, the case of no CSI-T is not considered here, as it is out of the scope of the design of wireless systems. However, there is an intermediate setup in between the absence of CSI-T and partial/perfect CSI-T, which can be denoted as low-resolution CSI-T. It describes the situation in which partial/perfect CSI-T is available only for a portion of the entire system bandwidth. This situation arises when practical implementation constraints are considered in the design of the transmission strategies. This is of particular importance for a good transmit-and-receive architectural development.

The IST project SURFACE considered the use of OFDMA for the downlink of a reconfigurable air interface. Studies on low-complexity and easy implementation were carried out on a large number of schemes for perfect and partial CSI.

Several transmission strategies were proposed for application in the downlink [40]. Along with their description, some remarks regarding its implementation, required signaling overhead, and suitability in case of non-full CSI-T were made as well. Finally, their performance was analytically characterized in the asymptotic regime of a large number of users.

From a system point of view, the packet scheduler and radio resource allocator will rely on these transmission strategies for scheduling transmissions based on higher layer information and assigning the limited network assets (power, rate, and bandwidth) based on their expected performance. Two features are hence considered key in the selection of the transmission techniques for system-wide integration:

* *Amenability*, from the perspective of simplifying the allocation of resources;

• *Performance* (in terms of probability of error), for meaningful ranges of the parameters of interest (number of users, bandwidth, and power).

When perfect CSI-R and CSI-T are assumed, the transmission techniques are able to adapt perfectly to the current link state. Although ideal, this setup can serve as a performance upper bound for more realistic scenarios. In particular, it can account for both channel estimation errors at the receiver (imperfect CSI-R) and the presence of a low-rate feedback channel (partial CSI-T) and study their separate and joint impact in the final performance in terms of throughput and symbol error probability. Figure 2.9 and Figure 2.10 show the results for the maximum achievable throughput and the diversity realized by varying the number of antennas for different transmission schemes employed in the downlink. In this case, the comparison in performance is made for zero-forcing beamforming (ZFBF), Dithered Zero-Forcing Tomlinson-Harashima Precoding (dZF-THP), and Vector Perturbation (VP).

dZF-THP outperforms ZFBF in terms of throughput, especially at low to moderate SNRs, where the gap between them is most pronounced. In contrast, VP only outperforms ZFBF at moderate to high SNR. Since VP has a modulo operation, it shrinks the decision regions of the symbols that are farthest from the origin. Therefore, whenever the effect of the noise is dominant with respect to the distortion of the channel, that shrinking will be dominant with respect to the alignment of the transmitted signal to the best directions of the channel matrix. As $n_T = K$ increases, the ill-conditioning of the channel matrix is more prominent, and so is the increase in the throughput of VP with respect to that of ZFBF at moderate to large SNRs. In fact, VP has larger diversity than either ZFBF or dZF-THP. However, by the time

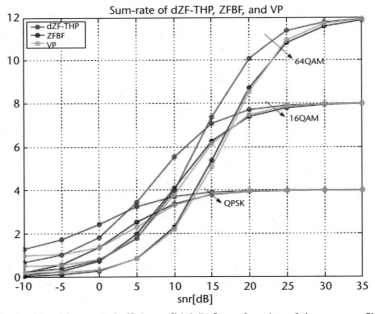

Figure 2.9   Total achievable spectral efficiency [bit/s/Hz] as a function of the per-user SNR [dB] for $n_T = K = 2$ and different modulations.

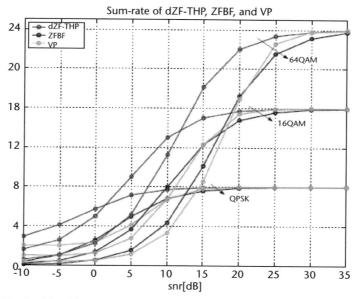

Figure 2.10 Total achievable spectral efficiency [bit/s/Hz] as a function of the per-user SNR [dB] for $n_T = K = 4$ and different modulations.

VP obtains the best results, most of the maximum achievable throughput has been already captured by dZF-THP.

Therefore, dZF-THP and ZFBF are the techniques that have a better complexity-performance tradeoff.

Figures 2.11 and 2.12 show that very little estimation error suffices to experience severe throughput degradation with any of ZFBF or dZF-THP.

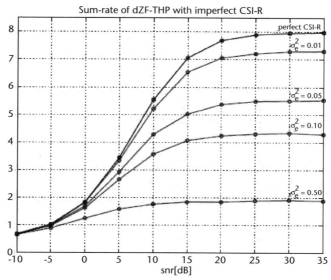

Figure 2.11 Total achievable spectral efficiency [bit/s/Hz] as a function of the per-user SNR [dB] for $n_T = K = 2$ users using dZF-THP and 16-QAM modulations for different values of the channel estimation error.

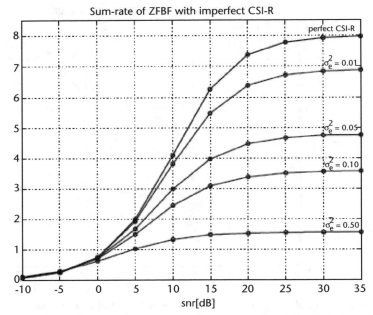

Figure 2.12   Total achievable spectral efficiency [bit/s/Hz] as a function of the per-user SNR [dB] for $n_T = K = 2$ users using ZFBF and 16-QAM modulations for different values of the channel estimation error power.

However, this degradation is not proportional to the power of the channel estimation error. In addition, although both techniques suffer from unacceptably high probabilities of symbol error [40], it is worth mentioning that dZF-THP is more robust to channel mismatches at the receiver than ZFBF.

When the receiver has perfect CSI but the feedback link to the transmitter is severely rate limited, the receiver must look for the closest channel vector in its codebook and feed back the index of the quantization vector. The quantization of the channel response due to limited feedback in the system introduces great losses in throughput (Figures 2.13 and 2.14) and probability of error (Figures 2.15 and 2.16). Moreover, the performance for a fixed number of feedback bits is strongly dependent on the codebook used for quantization.

However, the optimization of the codebook to minimize the probability of error is not a mathematically amenable problem, and, in fact, it has received little attention in the literature. The results reported here were obtained with arbitrary codebook designs.

Downlink transmission strategies capable of achieving high spectral efficiencies should be able to multiplex several users over the same bandwidth by means of spatial multiplexing and, at the same time, they should be able to provide the BS with accurate CSI. dZF-THP is a promising transmission strategy both from a complexity and performance standpoint, specially from low to moderate signal to noise ratios.

2.3.2   Uplink Scenario

One of the most important problems in networking is the allocation of limited resources among the users of the network. In traditional layered network architec-

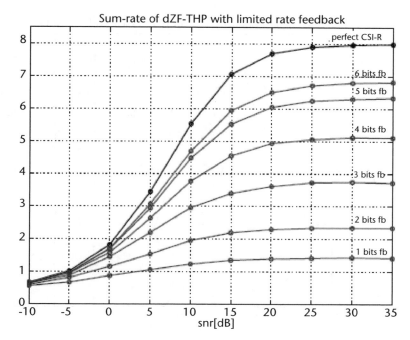

Figure 2.13 Total achievable spectral efficiency [bit/s/Hz] as a function of the per-user SNR [dB] for $n_T = K = 2$ users using dZF-THP and 16-QAM modulations for different values of feedback bits.

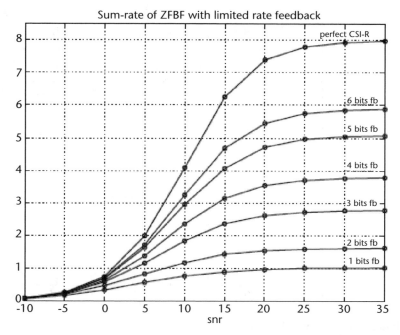

Figure 2.14 Total achievable spectral efficiency [bit/s/Hz] as a function of the per-user SNR [dB] for $n_T = K = 2$ users using ZFBF and 16-QAM modulations for different values of feedback bits.

ture, the resource to be allocated at the medium access control (MAC) and network layers is the communication links, which are considered "bit pipes" that deliver

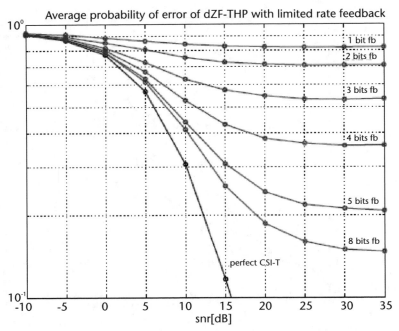

Figure 2.15 Average symbol error probability (SER) as a function of the per-user SNR [dB] for $n_T = K = 2$ users using dZF-THP and 16-QAM modulations for different values of feedback bits.

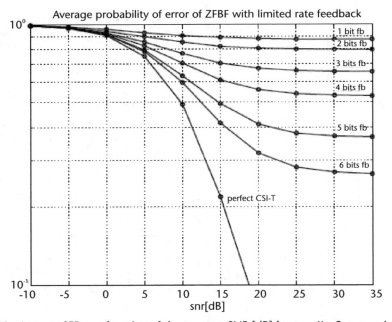

Figure 2.16 Average SER as a function of the per-user SNR [dB] for $n_T = K = 2$ users using ZFBF and 16-QAM modulations for different values of feedback bits.

data at a fixed rate with a certain error probability. This bit pipe is a simple abstraction of the underlying physical and data link layers. This abstraction has tended to

separate the treatments on the MAC layer and the PHY layer, generating two different areas of research in the field, which we shall refer to as the networking and communication groups. Research in the networking field has focused on allocating these bit pipes among different flows of randomly arriving traffic, using approaches such as packet scheduling and collision resolution. The main aim in networking is to use the bit pipes in the most efficient way possible while at the same time providing acceptable QoS in terms of delay and throughput to each user. The communication problem has focused on creating better bit pipes (i.e., improving the transmission rate or spectral efficiency for a given channel thanks to advanced detection, modulation, and coding). The random arrivals and departures of traffic are usually ignored, and delay is not considered.

Unlike the fixed wired network where the channel is time-invariant, the wireless channels are very dynamic and time varying. Hence, the conventional design approach of optimizing the physical layer and the MAC layer independently fails to exploit the dynamic nature of the physical layer and is a suboptimal in multiuser wireless channels. Therefore, despite the fact that the separation between networking and communication problems has many advantages, both practical and conceptual, there is an increasing awareness that this approach is not the proper one, in particular within the context of modern wireless data networks. This motivates the cross-layer approach, according to which physical layer and higher networking layers are jointly optimized.

In cross-layer optimization design, there are two great challenges. The MAC layer scheduling algorithm must adapt to the dynamics of the system, which in turn depend on the PHY-layer dynamics and the source statistics. In modern packet-based wireless data networks, the data traffic arrives randomly at transmitters in the form of variable-size packets. Transmitters then send the data over fading channels, generally in the presence of interference generated by other transmitters. In this scenario, the problem of resource allocation, in terms of power control and rate allocation, is of fundamental importance. Two issues are central to the resource allocation problem for multiuser communication. At the network and MAC layer, there are QoS issues regarding packet throughput and delay. At the PHY layer, issues of accurate channel modeling, detection, and coding are prominent. The cross-layer approach combines in an integrated framework all these issues.

In that respect, the trade-off between the fairness constraint and efficiency of an OFDM-MIMO wireless access has an important place. The approach consists of defining radio resource allocation (transmission power for each frequency sub-band) as an optimization problem, where fairness is obtained by introducing time-varying weights in the optimization target function. An information-theoretic point of view was adopted by the IST project SURFACE to simplify the description of the relationship between the channel model, transmission power levels, and obtained bit rates. The project established bounds by assuming perfect CSI-T. The focus was on the multiaccess (uplink) communication setting, where multiple transmitters send independent information to a single receiver, in the same time and frequency locality.

In the two-user case, the multiaccess capacity region C_{MAC} defines the set of all transmission rates r (in bits per second), at which reliable communication is possible under any coding and modulation scheme. This is shown in Figure 2.17. C_{MAC} is a pentagon shape.

An important observation is that to achieve all rates in the capacity region, joint multiuser decoding techniques must be exploited. In fact, CDMA-like strategies (in which the receiver decodes each user and treats the remaining users as noise) and simple time-sharing or scheduling strategies (whereby only one user transmits to the receiver at a time) are able to achieve only a proper subset of the rates in the capacity region.

To obtain all the feasible rates in $C_{MAC}(h, p)$, successive decoding can be used. For example, the corner point ra in Figure 2.17 is not achievable by pure time-sharing or a CDMA-like strategy, but it is achievable by first decoding user 1 (treating user 2's signal as interference in addition to background noise), and then decoding user 2 after subtracting the estimate for user 1 from the received signal. The other corner point, associated to rates r_b, can be achieved using the reverse order in the decoding of the two users. Successive decoding can easily be extended to the kth user case.

The ith data source can be modeled as generating packets according to an ergodic counting process $Ai(t)$, where $Ai(t)$ is the number of packet arrivals up to time t. The power and rate mechanism is shown in Figure 2.18.

An important hypothesis in this model is that all transmitters, as well as the receiver, have access to the global channel and queue information $H(t)$ and $U(t)$ (i.e., we assume that there exists a feedback channel). In a practical implementation, the base station collects, in every scheduling time-slot, the information regarding the queue lengths state and channels state.

The main question for the multiaccess queuing system is to determine the stability region, which is the set of all bit arrival rates, for which no queue "blows up." To understand the concept of stability, a number of defintions are available from the IST Project SURFACE [40]. The performance of some widely used resource allocation policies was also studied there. As a particular example, a multiaccess system of

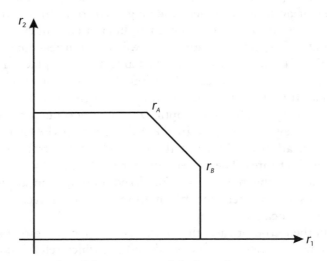

Figure 2.17 The capacity region of the two-user uplink channel.

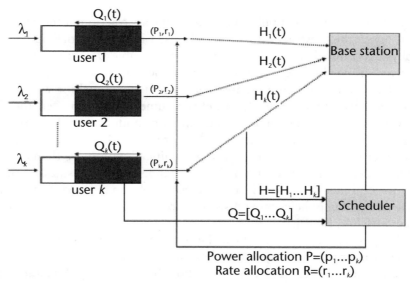

Figure 2.18   Power and rate allocation for multiaccess fading channels.

$K=2$ (two users), with each user seeing independent fading processes $Hi(t)$, was analyzed for the purpose.

The channel was modeled by an ON/OFF model, for which $|Hi(t)|2 = \{1$ good, $1/10$ bad} with probability 0.5. The fading state remains constant for T seconds and then changes to a new independent fading state. The arrival processes are independent Poisson with $\lambda_1 = \lambda_2 = \lambda$, and packets lengths are exponential with unit mean L.

Figure 2.19 shows the performance comparison of different scheduling rules applied to the scenario above; in particular, the total average queue size is plotted versus the bit arrival rate λ for each scheduling policy.

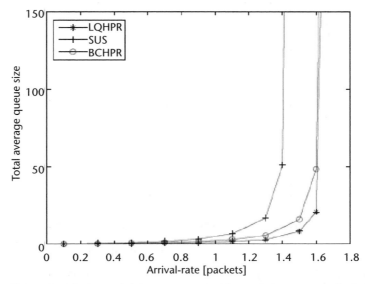

Figure 2.19   Comparison between total average queue size and average arrival rate of various scheduling strategies.

The ON/OFF channel model is a simple abstraction of the channel at the physical layer; this abstraction is generally used in the networking context, where the physical channel is modeled in the easiest possible way because the problems to be solved primarily concern higher layers in the OSI model. To give more concreteness to the example system presented here, a fading channel for which the statistics are characterized by a Rayleigh model is assumed. For each user, the channel, in any given coherence time interval, is given by a random complex circularly symmetric Gaussian variable with variance equal to 0.5. Moreover, it is assumed that the receiver (i.e., the base station) can be equipped with more than one antenna, while every user is equipped with a single antenna. The base station utilizes a maximal ratio combiner to decode the users' signals. Also in the SIMO case, one can consider the scheduling policies previously mentioned and described in detail by the IST Project SURFACE [40].

Figures 2.20 and 2.21 show the performance comparison between the cases of one, two, or four antennas at the base station, when the scheduling strategies implemented are Longest Queue Highest Possible Rate (LQHPR) and Semi-orthogonal User Selection/Single-User Scheduling (SUS), respectively.

When the number of antennas at the receiver is increased, the stability region becomes larger, as does the capacity region. In Figure 2.22, the performance comparison between the three scheduling strategies is shown, considering for each of them the cases of one and four antennas at the base station.

Stability in a queuing system implies that the queue sizes do not "blow up," but it does not say how large the queue size can grow. To minimize the average packet delay and other related QoS performance measures, it is necessary to keep the queue

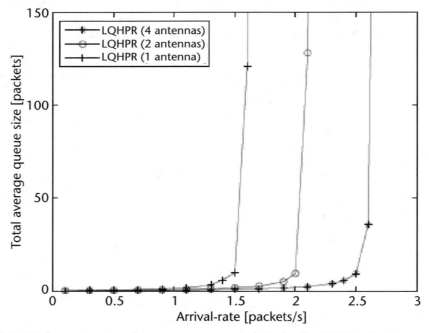

Figure 2.20 Total average queue size versus average arrival rate for the LQHPR policy in a two-user Rayleigh multiaccess channel with one, two, or four antennas at the receiver [46].

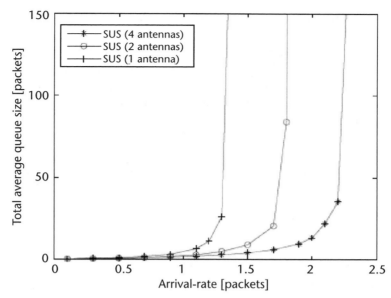

Figure 2.21   Total average queue size versus average arrival rate for the SUS policy in a two-user Rayleigh multiaccess channel with one, two, or four antennas at the receiver.

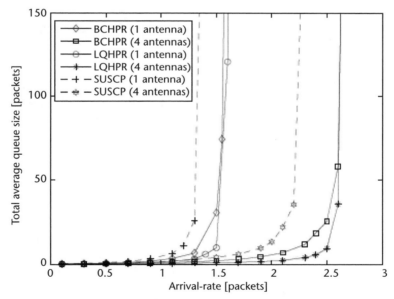

Figure 2.22   Comparison between total average queue size and average arrival rate of various scheduling strategies in the cases of one and four antennas at the receiver and Rayleigh model.

sizes as short as possible. The problem of finding joint power control and rate allocation policies able to minimize average delay in a queuing system has not yet been solved. The most important result in this research field is that in a symmetric multiaccess queuing system, the symmetric version of the Longest Weighted Queue

Highest Possible Rate (LWQHPR) rate allocation policy minimizes the average packet delay in a very strong sense [36].

Joint application of queuing theory on the data link layer and information theory on the physical layer can be used for channel aware scheduling for multiple-antenna-multiple-access channels. Further details are available from IST Project SURFACE [40]. It is possible to prove that for low SNR, only one user is allowed to transmit; solving an optimization problem of the "best" user (in other words, the user that transmits) is completely characterized. Compared to a case of a single antenna at the base station, using more antennas enables more users to transmit simultaneously at sufficiently high SNR. Thus, another question is to establish, for given queue sizes q and channel realization H, the SNR range in which only one user is allowed to transmit. The last problem to be considered is to determine whether the achievable region $S\pi(H, P)$ is convex, concave, or neither convex nor concave.

2.3.3   Exploiting the Diversity Gain

To show how diversity (time, frequency, and space) can be exploited, assume a scenario of a simplified receiver that does not use successive interference cancellation. Within each assigned frequency sub-band, the link is a single-user, flat-fading, SIMO link. The optimization approach is restricted to a single transmission format. The diversity gain provided by the SIMO system is exploited to reduce the transmission power without increasing the total transmission rate. The results are obtained with a multiantenna system and compared to those obtained with a single-antenna system.

An OFDMA multiple access system is considered, where the overall frequency bandwidth B is divided into N sub-bands. Radio resources can be assigned for an allocation time interval, called a frame. Radio resource allocation consists in assigning (part of) the N Basic Resource Units (BRUs) available in a frame to the K active links. For radio resource allocation purposes, it is assumed that each BRU can carry a fixed amount of information bits L, and that coded data frames are segmented into L-bit long chunks, referred to as "packets." The scheduler defines a list of packets from the active queues and the allocator assigns BRUs, each carrying just one single packet.

The simulation results (see Figure 2.23) consider a one-cell scenario, with $K = 20$ users randomly distributed in a cell with a radius of 400m. The traffic sources are assumed to be in saturation, i.e. each terminal always has packets to transmit. The bandwidth $B = 20$ MHz is divided into N = 8 sub-bands; the frame duration is $T_a =$ 10 ms. The channel is frequency selective, with a delay spread $\sigma_\tau = 0.1\ \mu$s, and thus a coherence bandwidth $W_c = 1/(2\sigma\tau) = 5$ MHz. A Doppler spread $f_D = 6.6$ Hz is assumed, thus resulting in a coherence time $T_c = 1/(4fD) = 38$ ms. The transmission format was chosen so that a data rate of 20 Mbps is achieved, corresponding to a BRU of 25 kbit. To evaluate the impact of multiple receive antennas in Figure 2.23, the total transmit power versus $\hat{C}_{max}$ (i.e., C_{max} normalized with respect to C_{req}) is plotted.

As expected the power decreases as $\hat{C}_{max}$ increases. With $\hat{C}_{max} = 1$, the allocator is forced to allocate all the requests selected by the scheduler, thus leading to a power-inefficient allocation; as $\hat{C}_{max}$ increases, the allocator has a higher degree of

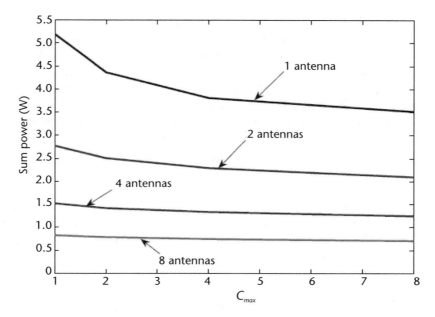

Figure 2.23   Total transmission power versus C_{max}/C_{req}.

freedom and can choose to allocate only the best users, resulting in a higher-power efficiency. As the number of receive antennas grows, the mean transmission power necessary to achieve the desired sum data rate decreases as expected (power and diversity gain). Note that, for more than one antenna, the average sum power is almost independent of the parameter C_{max}; thus, leaving more freedom to the allocator no longer results in a greater efficiency. Having a $\hat{C}_{max}$ higher than 1 (opportunistic scheduling) allows exploitation of the time diversity of the system (a user transmits only when it is experiencing a relatively good channel).

Exploiting space diversity (through multiple antennas) affects the potential for improvement brought about by opportunistic scheduling. With one antenna, a user may simply experience a bad channel in one frame, and allowing it to postpone transmission results in an efficiency improvement. With many antennas, the user will likely experience good channel conditions on some of the antennas, thus the transmission power required of that user to meet a target signal-to-interference-plus-noise ratio (SINR) will be less variable over time. This explains why adding a further degree of freedom through time scheduling results in a smaller performance improvement.

The statement of the radio resource allocation problem can be generalized by allowing more transmission formats [40].

2.3.4 Exploiting the Multiplexing Gain

This scenario assumes that the BS performs successive interference cancellation and thus the multiplexing gain provided by multiple antennas can be exploited. For a given channel realization and power allocation, the sum capacity can be achieved by successive interference cancellation (SIC), independent of the decoding order. The applied scheduling algorithm assumes first the case of the flat-fading channel, where in each frame the BS assigns transmission powers to the users by solving an optimization problem stated for the flat-fading SIMO multiple access link. This procedure is explained in detail in [40]. In the case of one receive antenna (a single-input-single-output, or SISO, channel), it is easy to show that, with the first probability, only one user is allowed to transmit in each allocation interval.

For the case of a frequency selective channel, OFDM with N sub-bands is applied, thus effectively transforming the frequency selective channel into N parallel subchannels. The optimization problem then must be solved in each allocation interval. The weights wk and the coefficients ck are computed as for the flat-fading case.

The simulation scenario proposed by the IST project SURFACE assumed a single-cell scenario. $K = 20$ users were randomly distributed in a cell with a radius of 400 m. The BS allocates the transmission power to (a subset of) the users for an allocation interval of duration $T_f = 10$ ms. These simulations were carried out under saturation traffic conditions. The Matlab Optimization Toolbox can be used to solve the convex optimization problems. Full, detailed information about the channel model is described in the IST SURFACE Deliverable 4.2, from June 2007 [40].

Figure 2.24 compares the performance compare the performance of a proportional fair (PF) scheduler with that of the maximum throughput (MT) scheduler for a flat-fading scenario. For the MT scheduler, the fairest decoding order could be determined by a greedy algorithm: on each step, the user who can achieve the highest rate is decoded, and the the users still to be decoded are treated as interferers [41]. Because the subset of users not allowed to transmit does not depend on the decoding order, for simplicity, a random fixed decoding order was selected.

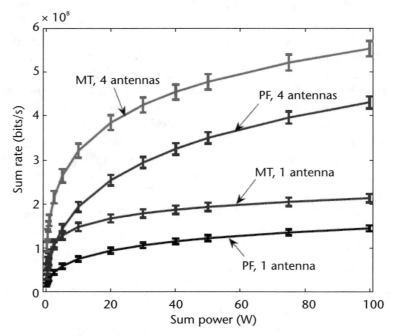

Figure 2.24   Sum of achieved transmission rates versus sum power.

The performance of the two scheduling algorithms is compared in terms of the total achieved data rate by varying the total sum power P.

The results were obtained for one and four antennas at the BS. It is clear that the use of multiple antennas brings about a substantial throughput enhancement. The PF scheduler, which aims at achieving some level of fairness among flows, suffers from a limited performance penalty when compared to the MT scheduler; thus we can see a tradeoff between fairness and sum capacity.

The results for the frequency selective scenario considered delay spread $\sigma\tau = 0.1$ μs, and thus a coherence bandwidth $W_c = 1/(2\sigma\tau) = 5$ MHz. A Doppler spread of $f_D = 6.6$ Hz was assumed, thus resulting in a coherence time of $T_c = 1/(4f_D)$ = 38 ms. The bandwidth $B = 20$ MHz was divided into N sub-bands. Figure 2.25 plots the sum data rates achieved with the two algorithms. Thanks to the frequency diversity of the channel, the achieved capacity has increased.

Figure 2.26 shows the average number of users served in an allocation interval. Figure 2.27 plots the sum data rates, obtained with a heuristic orthogonal allocator, in which only each sub-band is allocated to a single user. It is clear that there is a heavy penalty to pay for allocating only one user per channel.

Classical analysis of network resource allocation and packet scheduling tends to decouple the interactions between the physical layer and the multiple-access conflicts by using simplified models for packet transmission and reception. This usually leads to ignoring the time-varying dynamics of the wireless channel and, in essence, disregarding the soft nature of interference in wireless networks. On the other hand, the information theoretic analysis of the fundamental transmission limits of the multiple-access channels usually assumes the infinite backlog model for the user terminals. Hence, the bursty nature of practical packet arrival models is disregarded, and delay is not considered. Separate analysis and optimization of the MAC and the

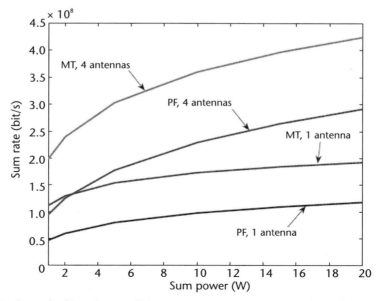

Figure 2.25   Sum of achieved transmission rates versus sum power, FS channel.

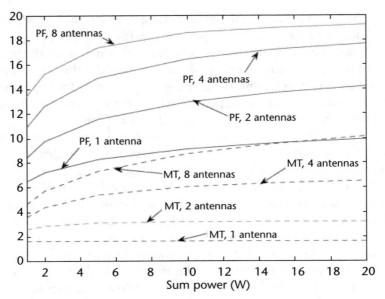

Figure 2.26 Average number of served users per allocation interval, FS channel.

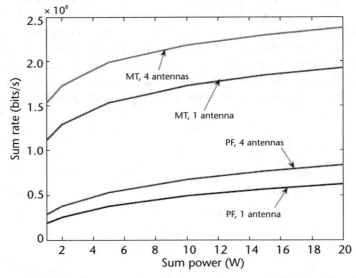

Figure 2.27 Sum of achieved transmission rates versus sum power, FS channel, orthogonal allocation.

physical layer leads to allocation policies and transmit strategies that are globally suboptimal.

A cross-layer approach in which both layers are jointly optimized can be very beneficial in view of the above. Two issues are central in a cross-layer optimization framework: network resource allocation for QoS achievement (in terms of throughput and/or delay) and optimal modulation and coding in the presence of realistic channel modeling.

Explicit resource allocation and packet scheduling algorithmic strategies that operate in a desirable point of the fairness-throughput tradeoff can be based on the proportional fair optimization criterion and can be applied to the case of single- or multirate signlling, single- or multiantenna reception at the BS, and frequency-flat or frequency-selective channels. The interplay between the use of multiple antennas and the diversity gain, as well as the effect of multiplexing gain in the final performance are also part of the strategies to enable a reconfigurable air interface.

Future high capacity communication systems will be optimized for packet data transfer only; thus, the packet scheduler functionality assumes a great significance. These systems, operating in bandwidths of 10 MHz to 100 MHz, facilitate advanced frequency-domain scheduling methods.

Similar to the time-domain (TD) only scheduling, the combined time and frequency-domain packet scheduling (FDPS) mechanisms have to be based on the knowledge of instantaneous radio conditions for the active users in the system. A promising cross-layer scheduling concept can be introduced for OFDM-MIMO transmissions: the space-frequency packet scheduling is based on opportunistic random single- or multibeamforming. These concepts can be further combined with queue-aware scheduling mechanisms, which try to optimize the match between the randomness of the traffic patterns and that of the radio environment while being subject to QoS constraints (queueing delay, etc.). Examples of such mechanisms are the LQHPR and Jointly Opportunistic Beamforming and Scheduling (JOBS).

A typical channel state indicator used for both DL and UL in OFDMA systems is the CQI, which represents, directly or indirectly, a measure of the estimated optimal transmission parameters (transport block size, modulation, and coding set) to be used in each physical resource unit. The design, measurement mechanism, and modeling of the CQI parameter has significant impact on the performance of multiuser transmit-receive (SISO/MIMO) schemes.

2.4   Practical Multistream Transmission Techniques

While OFDM is widely seen as the best technique to overcome the harsh wireless channel in downlink, several systems consider for uplink single-carrier frequency division multiplexing (SC-FDM), (e.g., UTRA LTE [42]) . Useful practical results were achieved and reported by the IST Project SURFACE.

2.4.1   Single-Carrier Versus OFDM

OFDM is a recognized multicarrier technique to combat the harsh environment caused by multipath. However, it has been shown that, when combined with frequency domain equalization (FDE), a single-carrier (SC) approach presents performance and complexity similar to OFDM, but at reduced peak-to-average power ratio (PAPR), a well-known drawback of OFDM [43]. In other words, SC transmission has less RF impairments, unlike OFDM, which requires an expensive and inefficient power amplifier in the transmitter.

For uplink transmissions, it is important is to allow for power-efficient user terminal transmissions by lowering the PAPR to maximize coverage. SC-FDM with dynamic bandwidth is therefore preferred. Slow power control, compensating for path loss and shadow fading, is sufficient, as no near-far problem is present due to the orthogonality of uplink transmissions.

2.4.2   Coexistence of SC-FDE and OFDM

The main difference between OFDM and SC-FDE is the placement of the IFFT operation at the transmitter or receiver, respectively. The coexistence in the same system of the two multiple access schemes for downlink and uplink, as shown in Figure 2.28, has several advantages. In a dual-mode system, the base station uses an OFDM transmitter and an SC receiver, thus concentrating most of the signal processing [43].

While the base station would have two IFFT and one FFT, the user equipment would have only one FFT. However, the UE transmitter consumes less power, due to the referred lower RF requirements of the SC scheme.

2.4.3   MIMO for Single Carrier

To fulfill the uplink requirements on coverage and capacity, multiantenna schemes must also be supported in uplink. In recent literature, different MIMO techniques have been evaluated for SC-FDE transmission. A variation of the Alamouti scheme is proposed for SC-FDE in Al-Dhahir [44] as an STBC scheme, designed for selective fading channels, that is applied to consecutive blocks of symbols. In the same direction, Reinhardt et al. [45] extend SC-FDE for STBC and spatial multiplexing with a linear MMSE receiver. Coon et al. [46] present a comparison between MIMO systems in OFDM and SC-FDE, concluding that SC-FDE is comparable to OFDM with respect to performance if MIMO is applied.

In several standards, multiple antenna schemes have been considered as open issues for research in the uplink channel. Although receive diversity in uplink has been more consensual, the feasibility of space-time and space-frequency transmit diversity schemes has also been studied. In the following, we picked up important research items that exemplify the opportunities posed in this field.

One of the key issues is the possibility of two users allocated the same resource block in uplink using spatial multiplex (SM) and transmit diversity (TD) simultaneously. Several standard specifications support the use of SM and TD, but do not

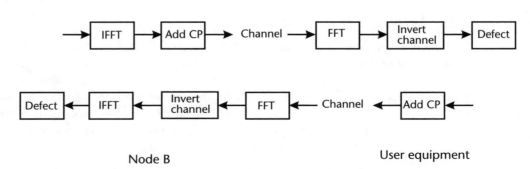

Node B   User equipment

Figure 2.28   Coexistence of OFDM (downlink, upper chain) and SC-FDE (uplink, lower chain).

allow two different users to use different MIMO modes at the same resource block, as it happens in 3GPP LTE. To cope with this issue, a proposal was presented based on a groupwise OSIC detector that uses the orthogonal property in STBC, which may increase the complexity of separating their signals [47]. On the other hand, multiuser scheduling issues need to be coped with carefully. A common approach for channel-based frequency scheduling and orthogonal multiuser MIMO scheduling was proposed in a white paper from Nortel [48]. A proportional fair scheduler based on channel sounding in time and frequency domain is assumed. In the same system, a virtual MIMO scheduler based on an orthogonal measure of users shall coexist. Two options—single-user and multiuser transmission—are considered.

Other research topics arise in this field, as the effect on PAPR of all techniques that involve a higher number of antennas in the uplink. Other key issues include control signaling (e.g., data control signaling, ACK/NACK, CQI report, resource request, etc.), channel estimation, virtual MIMO in uplink, localized or distributed subcarrier mappings, and hybrid automatic repeat request (HARQ).

2.4.4   Channel-Dependent Scheduling for SC-FDMA Uplink MIMO

Single Carrier (SC) FDMA for uplink transmission has attracted great attention due to its PAPR performance. The SC-FDMA signal can be obtained by using discrete Fourier transform (DFT) spread OFDMA, where DFT is applied to transform time domain input data symbols to frequency domains before feeding them into an OFDMA modulator.

From a user capacity point of view, MIMO is preferred due to its capacity enhancement ability. For wideband wireless transmission systems (e.g., LTE OFDMA downlink and SC FDMA uplink) to simplify scheduling, several consecutive subcarriers are usually grouped together in a resource block (RB), the basic scheduling unit. The scheduler in BS may hence assign single or multiple RBs to a MS.

Two MIMO schemes for SC-FDMA uplink transmission have been investigated for 3GPP LTE EUTRA, namely, multiuser MIMO (MU-MIMO) and single-user MIMO (SU-MIMO). For SU-MIMO, the BS schedules only one user per RB. For MU-MIMO, multiple single-antenna MSs are allowed to transmit simultaneously on the same RB.

For localized FDMA uplink MU-MIMO transmission, each RB can be further partitioned into several sub-RBs (SRBs) for the convenience of multiple user channel–dependent packet scheduling [49]. This is because for wideband SC-FDMA transmission systems (e.g., 3GPP LTE uplink), each RB usually contains a larger number of subcarriers in comparison with the number of users scheduled in each RB.

To optimally schedule users for each subcarrier is a very complicated task. For the SU-MIMO scheduling scheme, the scheduler in BS calculates the maximum achievable rate for each substream of each user and selects the one with the maximum utility function for each RB.

For the MU-MIMO scheduling scheme, the set of the SRBs should be optimized for each user within each RB.

Let $Q = L/L_s$ be the number of available SRBs of each RB. The optimization problem is thus: among KT number of users, K users are chosen and allocated to the Q available SRBs to maximize a utility function [40]. The optimal solution to the optimization problem involves high computational complexity, due to the nonlinear objective utility function with discrete constraint. Therefore, low-complexity suboptimal algorithms are needed for practical implementation.

2.4.5   SINR Distribution for SDM MIMO Schemes in DL

In a multiuser scenario, the scheduling algorithms take advantage of independent fading channel conditions observed by different users to increase the throughput of the system. This effect is regarded as multiuser diversity and has received considerable interest from research community.

Currently, two transmission schemes are supported in 3GPP for the downlink shared channel: localized transmission and distributed transmission. The basic scheduling unit in UTRA LTE is regarded as the physical resource block (PRB), which consists of a number of consecutive OFDM subcarriers reserved during a fixed number of OFDM symbols. One PRB of 12 contiguous subcarriers can be configured for either localized or distributed transmission in a subframe. If localized transmission is selected, only one user can be scheduled per PRB. If distributed transmission is selected, then multiple users can be frequency-multiplexed within the same PRB. For both transmission schemes, multiple PRBs can be allocated for transmission to a single user. However, the current working assumption is that only one transmission scheme is allowed per user and subframe on the shared downlink channel. Hence, transmission with both distributed and localized transmission in one subframe to the same user is prohibited. A more detailed description about the two transmission schemes can be found in [50].

With the localized transmission scheme, two spatial division multiplexing schemes are now under investigation: SU-MIMO and MU-MIMO schemes. They differ in terms of the freedom allowed to the scheduler in the spatial domain [51]. For the SU-MIMO scheme, only one single user can be scheduled per PRB; the data of the user can be transmitted by either single stream or two streams. For MU-MIMO scheme, multiple users can be scheduled per PRB, one user for each substream per PRB. Note that the SU-MIMO scheme is a special case of the MU-MIMO scheme (i.e., the same MS is allocated both MIMO streams on the same time-frequency resource block).

The FD scheduling algorithm used in this work is the maximum SINR scheduling algorithm. Under the assumption of equal average SINR of each user, the FD proportional fair [52] scheduling algorithm that is being investigated can be simplified to the maximum SINR scheduling algorithm.

Figure 2.29 shows the single-stream SINR distribution per PRB for SU-MIMO and for LTE local transmission using SDM MU-MIMO schemes. Here, two antennas are considered in both transmitter and receiver side. The postscheduling multiuser effective SINR distributions for the different scenarios are also shown. The number of users, that is, the user diversity order (UDO), is 20. It can be seen that the postscheduling SINR distribution per PRB for SU-MIMO scheme is worse than

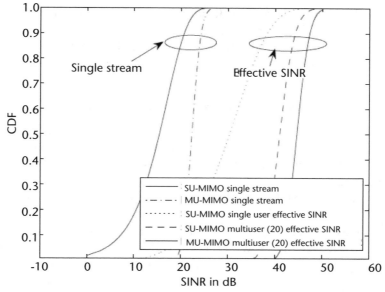

Figure 2.29   SINR distribution for SDM multiuser SU- and MU-MIMO schemes, UDO 20.

the one for MU-MIMO scheme. This is due to the fact that the multiuser diversity is exploited in SDM MU-MIMO schemes.

The average channel capacity for SDM multiuser SU- and MU-MIMO schemes with and without precoding is shown in Figure 2.30. It can be seen that the system using a linearly precoded MU-MIMO scheme has larger capacity than the one using an SU-MIMO scheme. For the SU-MIMO scheme, the precoded MIMO system always has higher average channel capacity than the one without precoding. For

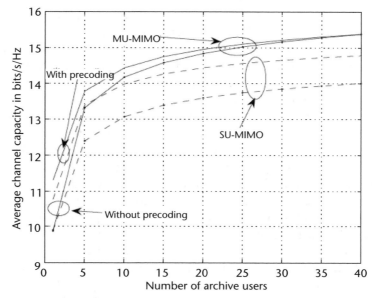

Figure 2.30   Average channel capacity for multiuser SU- and MU-MIMO schemes with and without linear precoding.

MU-MIMO, the above conclusion does not hold, particularly for systems with larger number of active users. This can be explained by the fact that the multiuser diversity gain has already been exploited by MU-MIMO schemes, additional diversity gain by using precoding does not contribute too much for this case.

2.4.6 MIMO Mode Selection for Multiuser Scheduling

The gains of MIMO communication systems can be realized by exploiting spatial multiplexing, spatial diversity, or both. However, in multiuser systems these gains may be limited by interference that can also introduce a loss in the performance of scheduling.

In multiuser SISO downlink systems, selecting the user with strongest SINR at each channel use attains maximum capacity.

The users need only to feed back the instantaneous SINR measure to achieve this goal in SISO systems. However, in MIMO systems, the performance of the scheduler is closely dependent not only on the instantaneous received SINR, but also on the channel matrix structure and on the receiver. Several issues are open when one considers scheduling algorithms coupled with multiple antenna systems, such as feedback channel quality and interaction of different sources of diversity [52]. Moreover, schemes that adapt the transmission parameters to the varying channel conditions considerably increase the spectral efficiency of the systemand have received particular attention from research recently [53].

The channel statistics of MIMO links depends on the spatial encoding technique applied. On the other hand, the success of scheduling depends on fluctuations of the channel (artificial or natural); that is, if the channel quality of all users is stable, the implementation of scheduling would not payoff.

It is therefore important to evaluate the interplay between multiuser diversity and MIMO schemes that attain spatial rate and/or spatial diversity. In this section, we consider that the transmitter does not have full CSI about the MIMO link, and open-loop MIMO schemes are employed. Partial CSI-T is used by the scheduling algorithm and/or MIMO mode selector algorithm.

Open-loop diversity schemes encode a stream of data into multiple transmitters to obtain spatial diversity without the need of CSI-T (e.g., STBC) [54]. This process reduces the fading dynamic range of the channel, producing a channel hardening effect or fading mitigation.

Although this effect provides robustness to the channel of one user, it degrades the performance of the multiuser scheduler [55]. In this way, not only are the deep fades of the link less frequent (robustness), but also the peaks of the channel are smoothed. Thus, the gains attainable by multiuser diversity are reduced. Hochwald et al. [55] show that the channel variation decreases rapidly as the number of antennas grow, hence extinguishing the gains due to scheduling.

In a MIMO open-loop system, spatial multiplexing can be used, transmitting independent streams of data over different antennas [56]. As optimal receivers have higher complexity and are difficult to implement, the focus is also on linear receivers (e.g., LMMSE). Heath et al. [57] argue that for a large number of users, linear receivers, as zero-forcing (ZF), approach the average performance of optimal receivers. This effect is due to multiuser diversity, which can compensate for poorly conditioned channel matrices.

Although spatial multiplexing offers high spectral efficiency, in multiuser environments there is a degradation of performance in scheduling all transmit antennas to a single user [57].

Thus, it results that independent multistream scheduling (i.e., letting all users compete independently for each antenna) achieves higher gains. This algorithm can be seen as a simple generalization of the SISO multiuser diversity case, multiplied by the number of antennas.

In a MIMO system, other techniques can also be considered that bridge the two previous schemes, achieving both rate and diversity gains. Examples of these are linear dispersion codes and UTOD.

In a single-user scenario, MIMO wireless communication systems can be used to exploit spatial diversity or increase throughput. At the transmitter, the selection of transmit diversity or spatial multiplexing is performed according to the system requirements. However, most of the practical systems require a tradeoff between robustness and throughput. Switching between spatial diversity and multiplexing is still unclear and subject to research. One obvious and simple approach is to evaluate the SINR and calculate the capacity achieved by diversity and multiplexing, selecting the scheme with highest value.

In a multiuser system, the MIMO selection problem becomes more complex. Not only does the choice lie between spatial rate and spatial diversity, but also the impact of multiuser diversity must be taken into account. In fact, spatial transmit diversity introduces the effect of channel hardening, thus limiting the gain of schedulers. Due to this fact, great attention has been turned to antenna selection, a source of diversity that is compatible with multiuser diversity.

However, scarce research has been devoted to the design of an integrated system that exploits the advantages of advanced ratio transmission technologies. Moon, Ko, and Lee [58] propose a framework that selects a set of transmit parameters based on user type of service and location within cell. The chosen transmit parameters incorporate various techniques such as link adaptation, opportunistic packet scheduling, channel coding, and multiantenna techniques (diversity, multiplexing, and beamforming).

Further work is necessary in this field. The IST Project SURFACE proposed a strategy for switching between different MIMO schemes according to a set of users' conditions. These conditions include the number of users and the scheduler employed, the strength of the channel, and the reliability of the feedback CSI. The potential of UTOD space frequency coding (SFC) was evaluated in a practical MIMO implementation, and results were compared with the use of other widely accepted multiantenna techniques to propose a selection of MIMO modes in an adaptive switching scheme.

The comparison of full-spatial-rate MIMO encoding schemes (SM and UTOD) with the rate-one Alamouti scheme under fair throughput is essential to understand the potential of UTOD.

Considering a MIMO 2×2 configuration (two transmitters and two receivers) as reference, a full-rate scheme encodes twice the number of bits as a rate-one scheme. Thus, for a fair throughput comparison, when UTOD and SM schemes use a two-bit modulation, QPSK, the Alamouti scheme needs to employ a four-bit modulation, 16-QAM. The following results were obtained using a fixed modulation order for each MIMO scheme. From the performance in terms of uncoded BER shown in Figure 2.31, it is evident that UTOD is superior to SFC above 10 dB of SNR. The spectral efficiency of the system is depicted in Figure 2.32. From these results, one can conclude that the performance obtained by employing UTOD with a linear receiver is superior to that obtained by SFC or SM.

However, the performance of the linear receiver with UTOD scheme suffers a degradation when increasing the order of the modulation implemented in the sys-

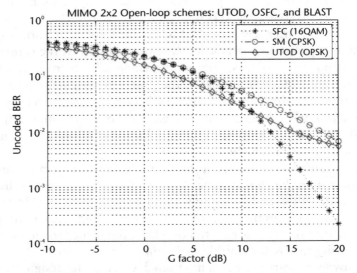

Figure 2.31 Probability of bit error of an uncoded system in MIMO 2×2 configuration. SFC uses 16QAM, while UTOD and SM use QPSK.

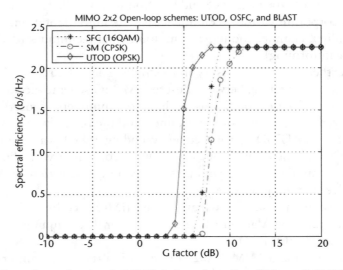

Figure 2.32 Throughput of system in MIMO 2×2 configuration. SFC uses 16QAM, while UTOD and SM use QPSK; both employ turbo coding with code rate of 2/3.

tem. Thus, the analysis of spectral efficiency and BER performance in a throughput fair comparison leads to conclusions that may differ from more practical scenarios. Hence, it is important to study the spectral efficiency when link adaptation is employed on top of the MIMO encoding schemes.

Figure 2.33 shows the spectral efficiency obtained for a MIMO 2×2 configuration. It verifies that SFC achieves better performance at lower SINR, but at higher SINR, SM clearly surpasses in terms of throughput due to the full-spatial-rate feature. One can see that UTOD achieves higher spectral efficiency in the lower range of SINR but fails at achieving reasonable performance in the mid- and higher ranges. In fact, in a system with equal number of receive and transmit antennas that employs a linear receiver, UTOD suffers a BER floor that in case of high modulation schemes compromises the throughput, since the target BLER is not achieved.

The effect of degradation in performance at high modulation orders can be controlled by increasing the number of receive antennas, thus allowing gains in the diversity freedom. The performance of a MIMO system 2×4 is shown in Figure 2.34. It shows the spectral advantage of an UTOD scheme at lower and midrange SINR; however, at high modulation orders, it still suffers losses that keep its performance below SM.

In conclusion, UTOD can be regarded as a tradeoff between SFC and SM. The linear receiver is optimized for low modulation orders, but performance can be dramatically improved by increasing the number of antennas at the receiver. In that case, MIMO mode selection can be performed between UTOD and SM to achieve the highest spectral efficiency.

For evaluation of the practical implementations, use of MIMO and scheduling in the final per-user SINR distribution is a key metric that determines the error probability and throughput in both the downlink (using a multicarrier approach) and the uplink (within a single-carrier framework).

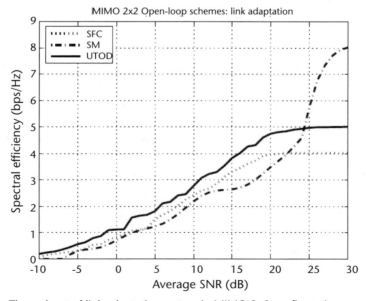

Figure 2.33   Throughput of link adaptation system in MIMO 2×2 configuration.

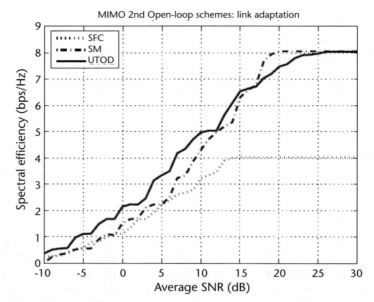

Figure 2.34 Throughput of link adaptation system in MIMO 2×4 configuration.

2.5   Air Interface Technologies for Short-Range Reconfigurability

The transition from the third to fourth generation radio systems is being driven by the demand and the deployment of WLANs and WPANs based on the IEEE 802 suite. Most of the short-range devices and networks mainly operate standalone in indoor environments, and their integration into the wireless wide-area infrastructure is difficult.

Some of the advantages of this technology have been summarized as follows [59]:

- *High data rate transmission:* It can support more than 500 Mbps within 10m.
- *Fading robustness:* It is immune to multipath fading and can resolve multipath components even in the dense environments that can be used to enhance system performance.
- *Security:* It is very difficult for users for whom it is not intended to detect the signal.
- *High precision ranging:* It has good time domain resolution.
- *Low loss penetration:* It can operate under both LOS and NLOS.
- *Low power spectral density:* It can coexist with all other systems such as GPS, WLAN, and so forth, as shown in Figure 2.35.
- *Single chip architecture:* It can be implemented fully digitally, which is especially good for handheld devices.
- *Scalability:* It is flexible and capable of dynamic trade-off of high data throughput for range.

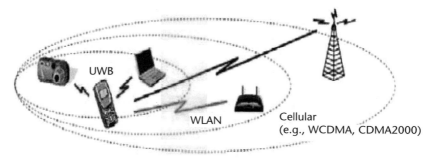

Figure 2.35   Coexistence of UWB with other systems.

The emerging UWB technology has the potential to provide solutions to problems related to spectrum management and radio system engineering. This technology is based on sharing the already-occupied spectrum by using the overlay principle. It will play a significant role in future pervasive and heterogeneous networking [60]. The spectral overlay principle is shown in Figure 2.36 [60].

A signal is said to be UWB if bandwidth is > 0.25 × carrier frequency. It is a developing technology that promises energy-efficient short-range communications (low power density, robust to narrowband interference). But the problem is that very large bandwidths on fading channels cannot be efficiently used by spreading power "uniformly" in both time and frequency, and the capacity for such schemes asymptotically approaches zero! To obtain high throughput on such channels, signals must be "peaky" in time and/or frequency—that is, localize power in time/frequency. Frequency-hopping wideband systems (e.g., FH-CDMA, multiband OFDM), if properly designed, do not have these limitations [61].

In February 2002, the FCC issued the First Report and Order, which permitted unlicensed UWB operation and commercial deployment. The various classes of UWB devices defined in the document range from imaging systems to vehicular radar systems, and include communication and measurement systems. The FCC assigned a specific spectral mask for each of the activities in the unlicensed radio spectrum, ranging from 3.1 to 10.6 GHz. The spectral masks are as shown in Figure 2.37 [60].

The FCC ruling was flexible and helped initiate a lot of interest, which led to the formation of industry alliances such as WiMedia and the UWB forum.

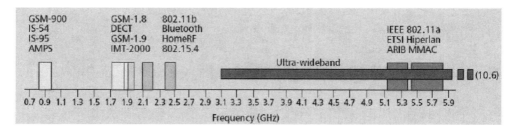

Figure 2.36   UWB spectral overlay principle.

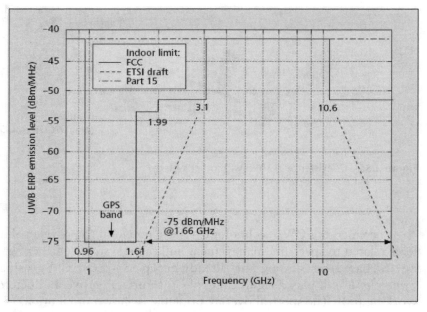

Figure 2.37 FCC First Report and Order and ETSI draft spectrum masks for transmissions by UWB communication devices in indoor situations.

Within Europe, ETSI and the European Conference of Postal and Telecommunications Administrations (CEPT) are working toward establishing a legal framework for UWB devices. Under ETSI, there are two task groups. ETSI TG31A is responsible for identifying a spectrum requirement and developing radio standards for short-range devices using UWB technologies; ETSI TG31B concentrates on developing standards and system reference documents for automotive UWB radar applications. CEPT SE24 focuses on regulatory issues and spectrum management [59].

In Japan, the Ministry of Internal Affair and Communications (MIC) issued a preliminary approval for UWB emissions policy in March 2006.

Figure 2.38 shows the UWB emission limits for the indoor communications systems and Figure 2.39 shows the UWB emission limits for outdoor communications systems in the United States, Europe, Japan, and Korea [59].

The following are some of the scenario system implementations based on UWB that would be helpful to the industry and service providers; they are also shown in Figure 2.40 [60]:

- High data rate wireless personal area network;
- Wireless ethernet interface link;
- Intelligent wireless area network;
- Outdoor peer-to-peer network;
- Sensor, positioning, and identification network.

The deployment of UWB systems has opened up many research areas, such as [59]:

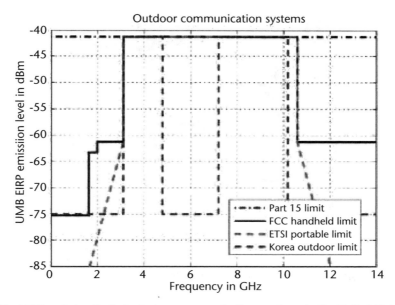

Figure 2.38 UWB emission limits for indoor communications systems in the United States, Europe, Japan, and Korea.

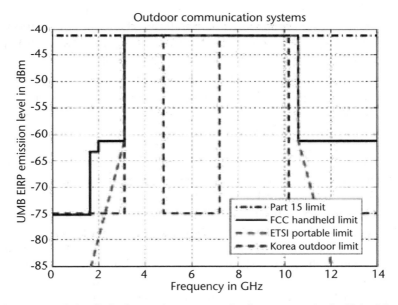

Figure 2.39 UWB emission limits for outdoor communications systems in the United States, Europe, Japan, and Korea.

- *Antenna design:* The design of the antenna should be efficient and small in size. It should be linear, to avoid any signal distortion, and at the same time be low in cost.
- *Propagation channel:* A deeper understanding of the propagation characteristics is required for more reliable detection.

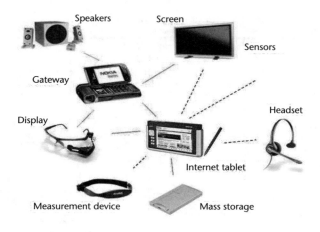

————— High Data Rate (HDR) provided by either UWB or MC-SS
- - - - - Low Data Rate (LDR) provided by UWB

Figure 2.40   Possible application scenarios for UWB technology.

- *Timing acquisition and synchronization:* Further research is required to reduce the searching space or time.
- *Coding and modulation:* More advanced coding and modulation schemes are required to design robust and reliable UWB communication systems.
- *Integrated circuit and digital signal processing:* The high bandwidth of UWB needs fast converters and high-speed sampling rates for DSP, which should also be available at low cost, small size, and less complex IC.
- *UWB networking:* More research work is required on UWB-based networks, such as MAC layer protocol, optimum routing algorithms, and resource and mobility management.

UWB and multicarrier spread spectrum (MC-SS) were two technologies upon which reconfigurability was achieved at physical layer by the IST projects MAGNET and MAGNET Beyond.

The specifics of the PAN radio environment, (i.e., user proximity, user dynamics, radio coexistence with legacy and emerging communication systems, terminal/device sizes and their use cases) affect the choice of a proper channel model and consequently the air interface configuration. Appropriate and accurate radio channel and radio interference models were investigated in the context of PNs with the objective of approximating the real-time varying PAN radio environment. The proposed MAGNET PAN radio access solutions related to the antenna, RF, BB, and MAC designs and were taken as a basis for the optimization of the air interfaces for typical PAN scenarios to ensure a favorable trade-off between user satisfaction (QoS) and system complexity.

MAGNET Beyond proposed air interfaces for high data rate (HDR) and low data rate (LDR) applications. The HDR applications are enabled by a MC-SS air interface solution. The only other available solution with similar capabilities at the time of writing of this book was WiMedia, a radio platform standard for high-speed UWB wireless connectivity. For LDR applications, a low-power, low-complexity frequency modulation–based UWB (FM-UWB) air-interface solution was proposed, compatible with standards such as Bluetooth, ZigBee, and WiBree. The MAC of these two is based on the IEEE 802.15.3 and IEEE 802.15.4 standards. The FM-UWB approach was adopted after being studied and compared with other solutions like ZigBee and Bluetooth. Accordingly, the MC-SS scheme was compared to the orthogonal frequency-division multiplexing (OFDM)–based UWB PHY scheme in a WiMedia system. Results were reported in detail [62] and show that the developed air interfaces fulfill the requirements for next-generation technologies.

Benchmarking at the physical layer was the approach used to compare the performance of two modulation schemes used in Bluetooth and FM-UWB. Under several operating scenarios, such as in Additive White Gaussian Noise channel (AWGN), in fading channel, and with and without interference, the robustness and drawbacks of the two systems were identified. The performance of these two systems in terms of the BER was evaluated at different Signal to Noise Ratio (SNR), Signal to Interference Ratio (SIR) level. The proposed benchmarking framework for FM-UWB and Bluetooth systems are shown in Figure 2.41 [63].

To evaluate the robustness of the FM-UWB system to interference, an MB-OFDM signal from the WiMedia system was used as interferer. The block diagram of the MB-OFDM UWB interference scenario is shown in Figure 2.42.

The interferers refer to the MB-OFDM signals that were summed to the signal of interest at the input of the receiver. Each interfering signal passes through an UWB multipath channel and therefore is added to the signal of interest (FM-UWB signal). The data rate of the simulated FM-UWB system was 100 Kbps.

Both signal and interference went through the UWB BAN channel. Fifty different channel realizations were generated for signals, and fifty unique channel realizations were also created for interferences. An average performance was obtained over all of the simulation results.

Complete details of this research are available in the books of the series entitled *Ad Hoc Networks* [64].

2.6   Adaptive Antenna Systems and Use of Localization

Adaptive antenna systems are also popularly known as smart antennas. Smart antennas can be described as an array of antenna elements that can dynamically change the radiation pattern to prevent noise, interference, and multipath fading.

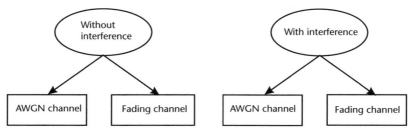

Figure 2.41   Benchmarking framework for Bluetooth and FM-UWB systems at the physical layer, proposed by the project MAGNET Beyond [63].

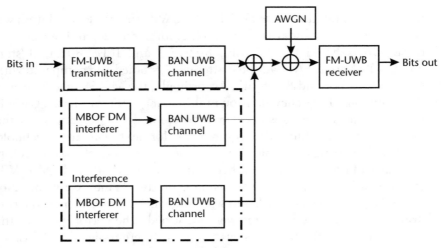

Figure 2.42 Block diagram of the MB-OFDM UWB interference scenario.

The idea is to reduce noise, increase the signal-to-noise ratio, and enhance the capacity of the system. In an adaptive type of beamforming, the gain and phase on each antenna element is adjusted until the desired pattern is obtained.

Reconfigurable antennas have an important role in communications, electronic surveillance, and countermeasures because they have the ability to adapt their properties to achieve selectivity in frequency, bandwidth, polarization, and gain. The attractive feature is that the user can roam any existing network using a single handset and has access to a variety of services, which is possible due to the switching property [65].

Further to the investigations carried out within the IST project SURFACE and presented in sections 2.3 and 2.4 of this chapter, the FP6 EU project RESOLUTION developed a wireless local positioning system with centimeter accuracy and real-time ability based on frequency-modulated continuous wave radar and common WLAN.

Figure 2.43 shows the comparison of the goals of project RESOLUTION with the other state-of-the-art approaches [6].

In a localization system, which is based on the time-of-arrival method, it is important to consider the existence of strong multipath components with very small delay in respect to the direct component. Such received components can severely degrade the accuracy of position determination. The main part of the work performed by the IST project RESOLUTION was related to the analysis of the existence of such components in the channel impulse response, considering their statistics of amplitudes, delays, and directions of arrival (DoAs).

Further, the project realized measurements in the frequency domain with 1 GHz bandwidth, which enables distinguishing multipath components with a delay difference of about 1 ns. The inverse Fourier transform was used to obtain the channel impulse response. To show the influence of the system bandwidth, measurements were realized with 140 MHz bandwidth.

The project developed a wireless local positioning system with centimeter accuracy and real-time ability, based on frequency-modulated continuous wave radar.

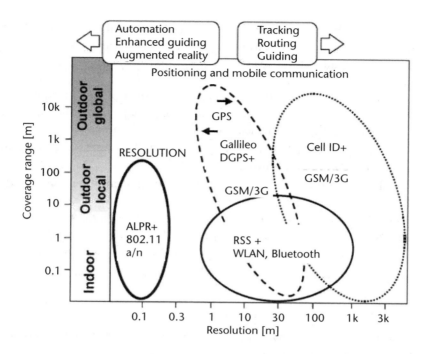

Figure 2.43 A comparison of the goals of project RESOLUTION with the other state-of-the-art approaches.

The baseband card was designed with the issue of scalability in mind. It could interface to a maximum of six antennas to minimize multipath phenomena, and the user could select one of seven possible FPGAs, depending on the application scenario and required costs. The card can also directly interface to the personal digital assistant (PDA) through the compact flash slot or to a laptop through an inexpensive CF-to-PC card adaptor, as shown in Figure 2.44 [22].

The possible integration of a microSD card slot that can accept either memory cards or WLAN radio modules in a microSD format makes the card very expandable and future-proof. When more WLAN standards become available, it would be possible to integrate the card with the positioning radar.

For the RESOLUTION positioning system, all efforts were made to design a custom baseband hardware platform with enough power to accomplish all the rele-

Figure 2.44 Compact flash-to-PC card adaptor.

vant tasks. The built system is flexible and powerful enough to implement both the mobile station and base station functions in a small space-saving size. The following important points should be noted:

- Trade-off had to be made between FPGA footprint size and flexibility, giving more importance to the latter. The FF668 package that was used gave access to seven different devices from 1 to 5 million gates.
- To minimize size and to simplify the PCB routing, an eight-channel A/D converter with serial outputs was used.
- Interfacing with transmitting antennas or analog sensors was done by using a dual-channel D/A converter.
- Extra memory was provided by using 2 Mbyte by 9 bits RAM.
- A microSD connector integrated in the design provides access to either inexpensive external nonvolatile memory cards or 802.11x WLAN modules implemented in the microSD format for hosts lacking wireless access capability.

The integration of all these subsystems in a relatively small size, combined with the ability to interface with the compact flash slot of a PDA directly or with a laptop using an inexpensive adaptor, accomplished a big challenge.

Table 2.1 summarizes the main application categories, emphasizing the reconfigurability of the design [66].

The results of the project are available in detail in the IST project RESOLUTION Deliverable D2 [66] and can be used as a base for different measurement campaigns and the design of indoor mobile communication systems.

2.7 Reconfigurable IA/MIMO Transceiver Algorithms

The IST Project MEMBRANE [67] established a framework for reconfigurable multiple-antenna studies. In the context of this work, the physical layer parameters for some channel models for various scenarios were proposed.

One of the main concerns of MEMBRANE was to enhance the presence of broadband in rural and remote zones, thus helping combat the "digital divide"

Table 2.1 Categorized Applications for Reconfigurable Design

Application	Short Description	Potential Customers	Location Accuracy/Range	Data Volume	Equipment	Building Type	Number of Concurrent Users
AGVs	Control of automated vehicles	Logistics companies	5cm/<500m	medium	Customer-specific	Factory halls	<5
Interactive guiding application	Real-time active mapping and guiding for advanced sightseeing	Museums, zoos, etc.	300cm/<50m	large	Specialized hand-held devices	Different room sizes	30–50

between these areas and the big urban centers, caused mainly by the inadequacy (or even absence) of backhaul infrastructure and the requirement for fast and cost-efficient deployment. Moreover, as the needs of users in the big urban centers become increasingly demanding, as a result of the large concentration of population around the major cities and the proliferation of a number of new data services, MEMBRANE addressed the problem of enhancing service provision in terms of coverage, capacity, and QoS in densely populated metropolitan areas. Figure 2.45 shows the scenarios adopted by the project.

One fundamental characteristic of the MEMBRANE system was that it was designed to backhaul the traffic of macrocellular, microcellular, and picocellular area access systems. Picocellular (hotspot in WiFi terminology) access devices (WiFi access points, UMTS femtonodes, or a similar WiMAX potential device) are often installed by the client. On the other hand, wider area coverage base stations (BSs), (WiMAX BSs or UMTS Node Bs) are carefully planned and installed by the operators.

In the picocellular case, an access device that incorporates MEMBRANE AN functionality is most likely to be mounted at below-the-rooftop heights—for example, on the side wall of a residence or a multiple-story building. Alternatively, the AN might be found on the rooftop of a building providing access to the backhaul for the in-building users through wired or wireless connections.

For wide area access, a BS with AN functionality is typically located on high towers or on the rooftops of buildings. However, as the capacity demands increase, the BS is moved closer to the end user; it could be placed at below-the-rooftop heights.

Furthermore, to reduce deployment costs, MEMBRANE envisions that the INs interconnecting the ANs with the EN are mostly placed at moderate heights (e.g., on

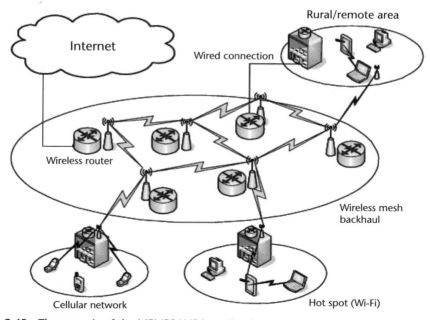

Figure 2.45   The scenario of the MEMBRANE investigations.

the order of 3–10m), mounted on electrical poles, telephone poles, or traffic lights. Still, it is possible that an IN could be placed at rooftop height. Specifically, the case of high-placed AN logical units that incorporate IN functionality is particularly attractive.

Overall, the aforementioned node deployment strategy is substantially different from what is seen in traditional backhauling solutions, in which nodes are exclusively mounted on high towers or at the rooftops of tall buildings and a clear line-of-sight (LOS) is required for reliable communication.

The implications of this type of deployment in channel modeling should be studied carefully. As a starting point, Table 2.2 classifies the various propagation scenarios that were met in the MEMBRANE urban target scenario. Prioritized cases are marked in bold. Reference to the nodes being involved in a link is not made since in general, any MEMBRANE node (EN/IN/AN) can potentially be placed either above or below the rooftop level (although an EN is typically mounted on a tower or at the rooftop of a high building).

It is noted that the listed propagation scenarios could be separated in two subcases, depending on whether the building density is classified as urban or suburban. However, due to the scarcity of channel measurements for this kind of deployment, it was chosen to simplify the studies by focusing on the radio environment of a typical city center. This case seems to be the most challenging one in terms of the capacity requirements of the backhaul network.

In the rural case, a large number of remote small population centers can be connected to a mesh backhaul network. Each center might be surrounded by large agricultural areas. Population centers are quite distant from each other. Land is assumed to be relatively flat. Various vegetation patterns can be met. Essentially, the EN and INs are placed on high towers. In practice, the tower heights are selected so that the first Fresnel zone is clear and LOS conditions are met for the required link distance. Furthermore, the actual delay spread and angle spread due to propagation environment are effectively reduced due to the use of directional antennas with high gain. With regards to the AN, two cases might be distinguished. Firstly, the AN could be placed on a high tower. This is more likely when a wide area access technology is used, such as UMTS or WiMAX. In that case, the IN-AN and EN-AN links have similar properties as the EN-IN and IN-IN links.

Secondly, the AN might be placed at some moderate height, for instance on an external wall or on the rooftop of a house. This is more likely where an AN incorporating a WiFi interface is installed by the customer without special care. In that case, LOS conditions are not guaranteed. Table 2.3 summarizes the aforementioned. Note that in scenario R2, the exact position of the AN (i.e., whether it is below the rooftop or at the rooftop of the residence/building) is not considered, since the building infrastructure is sparse and the major scatterers are trees.

Table 2.2 Propagation Scenarios in the Urban Case

#	Definition	LOS Conditions	Maximum Separation
U1	Rooftop to rooftop	LOS	8 km
U2	Below rooftop to below rooftop	LOS/NLOS	500 m
U3	Rooftop to below rooftop	LOS/NLOS	1 km

Table 2.3   Propagation Scenarios for the Rural Case

#	Definition	Nodes A-B	A Height	B Height	LOS Conditions	Maximum Separation
R1	Planned infrastructure	All cases	High	High	LOS	30 km
R2	Non-planned links	EN-AN, IN-AN	High	Moderate	LOS/NLOS	10 km/5 km

Spatiotemporal or MIMO channel modeling adds to the challenge of realistically representing the propagation environment by requiring the accurately incorporation of the spatial (angular) dimension. Channel models of this type can be classified broadly in two distinct categories:

- *Correlation-based channel models* are based on the statistical characterization of the matrix transfer function representation of the MIMO channel. The (*i,j*) entry in that matrix gives the SISO channel response from the *i*th transmit to the *j*th receive antenna element. As usual, this response can be represented by means of a tapped delay line that models the delay (frequency) domain characteristics of the corresponding SISO channel. To model the time (Doppler) domain properties of the channel, multiple independent realizations of each tap (corresponding to different time instances) are passed through a filter, with a frequency response that accounts for the shape of the Doppler spectrum. The aforementioned procedure is repeated for all the entries of the matrix transfer function. The spatial domain is modeled by applying the so-called correlation matrix, which models the correlation of the signal envelopes arriving at different elements of the receiver antenna array. This method relies on the statistical characterization of the delay, time, and spatial domains that are deduced by evaluating measurement data. For this reason, the term *stochastic* is frequently used in the literature for this type of model. However, this term is rather confusing and is avoided here because statistical means can be used in both categories.

- *Directional channel models* are based on a detailed reproduction of the channel ray structure. Each ray is characterized in terms of its complex amplitude, delay or time of arrival (ToA), direction of departure (DoD) and direction of arrival (DoA). To determine these parameters, various approaches exist. A deterministic approach can be followed when detailed environmental data is available (e.g., building geometry and construction materials, ground roughness and materials, etc.). Since the propagation environment is known to a sufficient degree, wireless propagation is a deterministic process that allows full description of its characteristics at any given point in space. Furthermore, a geometry-based stochastic approach has been traditionally used, according to which the probability density function (pdf) of the geometrical location of the scatterers is prescribed. For each channel realization, the scatterers' locations are taken at random from this pdf and then the ray delays, DoAs, DoDs, and complex fadings can be evaluated. The determination of pdf relies on empirical assumptions regarding the scattering environment at the vicinity of the transmitter and the receiver. Last but not least, a stochastic approach can be followed, using measurement data to characterize the statistical behavior of the required parameters. As more data becomes available, this approach is gaining ground. The terms *ray-based* and *physical* models are also used in the literature to name directional channel models. Moreover, the term *double-directional* is often used to stress that antenna arrays are used both at the transmitter and the receiver.

The most representative example of an analytical channel model is the well-known Stanford University Interim (SUI) model. This model has been adopted in the past by the 802.16 standardization body [68]. Analytic channel models for the indoor were used in 802.11. On the other hand, 3GPP/3GGP2 has adopted the directional channel approach to develop the spatial channel model (SCM) based on measurements (3GPP SCM). This model has also been adopted by IEEE 802.20.

Both methods possess advantages and disadvantages. In essence, correlation models employ a more abstract view of the channel. As a result, they are simpler to design and less computationally intensive, at the expense of some loss in accuracy. One practical advantage is that with this method, one can directly set the values of the correlation matrix. For these reasons, this method is very attractive for fast algorithm testing under several environments of various degrees of spatial selectivity and coherence. However, the lack of explicit parameterization of the angular dimension makes this method unsuitable for certain techniques, such as beamforming. Last but not least, apart form the propagation environment, the correlation values depend also on the array elements' separation. In practical terms, this means that specific correlation values observed in measurements are associated with the array geometry that was used in sounding.

The main advantage of directional channel models is that they are tailored to a certain propagation environment, and they can in principle accurately describe a certain propagation scenario at the expense of often prohibitively high numerical complexity. Moreover, they are independent of the array geometry, in the sense that the same realization of the ray parameters can be used to generate the channel coefficients of any array geometry. This is very useful when comparing the performance of different geometries under a specific environment.

The main disadvantage is that these models are scenario-dependent and have therefore little use in modeling other, even similar, propagation environments. Moreover, it is often difficult, if not impossible, to extract the statistics of the parameters from the measurements.

In MEMBRANE, one of the most crucial objectives was the interference management. To this end, multiuser techniques such as beamforming play a vital role. Based on the above, directional channel modeling was chosen as the basic framework for channel modeling. Furthermore, this decision was encouraged by the fact that the main source of information regarding spatiotemporal channel modeling for relay-based deployments is the IST WINNER project [69], which had adopted the same modeling approach.

Last but not least, when developing a channel model a typical distinction between link-level and system-level models has been made, in the sense that the former is focused on reflecting the parameters affecting a single link performance, whereas the latter is further concerned with the representation of parameters associated with the multiple-link, multiple-access point performance evaluation. As pointed out in the IST WINNER project [69], although this distinction is mostly convenient in that it simplifies modeling and decouples certain parameters, it may lead to inaccurate conclusions, as certain parameters' impacts at the link level do not necessarily translate to similar behavior at the system level, where intersector parameter correlations need to be taken into account.

In the IST WINNER project Web site [69], one can find a literature review of channel models that are suitable for the MEMBRANE propagation scenarios. Based on that, the project derived some useful channel models. In the same document channel, models that concern some other scenarios are reported as well. Although these scenarios have some differences from the MEMBRANE propagation scenarios, they can be used for initial simulations after some essential modifications.

As an example, the channel characteristics for the rooftop-to-rooftop channel model are shown in Table 2.4.

It is explicitly stressed that this selection was based on a literature review. However, this model can provide representative results and could be used as a starting point in simulations.

Based on the above results, the multipath component parameters are derived in Table 2.5. In total, 10 clusters are defined, from which the first is the direct one. Nondirect clusters are composed of 10 rays, while the direct cluster contains an additional ray. The total power is normalized to 0 dB. The ray offset angles are the same for all clusters; they are given in Table 2.6. These values have been chosen in such a way so as to approach a Laplacian intracluster power angle spectrum (PAS) with the desired intracluster angle spread.

Interference is a critical issue in the context of MEMBRANE, since it affects the maximum attainable performance in terms of BER and the achievable data rate of a certain MIMO link. Furthermore, it is a reconfiguration parameter, according to which robust transceivers are to be designed. In a MEMBRANE scenario, the number and propagation characteristics of strong interferers may vary, as a result of fading caused by moving scatterers and the scheduling and routing decision strategies deployed. Moreover, adequate interference modeling is critical, as scheduling, routing, and power control decisions may rely on interference characteristics.

Interference modeling is mainly a function of the propagation characteristics, the MIMO channel modeling, and the system deployment assumptions and dynam-

Table 2.4   Rooftop-to-Rooftop Channel Characteristics

Property	Value
Power delay profile	Exponential (nondirect paths)
Delay spread	40 μs
K-factor	10 dB
XPR	30 dB
Doppler	A peak centered around zero with most energy within 0.1 Hz
Angle spread of nondirect components	Gaussian distributed clusters with 0.5 degrees intra-cluster angel spread. Composite angle spread is 2 degrees. Same in both ends.

Table 2.5 CDL Model for the Rooftop-to-Rooftop LOS Case

Cluster #	Delay [ns]	Total power [dB]	Mean AoD(°)	Mean AoA(°)	Scatterer Doppler Frequency [mHz]	Number of Rays		Ray Power [dB]
1	0	-0.39	0.0	0.0	41.6	1 + 10	-0.42	-32.2
2	10	-20.6	0.9	0.2	-21.5	10		-30.6
3	20	-26.8	0.3	1.5	-65.2			-36.8
4	50	-24.2	-0.3	2.0	76.2			-34.2
5	90	-15.3	3.9	0.0	10.5			-25.3
6	95	-20.5	-0.8	3.6	-20.2			-30.5
7	100	-28.0	4.2	-0.7	1.3			-38.0
8	180	-18.8	-1.0	4.0	2.2			-28.8

Table 2.6 Ray Offset Angles

Ray Number	Basis Vector Offset Angles (BO)	Ray Offset Angles (OA)
1.2	± 0.0742	OA = intra-cluster angle
3.4	± 0.2532	(azimuth) spread x BO
5.6	± 0.4986	
7.8	± 0.8913	
9.10	± 1.9718	

ics. To capture the dynamics of both the propagation and system deployment aspects in an accurate interference model, computational complexity usually becomes a prohibitive factor. Interference modeling for the link-level studies is often simplified to overcome the computational complexity issues.

The simplest model for interference is the additive white Gaussian noise (AWGN). This is sufficient to assess transceiver performance as a function of the desirable channel characteristics and slow fading interference.

To reflect the impact of channel fading and dynamics, implementation of the interfering channels using the same model as the desirable channel is required. A common simplification in this case is to introduce at the link level only a small number of interferers—as the stronger ones—and abstract the remaining interferers by their "equivalent" AWGN component. This is equivalent to taking into account a number of interference "rings" around the cell of interest in a typical cellular setup.

In MEMBRANE, perfect synchronization was assumed for different nodes in the network, and some efficient algorithms can be developed to combat the interference.

Apart from the spatial model for interference, it is also important to model interference in time in the case when interfering and desired signal are asynchronous. The interference appears in an instant randomly chosen between the start of desired user transmission and its end. If the structure of the interference is not relevant for the algorithm performance, it can be simulated as spatially-colored temporally-white Gaussian noise. If the structure of interference is relevant (e.g., active interference cancellation techniques), then explicit representation of the interfering signal is necessary. Figure 2.46 shows the block diagram of an abstraction representation of the physical layer.

Striking the right balance on the trade-off between complexity and accuracy is the main challenge of system-level evaluation in a broadband, multiantenna deployment with a large number of resource elements to be managed (e.g., subcarriers, spatial channels) and the implementation of fast-scheduling, fast-link adaptation, in the form of fast power control and adaptive modulation and coding or other advanced schemes such as hybrid automatic repeat request (HARQ), to be performed.

Traditionally, the performance of the radio links has been evaluated in terms of the packet error rate (PER) as a function of SINR, averaged over all channel realizations of one specific channel model. PER versus average SINR performance has therefore been used as the interface between the link- and system-level simulators. This solution may be adequate, as long as every transmitted packet encounters similar channel statistics, which implies very large packet sizes/coding blocks with respect to the channel coherence time. But in the case of datacentric radio networks, this condition is generally not fulfilled. Consequently, assessing the performance of fast resource scheduling and fast link adaptation in system level simulations requires a more accurate link quality model that accounts for instantaneous channel conditions.

Since different bits can be transmitted on different spatial layers in different OFDM symbols and on different OFDM subcarriers, they may have different quality. Thus, we have a multistate channel. The total number of resource elements in frequency, time, and space is typically by far too large for direct use of associated quality measures (such as SINR) in mapping to PER. Therefore, the set of quality measures needs to be compressed to a much smaller set of typically only one or two scalar indicators. The selection of a suitable compression function (i.e., a link-to-system interface metric or physical layer abstraction model) is crucial for the accuracy of the system-level evaluation.

MIMO capacity for Gaussian signaling was shown to provide an adequate link-to-system interface metric. Under the assumption of AWGN and unstructured interference, the metric provides a good interface. However, in the presence of structured interference, and for small number of strong interferers, the metric cannot provide a suitable interface, due to considerable deviation from the Gaussian

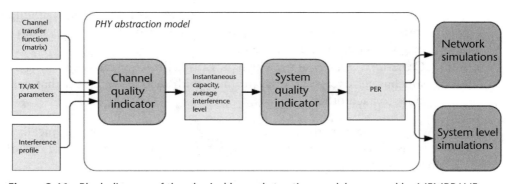

Figure 2.46   Block diagram of the physical layer abstraction model proposed by MEMBRANE.

assumption made for noise plus interference. For large numbers (greater than two) of interferers, the Gaussian assumption becomes more and more accurate, and the channel capacity metric can provide a suitable interface. Note that the channel capacity considered here inherently provides a sum across spatial modes, rendering the assessment of different receiver options impossible. Furthermore, a capacity metric will most likely work well for systems that exploit a large amount of the capacity (e.g., for open loop capacity and adaptive matrix modulation or closed-loop MIMO capacity and SVD-based MIMO), but might be less appropriate if an adaptation of the transmission mode is not applied.

To obtain a quality model of reduced complexity and training effort, a one-dimensional mapping between a so-called effective SINR and the PER can be considered. The compression function is of the form:

$$SINR_{eff} = \alpha_1 I^{-1}\left(\frac{1}{P}\sum_{p=1}^{P} I\left(\frac{1}{\alpha_2} SINR_p\right)\right)$$

where α_1 and α_2 are model parameters, the SINReff corresponds to the parameters resulting from the compression, and SINR p are the set of parameters to be compressed.

Four potential candidates for the selection of function $I()$ can be considered:

- Capacity effective SINR metric (CESM);
- Exponential effective SINR metric (EESM);
- Mutual information effective SINR metric (MIESM);
- Logarithmic effective SINR metric (LESM).

The complexity of the proposed model may be challenging, depending on the space-time processing and the propagation and system deployment; additionally, decimation in time and frequency need to be considered. Note that in project MEMBRANE, the instantaneous capacity–based compression function was considered the preferable approach.

When identifying interference modeling for the link- and system-level studies in MEMBRANE, interference in the backhaul multihop network was classified in two types:

- *Self-interference,* which prevents a single access point from simultaneously transmitting and receiving, or from simultaneously receiving from multiple neighbors on the same subchannel;
- *Cross-link interference,* which is caused by transmissions using the same subchannel over two separate links with distinct receivers that are located close to each other.

Access traffic from the users to their respective access points is typically assumed to be transmitted in a separate frequency band and does not interfere with the wireless backhaul traffic.

Separate links in the backhaul may interfere with each other, depending on the locations of their receiving end points and the antenna technology used for transmission and reception. Pairs of directed links that cannot be simultaneously active in the backhaul network can be precomputed.

Therefore, in principle, it can be assumed that interfering link pairs are specified as input. SDMA in MEMBRANE allowed nodes to have simultaneous reception or transmission of multiple packets. A spatial scheduling scheme complemented the SDMA transceiver based on a measure of separability between nodes, which is a function of parameters such as angle, correlation, and SINR.

One way to simplify interference modeling is to assume the interference comes from the strongest interferers located in the first adjacent ring that talk simultaneously.

The reconfigurable multiple antenna performance evaluations performed by MEMBRANE assessed the gains over conventional approaches that were considered the baseline. Based on that, the system-level performance was assessed, taking into account a realistic wireless backhaul node topology, traffic patterns, and delay constraints.

In a MIMO antenna system, such as the one shown in Figure 2.47, the data block to be transmitted is encoded and modulated to symbols of a complex constellation. Each symbol is then mapped to one of the transmit antennas (spatial multiplexing) after space-time weighting of the antenna elements.

After transmission through the wireless channel, demultiplexing, weighting, demodulation, and decoding are performed at the receiver to recover the transmitted data.

A large number of transmission schemes over MISO or MIMO channels have been proposed in the literature, designed to maximize spectral efficiency and link quality through the maximization of diversity, data rate, and SINR. Each of these schemes relies on a certain amount of CSI available at the transmitter and/or at the receiver side. CSI at the transmitter can be made available through feedback or can be obtained based on estimation of the receive channel. CSI at the receiver can be obtained using training-based or blind techniques, which exploit other properties of the received signal, such as constant envelope and the finite alphabet.

Novel reconfigurable designs can enhance the performance of the basic spatial processing techniques through the exploitation of the following:

- Channel parameters (e.g., antenna correlation, Demmel condition);
- Channel state information reliability (depending on feedback signaling and CSI);
- Network topology, resulting in varying interference both in terms of number of interferers and interfering channels.

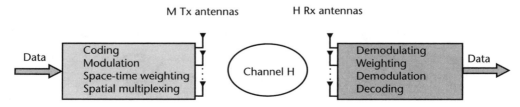

Figure 2.47   Multiple antenna transceiver architecture.

The basic deployment of intelligent antennas in a multihop backhaul network would consist of a set of predefined beams at each node. The beam to use is selected to maximize the signal toward a certain direction/node. The beams are assumed to have a narrow main lobe and small side lobes, as shown in Figure 2.48.

Adaptive beam systems consist of several antenna elements (arrays) whose signals are processed adaptively by a combining network. The signals received/transmitted from different antenna elements are multiplied with complex weights and then summed to create a steerable radiation pattern. The main lobe of the array pattern is steered towards the direction of desirable node, and nulls can be placed in the direction of the interfering nodes. Adaptive beam systems have the ability to change the antenna array pattern dynamically to adjust to noise, interference, and multipath, and to provide substantial gains (of the order of $10\log(M)$ dB, where M is number of array elements) as compared to the omnidirectional antenna system. Figure 2.49 shows an adaptive beam system.

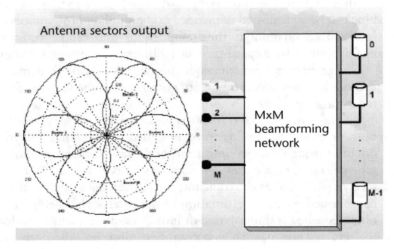

Figure 2.48   Switched beam system.

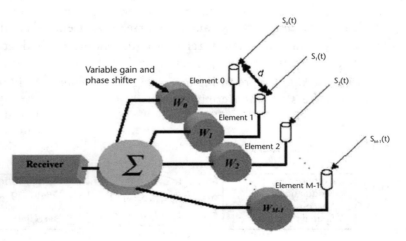

Figure 2.49   An adaptive beam system.

Switched beam systems may not offer the degree of performance improvement offered by adaptive systems, but they are often much less complex and are easier to integrate into existing wireless technologies. Figure 2.50 shows an adaptive versus a switched beam system.

Multihop networks with omnidirectional antennas use a RTS/CTS (request to send/clear to send) floor reservation scheme that wastes a large portion of the network capacity by reserving the wireless media over a large area. As a consequence, a number of nodes in the neighborhood of the transmitter and the receiver need to remain silent, waiting for the communication of the currently active nodes to be completed. The use of directional antennas (with fixed or adaptive beams) helps alleviate the problem by focusing—through the use of narrow beams—transmission and reception on the nodes of interest only. As shown in Figure 2.51, when omnidirectional antennas are employed, nodes A, B, E, and F have to be idle while nodes C and D communicate. The use of directional antennas removes this constraint, provided that the nodes can be separated in space by shaping narrow beams.

A transmitter equipped with multiple antennas can achieve transmit diversity by spreading the transmitted symbols over time and space (space-time coding). The amount of transmit diversity is equal to the smallest rank of codeword-error matrices, and the coding gain is related to the product of the nonzero singular values of

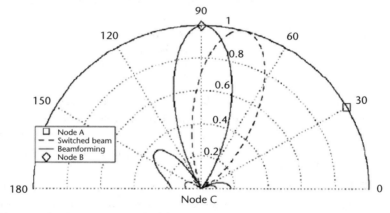

Figure 2.50   An adaptive versus switched beam system.

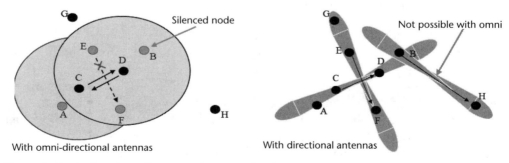

With omni-directional antennas   With directional antennas

Figure 2.51   Reduced interference and improved utilization of resources with directional antennas.

these matrices. The specific design and implementation of the spreading depends on the level of channel knowledge at the transmitter and receiver.

Recent work based on the capacity analysis shows that in the presence of high fading correlations, transmission on the eigenmodes of the transmit antenna correlation matrix [70] can be applied. The linear precoding technique proposed by Sampath and Paulraj [71], which assumes knowledge of the antenna correlations at the transmitter, improves performance of a space-time coded system by forcing transmission on the nonzero eigenmodes of the transmit antenna correlation matrix. The main advantage of this precoder is that it does not have to track fast fading, but only the slowly varying antenna correlations. The latter can be fed back to the transmitter using a low-rate feedback link or can be based on uplink channel estimation that exploits reciprocity. Furthermore, long-term channel knowledge contains only information on the antenna array response and ignores fast fading. The derivation of the linear precoding in Sampath and Paulraj [71] was based on the average pairwise error probability criterion. The design was then simplified by assuming that the worst pair dominates the performance.

The project MEMBRANE proved that this assumption is no longer valid when the worst pair is not unique (i.e., two or more error pairs with different eigenmodes and energy distribution may result in the worst performance). This situation arises in the case of nonorthogonal STBC.

The project built a generic framework for reconfigurability in STBC through a linear precoding scheme applicable to any STBC type (both orthogonal and nonorthogonal). The basic block diagram is shown in Figure 2.52.

Reconfigurability in terms of spatial correlation is achieved in the following sense:

- When the antenna correlation is zero (uncorrelated spatial modes), the eigenvalues of $\mathbf{R}_T^{1/2}$ are equal, and the matrix of the eigenvectors equals the identity matrix. In this case, the linear precoder becomes an orthogonal transformation and the scheme is equivalent to STBC, which is the optimal approach for zero correlation.
- When the antenna correlation is one (fully correlated spatial modes), only one eigenvalue of $\mathbf{R}_T^{1/2}$ is nonzero, and the precoder is equivalent to a beamformer, which is the optimal approach in this case.
- For partially correlated spatial modes, the water-filling solution adapts to the relative power of the eigenmodes.

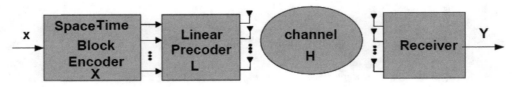

Figure 2.52 Transmission scheme combining space-time block codes and a linear precoder that exploits channel knowledge available at the transmitter.

In Figure 2.53, the performance is depicted in terms of frame error rate (FER). Clearly, linear precoding outperforms both beamforming and ABBA without precoding. The curve of old-linear precoding refers to linear precoding based on the criterion proposed by Sampath and Paulraj [71]. Constellation rotation is applied to the ABBA code according to Sharma and Papadias [72].

2.8 Conclusions

Transmission schemes, which do not require CSI at the transmitter, exploit the spatial dimension either by introducing coding on the spatial domain or by employing spatial multiplexing gain. The former approach, called space-time coding, increases redundancy over space and time, as each antenna transmits a differently encoded, fully redundant version of the same signal.

Transmission schemes, which require perfect CSI at the transmitter, optimize SINR by focusing energy into the desired directions, minimizing energy toward all other directions and satisfying transmit power constraints. Beamforming allows spatial access to the radio channel by means of different approaches, considering either short-term properties (e.g., directional parameters) or long-term properties (e.g., second order statistics) of the radio channel.

In the majority of cases, it is reasonable to assume that only partial CSI is available at the transmitter. Hybrid schemes (combining space-time coding and beamforming) which introduce precoding to exploit the available CSI when optimizing a certain criterion (e.g., pairwise error probability) have been proposed. Robust transmit beamforming schemes have also been proposed; these take into account CSI estimation errors and optimize transmission in terms of a performance criterion, such as SNR and mutual information.

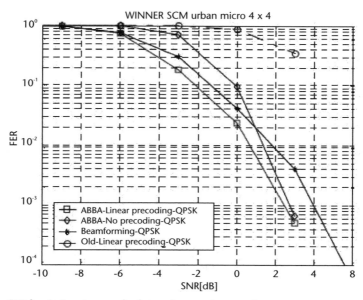

Figure 2.53 FER for 4×4 system and urban micro environment.

In the case of a frequency-nonselective SIMO channel, the optimal receiver strategy is to perform maximum ratio combining (MRC) and maximize the received SNR. For frequency-selective SIMO channels, an ML detector is an optimal receiver, but it is nonlinear with complexity that increases exponentially in the number of channel dimensions. Linear receivers can be used instead; they have lower complexity at the expense of lower performance. A zero-forcing (ZF) equalizer is designed to eliminate intersymbol interference (ISI) by employing channel inversion at the expense of noise enhancement. To overcome this drawback, the minimum mean square error (MMSE) receiver balances noise enhancement with ISI elimination. A suboptimal nonlinear scheme based on decision feedback (decision feedback equalizer) can be used to improve the performance of a linear equalizer by using the feedback filter to remove the portion of ISI from the present symbol caused by the previously detected symbols. Both ML and linear equalizers can be extended to the MIMO channel case. A new problem associated with MIMO receivers is the presence of multistream interference (MSI), resulting from the multiple data streams interfering with each other.

The use of intelligent antennas in ad hoc networks has recently attracted a great amount of attention as a means to optimize power transmission/reception.

Two basic types of intelligent antennas are considered in this context: switched beams (fixed beams) and adaptive antenna arrays. A switched beam antenna generates multiple predefined fixed beam patterns and applies one at a time toward the direction of interest. It is the simplest technique, essentially providing sectorization with the capability of illuminating the selected sector according to, for instance, an SNR-related metric. An adaptive (steerable) antenna array can formulate the beam structure based on a certain optimization criterion, such as maximizing the array gain towards the signal of interest and suppressing interfering signals.

The transmit/receive power optimization and interference suppression capabilities of intelligent antennas have encouraged the consideration of similar techniques in multihop networks that are, in general interference, limited. However, it was made evident that to be able to leverage on the prospective gains, appropriate higher layer protocols design is required.

References

[1]   http://www.e2r.motlabs.com.

[2]   http://b-bone.ptinovacao.pt.

[3]   http://www.ist-membrane.org.

[4]   ftp://ftp.cordis.europa.eu/pub/ist/docs/directorate_d/cnt/pulsers_en.pdf

[5]   http://www.imec.be/mascot.

[6]   http://www2.ife.ee.ethz.ch/RESOLUTION/index.html.

[7]   http://www.ist-surface.org.

[8]   https://ict-e3.eu/.

[9]   http://www.ict-aragorn.eu/.

[10] http://cordis.europa.eu/fp7/home_en.html.

[11] ICT project ARAGORN, "State of the Art," Deliverable D 2.1, April 2008, http://www.ict-aragorn.eu/deliverables.html.

[12] http://e2r2.motlabs.com/.

[13]   http://gnuradio.org.

[14]   http://networks.rice.edu/research.html.

[15]   http://www.etsi.org/WebSite/technologies/RRS.aspx.

[16]   http://www.sdrforum.org/.

[17]   http://www.omg.org/.

[18]   Lozano, A. et al., "Lifting the Limits on High Speed Wireless Data Access Using Antenna Arrays," *IEEE Communications Magazine*, vol. 39, no. 9, pp. 156–162, Sep. 2001.

[19]   Biglieri, E. et al., "Fading Channels: Information-Theoretic and Communications Aspects," *IEEE Trans. on Information Theory*, vol. IT-44, no. 6, pp. 2619–2692, Oct. 1998.

[20]   Ertel, R.B. et al., "Overview of Spatial Channel Models for Antenna Array Communication Systems," *IEEE Personal Communications*, vol. 5, no. 4, p. 10–22, Feb. 1998.

[21]   IST project MEMBRANE, *Joint Routing, Scheduling and Power Control Algorithms*, Deliverable 4.2.1, Dec 2006.

[22]   IST project RESOLUTION, "Development and Implementation of Algorithms for Local Positioning," Deliverable D11.

[23]   Yu, K. and B. Ottersten, "Models for MIMO Propagation Channels: a Review," *Wireless Communication Mobile Computing*, vol. 2, p. 653–666, Nov. 2002.

[24]   Foschini, G. and M. Gans, "On Limits of Wireless Communications in a Fading Environment when Using Multiple Antennas," *Wireless Personal Communication*, vol. 3, no. 5, pp. 311–335, 1998.

[25]   Telatar, I. E., "Capacity of Multi-Antenna Gaussian Channel," *European Trans. Telecomm.*, vol. 10, no. 6, pp. 585–595, Nov./Dec. 1999.

[26]   Shitz, S. S. and I. Bar-David, "The Capacity of Average and Peak-Power-Limited Quadrature Gaussian Channels," *IEEE Trans. on Information Theory*, vol. IT-41, no. 4, pp. 1060–1071, July 1997.

[27]   IST project SURFACE, Preliminary Design of the Multi-User Optimal Transmit and Receive Strategy, Deliverable 4.1, December 2006.

[28]   Cover, T. M. and J. A. Thomas, *Elements of Information Theory*. John Wiley & Sons, 1991.

[29]   Yang, J. and S. Roy, "On Joint Transmitter and Receiver Optimization for Multiple-Input Multiple-Output (MIMO) Transmission Systems," *IEEE Trans. on Communications*, vol. 42, no. 12, pp. 3221–3231, December 1994.

[30]   Scaglione, A. et al., "Redundant Filterbank Precoders and Equalizers, Part I: Unification and Optimal Designs," *IEEE Trans. on Signal Processing*, vol. 47, no. 5, pp. 1988–2006, July 1999.

[31]   Sampath, H. et al., "Generalized Linear Precoder and Decoder Design for MIMO Channels Using the Weighted MMSE Criterion," *IEEE Trans. on Communications*, vol. 49, no. 12, pp. 2198–2206, December 2001.

[32]   Yang, J. and S. Roy, "Joint Transmitter-Receiver Optimization for Multi-Input Multi-Output Systems with Decision Feedback," *IEEE Trans. on Information Theory*, vol. IT-40, no. 5, pp. 1334–1347, Sept. 1994.

[33]   Palomar, D. P. et al.,"Joint TX-RX Beamforming Design for Multicarrier MIMO Channels: A Unified Framework for Convex Optimization," *IEEE Trans. on Signal Processing*, vol. 51, no. 9, pp. 2381–2401, September 2003.

[34]   Love, D. J. and R. W. Heath , "Multimode Precoding for MIMO Wireless Systems," *IEEE Trans. on Signal Processing*, vol. 53, no. 10, pp. 3674–3687, Oct. 2005.

[35]   Palomar, D. P. et al., "Optimum Linear Joint Transmit-Receive Processing for MIMO Channels with QoS Constraints," *IEEE Trans. on Signal Processing*, vol. 52, no. 5, pp. 1179–1197, May 2004.

[36]   IST project SURFACE, Preliminary Design of the Multi-User Optimal Transmit and Receive Strategy, Deliverable 4.1, December 2006.

[37] IST project SURFACE, System Requirements on QoS. Definitions and Specifications, September 2006.

[38] http://www.ist-surface.org/deliverables.htm

[39] IST project SURFACE, "System Requirements on QoS. Definitions and Specifications," Deliverable 2.1, September 2006.

[40] IST project SURFACE, "System Requirements on QoS. Definitions and Specifications," Deliverable 4.2, June 2007.

[41] Maddah-Ali, M. A. et al.,"Fairness in Multiuser Systems with Polymatroid Capacity Region," *IEEE Trans. on Information Theory*, June 2006.

[42] http://www.3gpp.org/Highlights/LTE/lte.htm.

[43] Falconer, D. et al., "Frequency Domain Equalization for Single-Carrier Broadband Wireless Systems", *Communications Magazine*, IEEE, Vol. 40, No. 4, Apr 2002, pp. 58–66.

[44] Al-Dhahir, N. "Single-Carrier Frequency-Domain Equalization for Space-Time Block-Coded Transmissions Over Frequency-Selective Fading Transmissions," *IEEE Communications Letters*, vol. 5, no. 7, July 2001.

[45] Reinhardt, S. et al., "MIMO Extensions for SC/FDE Systems," *8th European Conference on Wireless Technology*, Paris, France, 2005.

[46] Coon, J., J. Siew, M. Beach, A. Nix, S. Armour, and J. McGeehan, "A Comparison of MIMO-OFDM and MIMO-SCFDE in WLAN Environments", Global Telecommunications Conference, 2003. GLOBECOM '03. IEEE, Vol. 6, December 2003, pp. 3296–3301.

[47] R1-063454, "Simultaneous Transmit Diversity and Spatial Multiplexing for E-UTRA Uplink Multiuser MIMO," ITRI, http://www.quintillion.co.jp/3GPP/TSG_RAN/TSG_RAN2006/TSG_RAN_WG1_RL1_2.html

[48] R1-060657, "Orthogonal Multi-User MIMO based Channel-Dependent Scheduling for SCFDMA on E-UTRA Uplink," Nortel, http://www.quintillion.co.jp/3GPP/TSG_RAN/TSG_RAN2006/TSG_RAN_WG1_RL1_2.html.

[49] Jianhong, C. et al., "Development of RF Subsystems for MIMO and Beyond 3G Systems," IEEE 2005.

[50] "Design of the Single-user Optimal Transmit and Receive," Number D3.1. IST-027187-SURFACE, Sept. 2006.

[51] 3GPP TR 25.814 V7.0.0, "Physical Layer Aspects for Evolved UTRA," *Tech. Rep.*, June 2006.

[52] Park. T. et al., "Proportional Fair Scheduling for Wireless Communication with Multiple Transmit and Receive Antennas," in *IEEE VTC*, Florida, U.S.A., October 2003, vol. 3, pp. 1573–1577.

[53] Catreux, S. et al., "Adaptive Modulation and MIMO Coding for Broadband Wireless Data Networks," *IEEE Communications Magazine*, June 2002.

[54] Alamouti, S. M., "A Simple Transmit Diversity Technique for Wireless Communications," *IEEE JSAC*, vol. 16, no. 8, October 1998.

[55] Hochwald, B. M. et al., "Multiple-Antenna Channel Hardening and Its Implications for Rate Feedback and Schedulling," *IEEE Trans. on Information Theory*, vol. IT-50, no. 9, pp. 1893–1909, September 2004.

[56] Foschini, G. J., "Layered Space-Time Architecture for Wireless Communication in a Fading Environment When Using Multiple Antennas," *Bell Labs Technical Journal*, pp. 41–59, Autumn 1996.

[57] Heath, R. et al., "Multiuser Diversity for MIMO Wireless Systems with Linear Receivers," *ISSS*, Asilomar, California, U.S.A., July 2001.

[58]   Moon, J., J. Ko, and Y.-H. Lee, "A Framework Design for the Next-Generation Radio Access System," *IEEE Journal on Selected Areas in Communication*, vol. 24, no. 3, pp. 554–564, March 2006.

[59]   Chia-Chen, C. et al., "Potential of UWB Technology for the Next Generation Wireless Communications," 9th International Symposium on Spread Spectrum Techniques and Applications, IEEE, August 2006, Manaus, Brazil, pp. 422–429.
Digital Object Identifier 10.1109/ISSSTA.2006.311807.

[60]   Domenico, P., and H. Walter, "Ultra-Wideband Radio Technology: Potential and Challenges Ahead," *IEEE Communications Magazine*, July 2003.

[61]   Geir,   E.,   "Flexible   and   Heterogeneous: Radio   Access   beyond   3G,"   at link..http://www.telenor.com/telektronikk/Oien_Beyond3G.pdf

[62]   http://www.ist-magnet.org/.

[63]   IST project MAGNET Beyond, Deliverable 3.2, December 2007.

[64]   http://www.artechhouse.com/.

[65]   Hamid, T., and D. Yasin, "Design of Dual-band Reconfigurable Smart Antenna," *Progress in Electromagnetics Research Symposium 2007*, Prague, Czech Republic, 27-30 August 2007.

[66]   IST project RESOLUTION, Deliverable D2.

[67]   http://www.imperial.ac.uk/membrane/.

[68]   http://ieeexplore.ieee.org/.

[69]   http://www.ist-winner.org/.

[70]   Brueninghaus, K. et al., "LinkPerformance Models for System Level Simulations of Broadband Radio Access Systems", *16th Annual IEEE International Symposium on Personal Indoor and Mobile Radio Communications*, Berlin, Germany, September 2005.

[71]   Sampath, H. and A. Paulraj, "Linear precoding for space-time coded systems with known fading correlations," *IEEE Communications Letters*, pp. 239–241, June 2002.

[72]   Sharma, N. and C.B. Papadias, "Improved Quasi-Orthogonal Codes Through Constellation Rotation," *IEEE Trans. on Communications*, Vol. 51, No.3, pp. 332–335, March 2003.

CHAPTER 3

Autonomic Communications

The convergence and interworking of present, emerging, and future radio systems and mobile networks has gained a lot of attention, and it is expected to boost revenues. Reconfigurability [1] can be understood to be the collection of software and cognitive radio technologies [2, 3] that aim to optimize network resources. For example, over-the-air software download has benefited mobile operators by enabling online upgrades of mobile devices' capabilities, as well as allowing remote fault management. The existing radio standards have achieved the "always-best-connected" objective by allowing dynamic network access over both the licensed and unlicensed spectrum. But frequent protocol updates demand that reconfigurability be performed in an autonomous way. Generic self-management and self-configuration mechanisms are needed to achieve protocol stack and component-based protocol layer reconfiguration.

This chapter describes advances in the state-of-the-art of autonomic communication systems enabling the adoption of reconfigurable system architectures. Section 3.1 introduces the main aspects related to autonomic communications. Section 3.2 describes self-configuration, self-management, and autonomic decision making as the main capabilities for the realization of autonomic systems. Sections 3.3, 3.4, and 3.5 describe the profile representation, ontology and context models, and device management, respectively. Section 3.6 talks about research advancements toward autonomous next generation networks (NGNs), and gives a perspective on the network evolution toward openness. Section 3.7 describes a unified scenario for use of autonomic communications to achieve a seamless user experience. Section 3.8 concludes the chapter.

3.1 Introduction

Autonomic and self-concepts have emerged as a new paradigm for the administration and management of complex information and systems. It helps to manage complex tasks at the business, system, and device level without human intervention [4]. The access to information is made transparent through self-managed and self-diagnostic distributed computing and networking facilities. The short-term benefits include cost savings, stability, and scalability; the long-term benefits, which will be an inherent functional part of devices, servers, and networks, will offer global tools for common tasks such as information retrieval. Telecom manufacturers and operators have been developing complex systems for mobile and wireless communications, including the operational infrastructure and support systems such as the Operation Support System (OSS). Recent efforts propose the adoption of the autonomic computing principle in communication systems, focusing on self-governance aspects and dynamic assembly and adaptation of policy rules [5].

Incorporated in the system architecture framework, autonomic and self-concepts allow the system architecture to integrate self-managed and self-organizing systems' governance in terms of reacting to changes in its environment in a dynamic and efficient way, demonstrating the cognition loop functionality. Depending on the context, the intention is to apply collaborative or autonomous distributed decision making–based approaches, considering both single-operator and multioperator scenarios. On this basis, the specification of behavioral and functional aspects, as well as protocols/signaling and functional incorporation issues, must be studied together with the impact such concepts might have on legacy network architectures and emerging cognitive networks and systems.

Autonomic and self-functionalities on all levels of the system architecture are seen as a key solution to the efficient management of cognitive, heterogeneous wireless systems. In this context, *self-* refers to aspects such as configuration, planning, optimization, healing, retuning, and resource reallocation, which may be applied to the edges of the network and/or the user terminals [6]. Solving this challenge could lead to an increased degree of automation in the operation of radio network elements. Fast-automated adaptation of radio parameters could be achieved as function of measurements/parameters like traffic load, user profile, available services, and so forth.

The evolution of system design can be understood as three main levels. At the first level, the system is designed for a specific context, and the designers exploit the features of the context to optimize the architecture. At the second level, the system is designed to operate in multicontext scenarios, which are known *a priori*. At the third level, the design is more sophisticated, since the multiple contexts are not known beforehand. In this case, mechanisms in the architecture to infer contexts and to learn the pattern of context changes, along with mechanisms for self-reconfiguration, are needed to realize autonomic capabilities.

The major contributor for this chapter is the IST FP6 project E2R II [1]. It has addressed the design evolutions in the context of telecommunication infrastructures [7]. The basic idea was to build a unified scenario that incorporates features that cover all aspects of reconfigurability. A functional analysis of the scenario brought out the set of capabilities required from the architecture to support it, which were further integrated into the functional entities comprising the reconfiguration management plane (RMP). Since the different currently available systems have been designed under different contexts, standardization is key for interoperability in a fully autonomic system. One approach is to make the features that prevent interoperability part of the context so that the system is able to adapt to them.

3.2   Capabilities of Autonomic Communications

Research activities in the past years have been focused on autonomic computing and communications. These activities include defining, designing, and deploying autonomic features in emerging communication systems and devices [8]; enhancing the standards functions of existing systems; and posing new requirements regarding their functionality.

According to a model proposed by IBM [4], the four main attributes that an autonomic system should incorporate, which are also known as self-CHOP features [9], are as follows:

- *Self-configuration* includes the systems' capabilities to configure, reconfigure, and adapt themselves according to the defined policy rules to meet the requirements posed by environment changes and evolving users' needs.
- *Self-healing* means that the system incorporates diagnosis and irregular behavior detection procedures with the ability for automatic remedy.
- *Self-optimization* means that the system should be able to perform performance tuning and continuously monitor the environment to optimize its performance and behavior [10].
- *Self-protection* means that the system should also be capable of protecting itself from attacks and of maintaining the system security and privacy.

Figure 3.1 shows a fundamental reference model for autonomic computing systems as proposed in [8].

An autonomic manager monitors the managed elements and their environment, and, based on the analysis of retrieved information, it plans and executes the necessary actions associated with the autonomic elements. The number of managed components could be one or more and include all types of components, such as software components, hardware components, and so forth.

The incorporation of autonomic capabilities into computing and communications systems is shown in Figure 3.2 as a process that has gradually been realized over different phases and enables the following capabilities [11]:

- *Basic and Managed Levels:* System management technologies are used for the collection and synthesis of information.
- *Predictive Level:* The system monitors the environment, predicts the optimal configuration, and provides recommended actions to the IT staff.

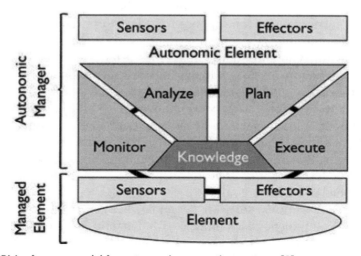

Figure 3.1   IBM reference model for autonomic computing systems [8].

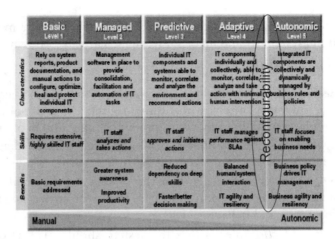

Figure 3.2   Phases of autonomic evolution [11].

- *Adaptive Level:* The system can make certain decisions with minimal human intervention, based on the analysis of the available information.
- *Autonomic Level:* Business policies are used for managing the IT infrastructure operation.

As shown in Figure 3.2, reconfigurability forms a key enabler for the evolution toward an autonomic communication environment and may be considered an intermediate phase between the adaptive and the autonomic level.

3.2.1   Self-Configuration and Self-Management

Two important aspects of autonomic systems' behavior are self-configuration and self-management. A lot of work has been done in the context of deploying such features.

For example, an innovative approach that is tightly coupled to IBM software and applications has been introduced by the IBM Computing Toolkit [12] for the development of autonomic applications. It incorporates sample implementations of the autonomic manager concept with necessary mechanisms to achieve different levels of maturity in autonomic capabilities. An object-oriented system that provides the infrastructure for hot-swapping and meets autonomic challenges by dynamically monitoring and modifying the operating system is proposed by Appavoo et al. [13]. It uses a research operating system known as K42 [13] for demonstrating the implementation. Semantics-based dynamic service composition architecture was proposed by IBM [14]; it assumes that a user requests a service in an intuitive manner (e.g,. using a natural language) and dynamically composes the requested service based on its semantics. The self-deployment and self-configuration aspects are mentioned in Mullany et al. [15].

The conventional software design has been extended through many works that propose component-based models with dynamic configuration or autonomic capabilities. A cross-layer protocol stack design for autonomic communication based on the performance-oriented reference model (POEM) has been analyzed by Gu et al.

[16] that; it focuses on the advantages of layering and the benefits of holistic and systematic cross-layer optimization.

In an autonomic environment, the traditional centralized decision process is split between the terminal and network in such a way that it allows the network to federate coexisting users, thus enhancing scalability. The terminal satisfies its request in terms of QoS by applying simple behavioral rules of its own to reach an efficient operating point, whereas the network broadcasts policies to highlight the action space allowed for users.

3.2.2 Autonomic Decision Making

Autonomic decision making is important, as it reduces the human intervention and thus reduces complexity in management. The concept of closed control loops, which is derived from process control theory, can be used here. Resources should be monitored to keep the parameters within the desired range determined by the policies.

Figure 3.3 shows the use of control theory in self-management of the computing systems to implement the constraint policy rules.

A hierarchical policy model captures the higher goals of users and administrators into network level objectives. The design of an autonomic computing model was described by Fei and Fan-Zhang [18] together with an algorithm for autonomic decision-making based on behavioral rules. PROTON [19] is a novel solution that assists mobile users in the decision-making process concerning roaming between heterogeneous technologies. It uses a formal policy representation model, based on finite state transducers, that evaluates policies using contextual information to manage the behavior of mobile users in a transparent way.

3.3 Profile Representation

It has become increasingly important for services, applications, operating systems, and devices to be configured or reconfigured dynamically depending on a number of factors, such as context information, user preferences, and policies that are defined by manufacturers, operators, developers, and end users. Due to the increased com-

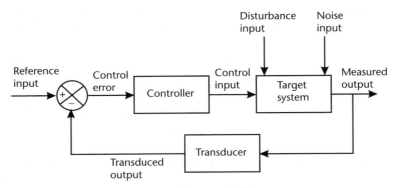

Figure 3.3 Block diagram of feedback control system [17].

plexity that this brings, it is important to achieve reconfigurability in an automatic fashion. The initial efforts made would help to make the concepts machine-understandable.

The concept of the user profile refers to the relationship between a given user and one service that the user is going to use. A device profile is needed that will make the service provisioning features' procedures aware of details about the device currently being used. A network profile would help to understand the characteristics of a networking environment when providing a given service. The term *profile* encompasses personal data, users' preferences, service and application profiles, and device and network profiles.

The following subsections discuss briefly the various types of user, device, and network profiles.

3.3.1 User Profiles

The user profiles can be classified as follows [7]:

- *User Profiles and Single Sign-On (SSO) Architecture:* A SSO is a session/user authentication process that allows a user to enter one name and one password to access for using multiple applications and services. Microsoft's Passport [20] and Liberty Alliance [21] are some examples of SSO services that use centralized and distributed ways, respectively, to share personal data between different Internet sites.

- *3GPP Generic User Profile (GUP):* The 3GPP GUP is a solution to overcome the variety of capabilities associated with the introduction of sophisticated user terminals, the hybrid combinations of mobile network domains, the increasing use of downloadable applications, and the desire of customers to customize services to individual preferences and needs [22]. It provides architecture, data description, and interface with a generic mechanism to access and manipulate user-related data for suppliers and consumers. The structural relationship between the main parts of 3GPP GUP are shown in Figure 3.4. There are a number of independent GUP components in a generic user profile. The composite data type defines the structure of the whole GUP component. The structure includes the definition of the types of data element groups and/or which data element groups belong to which defined GUP component, along with the data types and the valid values for the data. The data description method (DDM) is used to describe how different profile components are specified, based on an XML schema. It is a set of common rules that specifies the data components and also acts as a template for constructing the data description to avoid incompatibilities and inconsistencies between different profile components.

- *The ETSI Universal Communications Identifier (UCI):* It is a unique identifier, stable and immutable over time, which can be used for all kinds of communication with a given user. The user may hold several communication devices, each with different characteristics, network bearers, capabilities, and so forth, but he or she may also wish to receive certain communications in particular forms and on particular devices.

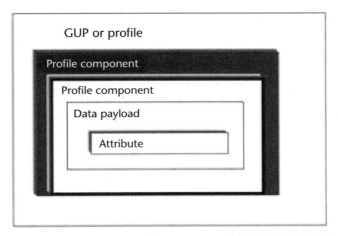

Figure 3.4 Structural relationship between the main parts of the 3GPP GUP [17].

• The important concepts are shown in Figure 3.5. The personal user agent (PUA) is an autonomous agent that is responsible for the adaptations. In this capacity it is an external entity to the main communications networks with a one-to-one relationship to a specific UCI. The adaptations are possible depending on the data accessed and a policy-based approach is followed. The service agent (SA) interfaces with the PUA, and a given communication service (or network) is basically responsible for the policy enforcement. It represents the link between the UCI system and the legacy network and services, and is typically provided by the network or service provider.

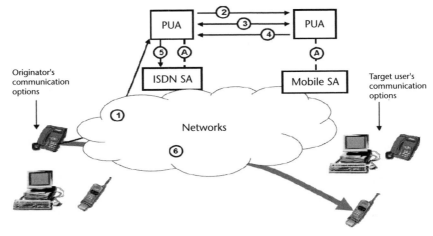

The originator requests a voice call to the target user:

A–Each PUA exchanges information with the SAs of its user's networks/services before, during, and after communication attempts take place. The target user's PUA knows that the user's mobile phone is able to receive voice calls.

1–The originating user enters the UCI of the taerget user.
2–The originating PUA makes a request to the PUA of the target user.
3–The PUAs negotiate communication options if necessary.
4–The target user's PUA takes account of its user's preference and proposes the user's mobile phone to receive the call.
5–The originator's PUA instructs the originator's network to set up the call.
6–A voice call between the originator's ISDN phone and the target user's mobile phone is established.

Figure 3.5 ETSI UCI: Example of PUA-SA interaction [23].

OASIS I-names is a concept that, unlike the conventional addresses, provides binding to legacy addresses and is not tied to a specific location. It is based on two complementary technologies which are under development at OASIS. One is extensible resource identifier (XRI) [24], which is an extension of the actual system based on URLs, and the other one is XRI Data Interchange (XDI), which defined an interoperable XML metaschema in which every element of the data is identified with one or more XRIs. It is inspired by the model provided by the Web, but legacy Web pages are different, in the sense that the consumer is no more represented by a human but by an automatic process. An important feature is that Dataweb pages provide a single format in which any XML-encoded data can be shared, independent of the application or domain from which they originated. A model to allow semantic description of every Dataweb page, involving XRI and XDI, is a promising technology in the field of profile definition because it allows sharing and linking of existing data in a very intuitive way. In principle, it is possible to address the semantics of every new data format that is going to be defined without having to modify the architecture and the data model of the legacy system.

3.3.2 Device Profiles

The emergence of new types of mobile and embedded computing devices and the advances in wireless networking imply that services will be accessed on a wide set of different terminals, creating a need for the applications and services to become more sensitive to context and for different devices for exchanging context information. Two types of device profiles are mentioned as follows:

- *W3C CC/PP:* The Composite Capabilities/Preferences Profile (CC/PP) [25] defines a high-level structured framework for describing information using the resource description framework (RDF), is a language used to represent metadata about Web resources in a machine-understandable format. To enable the adaptation to take place on the server side and render the content to the device, the framework communicates the information about the device's capabilities and characteristics, as well as communicating information about the network bearer to the Web servers and gateways/proxies. CC/PP does not include any specific attributes about the device capabilities and user preferences. It provides the rules to help construct a vocabulary that describes capabilities and preferences.
- *OMA UAProf:* The User Agent Profile (UAProf) [26] is an application of CC/PP proposed by the OMA [26]. It is concerned with capturing classes of device capabilities and preference information, such as the hardware and software characteristics of the device, as well as information about the network to which the device is connected.

3.3.3 Network Models

Two types of network models are described as follows:

- *DMTF CIM:* The Distributed Management Task Force (DMTF)'s common information model (CIM) [27] describes and manages communications connectivity and the network "cloud," as well as the individual services and protocols in the network. It also provides a way to hook the logical entities to physical components, which can be modeled in turn by the DMTF core, physical, and device common models. Since it has been assumed that the network systems fit the same pattern as computer systems, each network element maybe provided with a certain degree of "intelligence" and reasoning capabilities which makes the model suitable for modelling autonomic network and elements.
- *TMF SID Model:* An advanced model called shared information/data (SID) model [28] has been defined by the TeleManagement Forum (TMF) for the telecommunication domain. It serves as an information model and a common vocabulary that acts as a bridge between the business and information technology people, providing definitions that are understandable by the business and are also rigorous enough to be used for software development.

3.3.4   General Concept of Profiles

The views concerning the user's personal profile, the device profile, and the network models are complementary, considering that the first refers to the users themselves, the second to the devices that the users may use, and the last to the network users may access. It must be noted that various kinds of profile information and incompatibility among these, due to different syntax, semantics, and ways of data access, can increase the complexity and, therefore, are not practical for use in automatic computations.

The concept of one-to-many relationship between the user and his or her devices and access networks means that a user may own or use a number of devices and may wish to define rules for each of them, or rules involving two or more terminal devices and their underlying technologies. But a one-to-one relationship between the user profile and the user's identity would mean that the user profile is in a way "dual" to the user's identity, providing complementary information, and this concept has been envisaged in [23].

3.4   Ontology and Context Models

Ontology and context are complementary aspects of the descriptions of reality, or the state of the world. An explicit requirement for autonomic communications is that this world should be represented in a machine-processable form.

3.4.1   Ontology

Artifacts such as databases and programs implicitly assume a certain model, with the aim of recording the facts and solving the problems. The term *implicitly* is used in the sense that the information corresponding to this model is scattered within these artifacts in a rather nontransparent manner. This would be no problem in a standalone mode. However, when two databases or programs are required to communicate and interoperate, the implicitness can cause misinterpretation of exchanged information, with undesirable consequences. This problem is called semantic interoperability and could be solved case-by-case for every distinct pair of interacting artifacts. A more scalable solution would be to define a shared model of reality interpretable by all the artifacts; this corresponds to the term *ontology* as used in information sciences (distinct from its use in philosophy). Gruber [29] has defined ontology as "the specification of a conceptualization"[30].

3.4.2   Context

In common language, the term *context* refers to situations of an (external) reality that can vary. In the IST FP6 project E2R, this term was used to refer to information that surrounded and gave semantic meaning to an entity [31].

3.4.3   Relation Between Ontology and Context

The term *ontology* is reserved for referring to the relatively invariant part of reality, whereas the term *context* refers to the variant part. But ambiguities could arise due to the use of phrases such as *relatively invariant*. To avoid this, a clarification can be made about the distinction using a topological framework [32]. Contexts can be understood to be open sets and concepts belonging to ontology as accumulation points; in other words, whatever is global is ontology, and whatever is local is context.

3.4.4   Representation for Ontology and Context

Formal notations and languages with well-defined semantics are needed for automatic interpretation and use of ontological and contextual information.

3.4.4.1   Ontology Languages

The Knowledge Interchange Format (KIF) [33] was developed as part of the DARPA Knowledge Sharing project and is based on first order logic, which is quite expressive but difficult to reason with. A widely used popular ontology is OWL (Web ontology language) [34] that was developed as part of the Semantic Web Project and is an extension of the earlier RDF [35] in which resources are described as subject-predicate-object expressions. Other parts of OWL are derived from the DARPA Agent Mark-up Language (DAML) [36].

OWL Lite, OWL DL, and OWL Full are three sublanguages of OWL.

3.4.4.2   Expressiveness and Inference

With respect to powerful logics, there is a model space and an expression of the logic that defines the subsets of the space. Based on intuition, one logic may seem more expressive than another if it can define a larger family of subsets. The ability to be more precise means the ability to make more distinctions; a more expressive logic enables one to say things more precisely.

A balance needs to be maintained between enhancing expressiveness that increases the difficulty of reasoning with the logic. First order logic is very expressive, but this expressiveness causes the reasoning algorithm to get into an unending loop such as in OWL. Below, the relation between the sublanguages of OWL has been given in terms of degree of expressiveness:

$$OWL\ Lite < OWL\ DL < OWL\ Full$$

OWL Lite is efficient in terms of computational complexity of reasoning, but is least expressive. OWL Full is not even decidable, though it has the expressive power of first order logic. OWL DL is computable.

Reconfiguration decisions need to be taken quickly; therefore, the first two are used for representing ontologies, unless time-out mechanisms are needed during reasoning.

3.4.4.3   Context Models

The localized nature of contexts means that the representations could be optimized for different purposes. They can be classified into the following types of models:

- *Key-value models:* These model the context by providing the value of context information as the value of an environment variable.
- *Mark-up scheme models:* These allow for hierarchical data structures using mark-up tags with attributes and content.
- *Graphical models:* Models such as Unified Modeling Language (UML) can also be used to model contexts.
- *Object-oriented models:* These allow encapsulation and reusability of context models [37].

3.4.5   Role in Communications and System Aspects

Connectivity, QoS, and services are some of the primary drivers for adaptation for autonomic communication systems. The adaptation process involves certain decisions regarding reconfigurations with the decisions based on local and global knowledge about the network, the devices and the user where global knowledge corresponds to the ontological part and the local to the contexts. Table 3.1 shows the examples.

Regarding system aspects, Chun et al. [38] have identified some basic functionalities of filtering, aggregation, and correlations which mandates the existence of an extensible data flow framework for supporting these functionalities for ideal information-plane architectures [38]. It basically defines the concept of information plane for large decentralized systems that aspire to be self-managing.

Table 3.1 Examples of Use of Ontology and Contexts in Communications [7]

	Ontology	*Context*
Connectivity	Facts about standards, radio propagation characteristics, theorems of graph theory, design rules from communication theory.	Transmission power level, interference temperature, RAT, link status.
Quality of Service	QoS mapping, trust, and reputation rules, properties of crypto schemes.	Bit error rate (BER), packet error rate (PER), delay, traffic pattern.
Services	Service classes and their mapping to network resources and standards, composition rules, and constraints.	Service availability, usage statistics.

3.4.6 Applications of Autonomics

The system for autonomic communication consists of adaptive networks that are governed by human-specified goals and constraints on the network behavior. In turn, these goals are mapped to appropriate policies at a system level and are subsequently enforced.

The self-management of network elements needs dynamic mapping of human management goals to enforceable policies across a system, based on a context change. It is a big challenge to map the high-level policies or governance directives down to low-level adaptation and control policies for individual heterogeneous functional elements accurately in an autonomous way. Keeney et al. [39] explain a service-oriented approach using a model of constrainable adaptivity for heterogeneous network management functions in which the resources are managed as composeable services, called Adaptive Service Elements (ASE), that have inbuilt application-specific adaptivity in their use of subservices and their subscription to relevant context information streams. This would enable services to be composed into value chains and workflows while also exposing an elemental resource management view that can form part of end-to-end resource management activity. Issues pertaining to heterogeneity between services and context can be addressed using ontology-based semantics that also provide a reasoning framework for policy refinement.

Considering the fact that service-oriented architectures were influencing telecommunication systems, the TMF introduced the NGOSS initiative to address management modeling [40]. This initiative was meant to provide a framework for management information exchange between business processes and service definitions, enabling an open market for telecom software components.

Positioning, tracking, and seamless service access to personalized services have been demonstrated in various environments, but future services need much more than connectivity. Seamless integration of heterogeneous external services has become a critical issue. Some examples are the Web Service Modeling Ontology (WSMO) [41], the use of OWL to model the location ontology as discussed by Flury et al. [42], and the use of ontologies within the e-Automation correlation engine to deal with policy-based decisions at a higher abstraction level as presented by Lewis et al. [43].

3.4.6.1 Network Ontology

Ontologies are extremely important for processing autonomic decision making. An ontology is a data model that represents a domain, which could be used to reason about the objects in the domain and the relationship between them, in order to allow the representation of knowledge in a standard way. Lewis et al. [43] showed that ontologies provide a strong mechanism for heterogeneity in user task requirements, managed resources, services, and context. Two complementary approaches were presented that exploit ontology-based knowledge in support of autonomic communications for service-oriented models in policy engineering and for dynamic semantic queries using content-based networks.

The role of semantic Web technologies in context-aware systems has been explained by Laukken [44]. Also in that source, the role of context ontologies for service adaptation was described by simplified network ontology and a network QoS ontology example. A network ontology block diagram is shown in Figure 3.6.

3.5 Device Management

Device management (DM) is the management of device configuration and other managed objects of devices performed by various management authorities. For

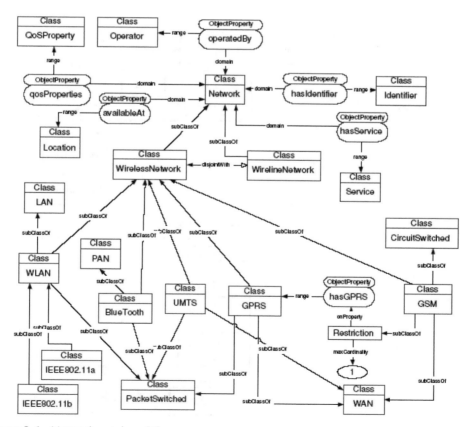

Figure 3.6 Network ontology [7].

example, it includes the setting of the initial configuration information in devices, subsequent updates of persistent information, the retrieval of management information from devices and processing events, and alarms generated by devices. Third parties (i.e., wireless operators, service providers, corporate information management departments, etc.) can carry out the procedures of configuring the mobile devices for the end user.

3.5.1 OMA DM Architecture and Enablers

The OMA DM Workgroup has defined a Device Management Protocol [45]. It enables the device to present its resources in a standardised way, in the form of a management object (MO) on the DM tree [46], an example of which is shown in Figure 3.7. A DM tree is used to represent the information and parameters stored in the device in a bearer-agnostic way.

The device management server (DMS) can modify, retrieve, and configure the parameters inside the different MO. The architecture is shown in Figure 3.8.

The list of enablers for different features of management that are being worked on currently are:

- *Firmware Update (FUMO)* [47]: It provides information on managed objects associated with firmware in OMA DM and allows over-the-air firmware

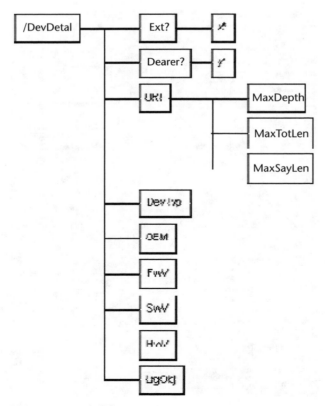

Figure 3.7   OMA DM tree example [7].

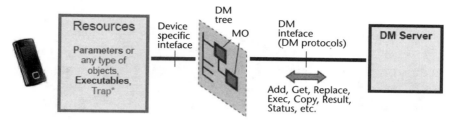

Figure 3.8   OMA DM architecture [7].

updates. Figure 3.9 shows the management object for firmware update with the description of each of the comprising nodes.

- Figure 3.10 shows an example of a scenario, in which OMA DM is directly used to move a firmware update package to the device, using a DM "replace" command to access a management object representing the actual firmware binary package data.

- *Software Management (SCOMO)* [48]: it enables remote operations for software components in the device. The specifications provide the ability to process management actions, such as installation, upgrade and removal of software, over-the-air.

- *Diagnostics and Monitoring (DiagMon)* [49]: it enables the management authorities to proactively detect and repair errors before the users are impacted and helps to determine the actual or potential problems with a device when required. The management authorities should be able to

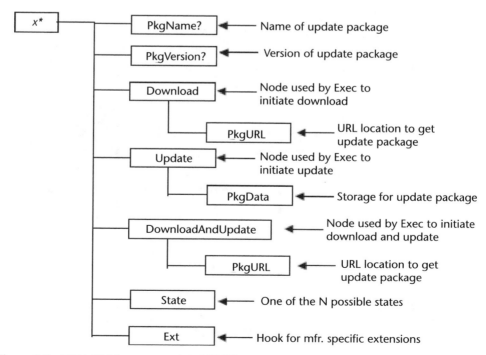

Figure 3.9   OMA DM firmware update MO [7].

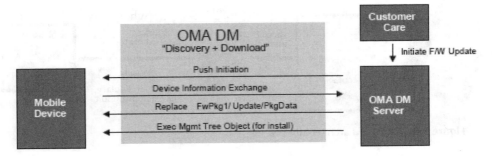

Figure 3.10 Example illustrating use of OMA DM replace command [7].

remotely interrogate the device for trouble isolation. The main areas focused on are diagnostics policies management, performance monitoring, device interrogation, remote diagnostics procedure invocation, and remote device repairing. The management authority can set up monitoring traps to monitor the device.

- *Device Management Scheduling* [50] helps to further reduce network operation costs by processing the scheduled management commands offline. It enhances the user experience by allowing an earlier response time to the local events of the device.

- *DM Scheduling*, which is both time- and event-based, is the ability of the client to perform management actions requested by a DMS, while not in session by that DMS. This is a significant step towards self-management and configuration, since the device can perform management actions in an autonomic fashion. The scheduling operation is shown in Figure 3.11.

DM Scheduling would enable management authorities to schedule tasks on the device and to define policy rules specifying what event should trigger those tasks or when to do those tasks. An example of the use of DM scheduling

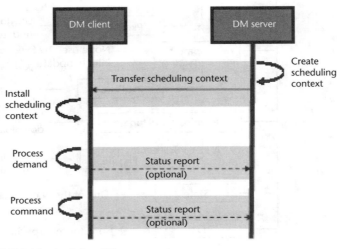

Figure 3.11 OMA DM scheduling [7].

is as follows: the DM server creates schedules (which have information about when to start and stop the tool, what arguments to be used for each invocation, etc.) to control the invocation of an on-device tool. After the schedules have been set, the tool is invoked automatically with proper arguments based on the schedules.

3.5.2 Role of Device Management in Self-Management and Autonomic Decision Making

Device management plays an extremely important role in facilitating self-management and autonomic decision making, which can be understood as follows.

The various DM enablers allow the DMS to configure the device over the air, to update firmware, to manage software on the device, to monitor the device, and eventually to reconfigure it after the monitoring. All these features, along with DM scheduling, realize the concept of self-management and configuration capabilities.

Autonomic decision making is made possible by the use of traps and schedules that alert the device that some conditions have been met and, as a result, some parameters in the device should be reconfigured, therefore triggering that configuration. Figure 3.12 is an example of the use of schedule and trap functionalities.

The self-configuration and management capabilities offered by the OMA DM can be used by the network to send reconfigurable terminals programs on how to behave in certain conditions, to analyze some conditions, and to use the DM subsystem to send orders or new configurations to the device.

3.6 Operation Support Systems

The NGNs require a more evolved OSS, as the traditional system may not support autonomously reconfigured equipment and may not be able to manage the network

1. Management authority sends a TRAP and schedule to the device.

1.1   DMI server   Management authority sends a TRAP (e.g., strength of the signal for a specific bearer) to the device and a schedule including what to do in case the TRAP is triggered (e.g., connection to s different bearer).

1.2   Device verifies and accepts the schedule including user preferences

2. Device starts monitoring. Schedule is triggered and the device is reconfigured.

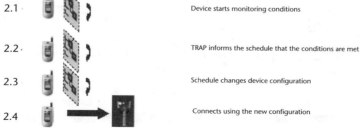

2.1   Device starts monitoring conditions

2.2   TRAP informs the schedule that the conditions are met

2.3   Schedule changes device configuration

2.4   Connects using the new configuration

Figure 3.12 Example: OMA DM schedule and trap functionalities [7].

elements. The OSS architectures, therefore, need to support adaptability and autonomy.

3.6.1 Autonomous Next-Generation Networks

The monolithic solutions proposed in the 1960s and 1970s, the Telecommunication Management Network (TMN) proposed in the 1980s [51], the Telecommunication Information Networking Architecture (TINA) of the 1990s [52], the Tele Management Forum (TMF) proposal in 2000 [53], and the M.3060 proposed ETSI TISPAN Service Oriented Architecture (SOA) [54] endorsed by ITU in April 2006 are some of the legacy examples of the evolution of OSS architectures.

Even though SOA successfully manages the complexity of NGN, the business and operations support system (BSS/OSS) is not yet ready to support the NGN standards. Figure 3.13 shows a possible evolution path where management solutions take into account the limitations, keeping in focus the relationship and interfaces between the Network Element (NE), the Element Manager (EM), and the Network Manager (NM).

In addition, a generic OSS-network mediation layer was proposed, as shown in Figure 3.14.

The proposal targeted the following goals:

• *Establishment of a single multitechnology multivendor layer between the OSS environment and the network to virtualize the network:* The layer hosts an image of each network element, called the virtual NE, on which the NE-relevant information is uploaded from the network, stored, and made available to the OSS through the supported functionalities. NE data is aligned to the real network via both direct polling and real-time updates, due to the executed fault, configuration, accounting, performance, security (FCAPS) functionalities.

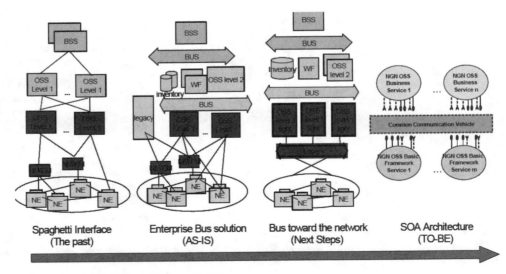

Figure 3.13 NGN OSS evolution path [7].

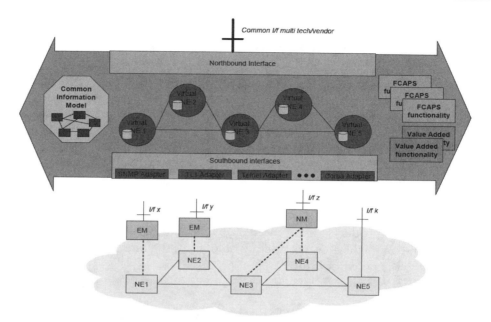

Figure 3.14   OSS NE/EM interface requirements [7].

- *Offer of a common, multipurpose, northbound interface:* For this, all the virtual NE information is mapped into a common information model (CIM), and all the functionalities work on this data. A common multitechnology and multivendor interface is used by the OSS to interoperate with the layer. Propriety interfaces exposed by the different EMs and NEs are used between the layer and the network. The CIM could be either a propriety model or designed starting from a standard information-data model, the latter being preferred as it allows interoperability with the OSS.
- *Support of proprietary southbound interfaces:* The layer manages the set of protocol adaptors, thus guaranteeing connectivity between pairs.

The framework could be equipped further with a wide set of capabilities, ranging from basic, interdomain functionalities to more sophisticated functionalities, such as the capability to automatically capture and react to network behavior. Although there is no standardization body or forum that supports this solution, there are commercial products in the market (e.g., the Nakina Systems' Multi-Vendor Element Management Solution [55] and the Sheer Networks Active Network Abstraction (ANA) [56].

3.6.2   Network Evolution Toward Openness

The Open Communication Architecture Forum (OCAF) [57], which is a focus group of ITU-T SG13 NGN group [58], plays a significant role in driving a revolution in the classical OSS architecture. The network node acts like a computer, such that the network capabilities and management functions interoperate directly on the node. This has been strongly supported by the IT vendors, who view the lack of openness in telecommunication solutions as a main problem for network/service providers and system integrators.

A set of hardware and software customized off-the-shelf components, which enable the creation of network elements, platforms, and applications using a common reference model have been defined. The three types of providers envisaged are:

- *Service provider:* It delivers services to the end users.
- *Solution provider:* It delivers solution building blocks.
- *Technology provider:* It delivers functional components.

Their roles are summarized in Figure 3.15.

Currently, no commercial OCAF-compliant solutions for routers and switches exist, but products are expected in the future. Figure 3.16 shows how the OCAF-compliant network nodes would move the functionalities from the OSS/EM layer to the network layer.

3.7 Policy Framework for Opportunistic Communication

One of the challenging aspects in opportunistic radio (OR) is the selection of policy/policies to be applied to a certain communication scenario. An overview of the relations between the components used in IST FP6 project ORACLE [59] is shown in Figure 3.17. The context management system forms the basis of policy framework.

Some of the benefits of using ontology-based structuring are as follows:

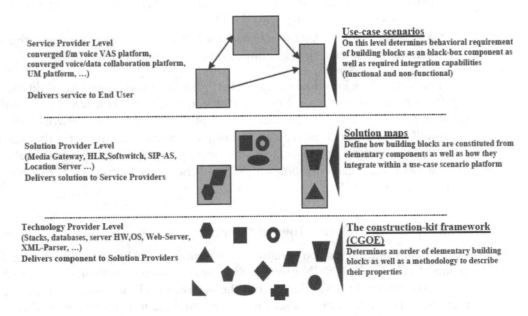

Figure 3.15 OCAF providers and their roles [7].

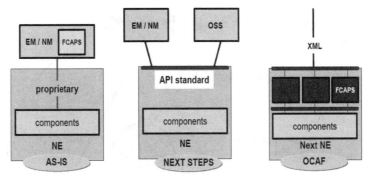

Figure 3.16 Management architecture based on OCAF recommendations [7].

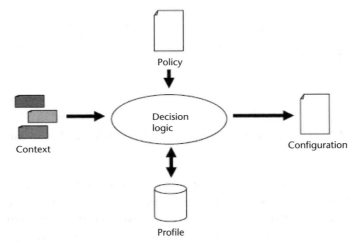

Figure 3.17 Relationship between context, profiles, policies, and configuration.

- Enhancing compatibility and consistency;
- Frequency changes of policies during run time is supported;
- Learning process in decision-making framework, which leads to continual growth and increasing complexity, is supported.

The state-of-the-art for policy syntax and format is explained in detail in IST project ORACLE Deliverable D4.2 [59]. Within the scope of the IST FP6 project ORACLE, policy can be understood as "a plan of action to guide decisions and actions. A policy selects actions to be taken and establishes rules for these actions to reach an explicit goal. A policy is defined by a set of rules and a rules-based reasoning algorithm." A number of policy languages are described in the ORACLE Deliverable [59] with special emphasis on policy description language (PDL) [60], the Ponder policy specification language [61], and the DARPA-XG Policy Language [62]. In general, they can be categorized into those that address access control and those that address resource management.

Figure 3.18 shows the basic information flow in context and policy processing within the ORACLE policy framework.

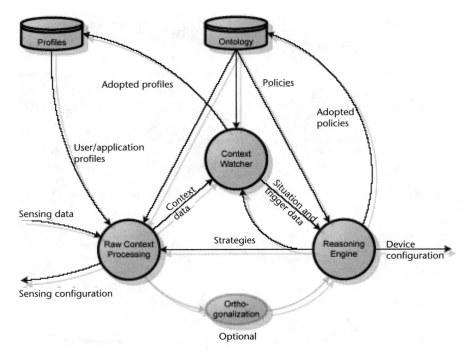

Figure 3.18   Overview of basic information flow in context and policy processing.

The *raw context processing* entity is responsible for obtaining the sensing data from any available sensor device or protocol layer service, which may include the lower layer, network layer, application layer sensing, and metering services. These profiles provide metadata, which is used to determine which type of information can be obtained from which source available in the architecture. It directly interfaces with the local sensing devices and is also responsible for configuring them. A unified vector context vector is the output of this entity and is provided to the *context watcher* entity. It implements a rule-based logic, which determines which change of context will trigger the policy processing by the *reasoning engine*. The behavior of the raw context processing entity is controlled by policy rules provided by ontology and by strategies from the reasoning engine. It obtains the required information from the context watcher to trigger a matching policy rule for the opportunity specified by the current situation. The context watcher may generate new profiles or update existing ones. The reasoning engine is capable of creating or updating the policies within the ontology. A key point of the ORACLE policy framework is that the collaboration not only supports the exchange of sensing data between the networked OR terminals but also allows the exchange of policies to be applied to all of the context processing, environment observation, and opportunity detection and utilizatio.

The envisaged ORACLE ontology consists of facts and rules required to define the behavior of collaborative OR terminals and base stations. It is assumed that the semantics of this ontology are based on extensible markup language (XML) [63], RDF [35] and OWL [34]. It is comprised of the knowledge base as well as its content and its OWL-based semantics. The approach for controlling the communication of sensing parameters and policies in a networked environment of collaborating OR terminals is described briefly as follows:

- The reflection of remote sensing parameters towards the ontology requires an appropriate context processing that is able to collapse single values into a context data vector, introducing an abstraction level that isolates context watching and reasoning from the need to be informed about each neighboring terminal in full detail. Further study can be conducted to achieve this by introducing opportunity map objects as an outcome of the raw context processing into the XG ontology.
- To exchange policies between collaborating OR terminals, a process needs to accept the policy rules as an input parameter, and it must be possible to bind the parameters to values that represent policy rules received from remote OR terminals.

The ORACLE policies for a WLAN scenario are covered in [63].

Figure 3.19 shows the various types of profiles considered in IST FP6 project ORACLE, which are categorized as follows:

- *User profiles* define the settings for applications and OR terminal associated with a single user. It is further categorized into classes such as premium users and regular users. User preferences and person service descriptions, such as QoS level, tariff preferences, service personalization, and subscription requirement, are included. The profiles may be dynamically associated with the user under consideration. Further study is needed regarding the storage of a human user's information in the user profile for personalization purposes.
- *Application profiles* store configuration settings for a distinct application or group of applications with common attributes. They assist the decision process because they define application requirements such as bandwidth request, minimum requirement for the transmission speed, minimum throughput, and

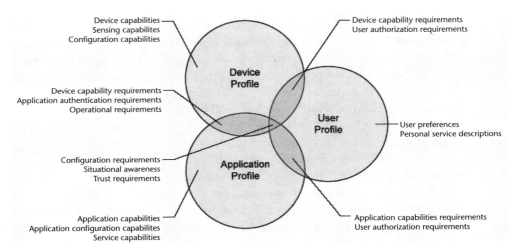

Figure 3.19 Overview of ORACLE profiles.

allowed/maximum delay and jitter. They also include the operation profile, which bridges the gap between device configuration and service/application configuration.

- *Terminal capability profiles* are a subset of the device capability profiles and contain the capability settings for the OR terminal. They include parameters that can be configured, such as radio access options, transmission power range, terminal resources, protocol environment, and so forth, as well as sensing services available to the neighboring OR terminals.

The cognitive engine analyzes these profiles so as to configure its decision-making process by considering the degrees of freedom and the terminal constraints, which helps to identify authorized decisions. These profiles provide the rules (in a predefined format handled by ontology) that are used by the cognitive engine to make coherent decisions taking possible interactions into account.

Currently in communication devices, the decision-making processes takes place throughout the protocol stack. While some of the decision processes are visible to the human user, the rest are hidden within the system operation. These hidden decision processes mainly consist of built-in initialization procedures and run-time control loops. They are based on simple interactions with the user, who initiates communication activity in a rather constrained manner, or on simple organization procedures at the system level, based on wide margins and simple rules that permit convergence at a good operating point. The requirements of opportunistic access are as follows

- More extensive decision making;
- More often;
- Based on much more interrelated parameters.

The decision-making problem is exacerbated by mobility, system coexistence, and the need for increased transparency for the user. A complete decision-making framework for OR communication devices should support any kind of decision making influencing the operation of a communication device that is capable of engaging in opportunistic communication. Some of the important assumptions made are as follows: The radio terminals should be multimode to adapt to different radio norms. The base station devices should support multiband communications with different users. Also, wideband sensing should be possible so as to decrease the sensing time. It is important that the radio devices support and provide explicitly different configuration options, which should be available for allowing dynamic reconfiguration satisfying policy components such as satisfaction of QoS constraints while minimizing energy consumption.

The decision engine framework is shown in Figure 3.20. The three main elements are context definition, context modeling, and opportunity detection and reconfiguration decision for opportunity exploitation. The cognition and reconfiguration manager supervises the detection process. The decision cycle starts by the detection of the potential opportunities resulting from the analysis of the current environment metrics. The opportunity identification is assured due to specific knowledge regarding the opportunity possibilities in space, frequency, time, and

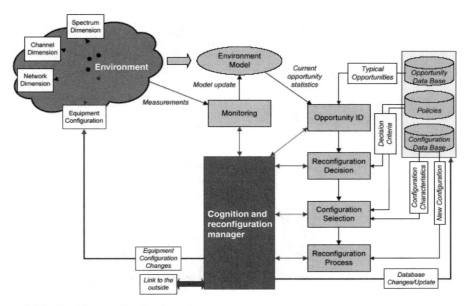

Figure 3.20   Decision engine framework.

code. An opportunity once detected, should be exploited and this phase can be decomposed into the following tasks:

- Decide whether reconfiguration must occur and appropriate decision-making policies should be used to minimize the number of reconfigurations;
- If necessary, the decision unit should select appropriate configurations and parameters to adapt the OR the environment according to some policy. The configuration space is limited by the terminal capabilities stored in the configuration data base;
- Reconfiguration is executed based on the decisions of the cognitive process.

Fuzzy logic is a reasoning mechanism associated with fuzzy set theory; it is a well-founded logic responsible for approximate reasoning. It is one of the soft computing tools, a category that includes other tools which may be of interest for working out the cognitive part of cognitive radios. Expert knowledge is encoded in fuzzy rules close to natural language expression. The number of rules is decreased significantly compared to conventional expert systems, thanks to fuzzy logic's capacity to sum up knowledge. The inference speed is then increased, allowing real-time decision making. Figure 3.21 shows the decision making process using fuzzy logic.

The decision process is broken up into several decision levels. First, the need for reconfiguration is assessed. The cognitive engine needs to determine which spectrum portion to use and how to share it. These decisions are based on spectral occupancy. Finally, the needed signal power is computed. Since these decisions are not independent, the cognitive engine needs to integrate a mechanism to represent the influence of one decision on the others. A distinct separation of the decision level is preferable, but it implies finding a way to determine on which level to play while maintaining knowledge of the interactions. Each decision-making unit is repre-

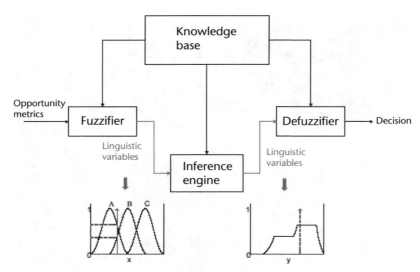

Figure 3.21   Decision-making unit architecture using fuzzy logic.

sented by a fuzzy expert system with specific knowledge encoded in fuzzy rules, which represent the expertise governing the decision process and link the sensing space to the decision space. Metrics that are *fuzzified* (i.e., converted to fuzzy sets) are fed to the fuzzy expert systems. After fuzzification, the fuzzy rules and fuzzy inference mechanisms are used to map the metric space to the decision space. Then a *defuzzifier* is applied to obtain a crisp value which corresponds to the degree of satisfaction for each possible alternative and the alternative with maximum satisfaction degree is selected.

The genetic algorithms (GAs) can be used to optimize the decision-making framework. In OR, decision making involves detecting spectrum opportunities based on the current environment and then reconfiguring the radio device to exploit them and make opportunistic allocation decisions. Several contexts, including profiles, policies, and configuration, are used to achieve various targets that could be in conflict. Selecting the most appropriate contexts in opportunity exploitation is a complex task that is generally solved by multiobjective optimization techniques. Due to the complexity of this problem, the nonlinear behavior is such that explicit functions modeling the system evolution are not readily available. These difficulties pose severe limitations to the application of classical analytical and semianalytical multiobjective optimization methods, including those based on an evaluation of the gradient of the system. However, it has been reported that evolutionary algorithms, especially GA, are good candidates for solving these kinds of problems. These are stochastic search techniques that are computationally based on a method of adaptation similar to the process of evolution in living organisms. The techniques are programmed to work the way the population does, solving problems by changing and reorganizing their component parts via processes similar to reproduction, mutation, and natural selection. The multiobjective GA defines fitness functions according to the specific objective functions, and the algorithms evolve by using genetic selection techniques to produce the optimal solution. These techniques include:

- *Crossover:* Based on roulette-wheel selection, two new offspring are created by recombining portion of good individuals from the parental chromosomes, thus developing a higher value of fitness or survivability than the parents.
- *Mutation:* This process brings new individuals into the gene pool through random modification on an individual. Thus, a better solution than what was previously possible can be obtained.
- *Elitism:* This process guarantees the transfer of best chromosomes from one population to the next, reducing the risk of eliminating the best-fit chromosomes at the early stage of the algorithm.

Figure 3.22 shows a GA optimization technique in which the initial population of size *N* chromosomes was generated. The fitness function of each candidate was calculated, after which the best pair of chromosomes with best fitness value was selected and passed through the processes of crossover, mutation, and elitism to obtain the optimum solution. More information on the basic concepts of GA is available in [59].

For OR, the application begins with mapping the problem to artificial chromosomes, a process called decoding. Several fitness functions have been defined, considering different objectives using the specific knowledge of the OR framework. Figure 3.23 shows the GA-based decision-making tool used in OR. These algorithms are essentially a type of metastrategy for solutions. When applying these algorithms to solve multiple objective decision-making problems, it is necessary to define each of their major components, such as encoding methods, recombination operators, fitness assignment, selection and constraint handling, and so forth.

First, the initial population is generated and then subjected to the genetic search, which consists of fitness evaluation, selection, and recombination, which is carried out until the termination criterion is met. Finally, the best chromosome is selected and decoded to obtain the preferred solution.

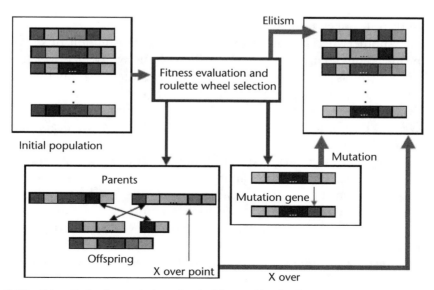

Figure 3.22   GA optimization technique for decision making [64].

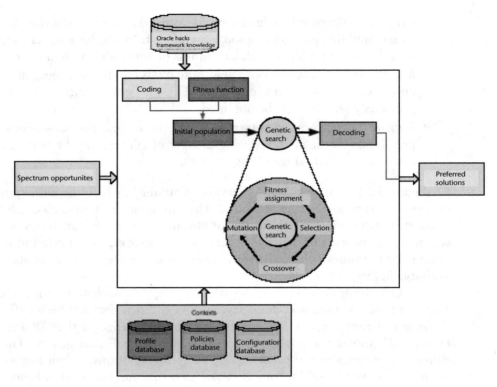

Figure 3.23 GA-based decision-making tool for OR.

Determining the fitness value of chromosomes according to multiple objectives is a challenging issue. Some of the methods used can be classified into one of the following approaches: vector evaluation approach, weighted-sum approach, pareto-based approach, compromise approach, and goal programming approach. It is important to identify which technique is appropriate for satisfying the objectives of the OR policy framework.

Another special issue is the encoding of the solution of the decision-making problem into a chromosome. Depending on the symbol used as alleles of the gene, these methods can be classified into binary encoding, real-number encoding, integer or literal permutation encoding, and general data structure encoding. A combination of binary and real-number encoding technique was studied in IST project ORACLE [59], in which the spectrum opportunities and decision-making policies were mapped using binary encoding technique and the terminal and base station configurations were mapped using real number encoding methods.

The policies used in this approach consist of optimization statements and constraint statements. Optimization statements are used to guide the design space exploration to identify the best decision. Complex policies that require several optimization statements to fulfill compromises between conflicting criteria require the introduction of multicriteria decision techniques in the decision engine. The constraint statements help to narrow down the decision space by eliminating the potential solutions in violation with the constraints imposed. These constraints are imposed by the regulation agencies, who dictate how the spectrum should be accessed and also imposed by the terminal capabilities profile, preventing some configuration from be used. Therefore, the policy framework adopted in this case represented the optimization statements as optimization criteria, weighted by their degree of importance and constraint statements as expert rules applying to the decision space before or during the decision process.

With respect to the architecture, the decision engine has a modular architecture including a module that specializes in radio environment modeling, several estimation agents responsible for the opportunities quality assessment, and a supervision expert system for coordinating the actions of the different entities.

To summarize, the policy and decision-making framework were based on self-computing methods. The fuzzy logic approach was articulated around fuzzy logic expressions of policies; it is a good tool for control, expert systems, and decision making, as it provides a pragmatic, efficient, and generic approach to problem solving. The advantages of using fuzzy logic are as follows:

- It is a low-cost, simple, and efficient technology, which is very important for hardware implementation in mobile terminals.
- Since it is flexible, the tradeoff between the description precision and the complexity may be easily adjusted.
- It is tolerant of imprecise and uncertain data, and sensing information uncertainties may be taken into account during the decision process.
- It may solve conflicting objectives.
- Since it is based on natural language, human policies maybe more easily implemented.

The advantages of using the GA approach are as follows:

- It simplifies the search of a large space with complicated problems.
- It offers a fast and low-cost search function, which supports real-time application.
- It provides optimal solution in a shorter time.

Thus, the decision-making framework can be easily adapted and extended to tackle the increasing decision-making needs of future intelligent communication devices and networks. Since it is possible to handle all separate problems within a single framework, joint optimization can be facilitated. Also, the chosen modular architecture allows easy extensions to handle unanticipated aspects of decision making, which are still open issues in the field of cognitive radio research.

3.8   Unified Scenario on Autonomic Communications Systems for Seamless Experience

A unified scenario on autonomic communications system for seamless experience was described by the IST project E2R II to highlight the main technical issues related to autonomic device management and network self-organization and self-management. The scenario was an extension of a previous proposal describing how pervasive services can exploit reconfiguration for advanced concepts and, specifically, for autonomic communication environments [65]. Four basic categories of capabilities were included in the scenario, as follows:

- A *self-management capability* for profile interpretation, policy-based management, autonomic decision making, formation of network compartments, and self-protection;
- A *self-knowledge capability* for resource management, profile management, RAT discovery and selection, cognitive service provision/discovery, traffic load prediction, and network monitoring;
- A *self-configuration capability* for self-healing, self-protection, self-configuring protocols, software download, and base-station reconfiguration;
- A *self-optimization capability* for resource allocation, traffic load balancing, dynamic spectrum allocation, flexible spectrum management, and the dynamic extension of network coverage.

Consider the following example scenario: a company group is traveling by train from a small city outside Beijing to Beijing to join a telecommunications exhibition. The workers have reconfigurable autonomic devices, which they can use to access different technologies and the company groupware software while the equipment is communicating with the company intranet. The complex and continuous change may require vertical handovers, capability negotiation, self-awareness, and the possibility of adapting the services based on the telecommunication system context, profiles, and policy rules; it also the ability to reconfigure the equipment with new protocols, codecs, and cipher algorithms. On reaching the destination, multimedia location-based services related to tourist content and information are available to the users by means of intelligent service provisioning. For example, a large number of fans arriving at the Olympic Center cause a burst for the network traffic load. Service provisioning is very complex. The events are broadcast nationwide, using a wide range of transmission technologies. Additional coverage and service provision facilities may also be needed. The service environment can become very complex, since the center is divided into different sectors and the attendees have different demands, requiring the convergence of heterogeneous wireless technologies, products, and services.

For the previous scenario, use of the state-of-the-art network management mechanisms based on manual configuration would lead to inefficient and costly results. An intelligent autonomic networking technology, on the other hand, could help solve the problems of a complex communication environment, as in the following example:

The employees leave by train to Beijing from a small city outside Beijing to join a telecommunications exhibition. They use their company groupware to access the company intranet with their reconfigurable autonomic devices. A secure WLAN ad-hoc mode between the devices is used for private communications and can provide an "always-on" communication link. The autonomic concept envisages a hierarchy of systems that are flexible, accessible, and transparent to the user for a platform- and device-agnostic approach for systems with always-on capabilities and on-the-fly seamless adaptation to the user needs. The laptop can act as a gateway for the station's WLAN facility. But if this connectivity breaks down due to an out-of-range problem, the equipment discovers an alternative wide-range radio access technology (e.g., UMTS). The autonomic device continuously monitors and interprets the context information and, based on this, selects a system to connect to in a personalized way. Service adaptation takes place to compensate for the QoS degradation when the handover takes place from WLAN to UMTS. The autonomic device may also trigger protocol/software downloading and reconfiguration to adapt to the environment. Reconfiguration of hardware resources may be triggered. These aspects come under the self-configuration capability of the autonomic device. It is assumed that from this step and beyond, there is no private communication.

Autonomic decision-making mechanisms embedded within the user terminal allow the RAT selection and software downloading procedures. In this way, the autonomic device can minimize human intervention, which maybe required only for privacy issues.

The decision for the intersystem handover falls within the policy-based self-management mechanisms of the autonomic device. The autonomic decision-making mechanisms are based on policy hierarchies and coordinate the self-configuration procedures locally in a context-aware manner [5]. The mechanisms considered extend the traditional protocol component functionality to autonomic elements that enable the self-configuration of protocols.

When the train approaches the next station where WLAN BSs are available, the WLAN BSs recognize the high traffic load due to the numerous devices connected simultaneously and already requesting service, and they predict the rise in requests with the arrival of a new train. The BSs negotiate with their respective UMTS counterparts close to the station for the formation of a broader WLAN compartment cluster around the station.

Autonomic network capabilities and characteristics involve functional and algorithmic compositions of components in the autonomic network nodes and also the cluster formation of these nodes. The BSs may use the prediction- and policy-based algorithms. On reaching the destination, the autonomic device continuously monitors the environment, seeking ways to tune performance and self-optimize; it performs updates depending on the preferences of the users. After setting up a secure configuration channel, a new ciphering algorithm is downloaded and installed, since the service and terminal use different ciphering algorithms.

A suitable patch is downloaded and installed, but the newly installed module crashes. In this situation, the autonomic device makes use of its self-healing capability by detecting the ongoing failure and rolls back to its previous configuration. The manufacturer and software provider are informed about the malfunction; they then develop a patch to fix the compatibility with the specific software and trigger a mass-upgrade for a designated group of terminals.

The workers arrive at the Olympic Center, which is divided into different parts. There is unacceptable interference, and the operating companies start to build a temporary radio network. The radio network uses its cognitive capability and negotiation schemes to map the spectrum distribution, the coverage of different networks, and the interference level, and reconfigures the parameters of the network deployment. The system can also classify the users into high-density and slowly-moving groups and multicast a reconfiguration decision to these terminals according to the location and the network deployment.

On the following day, the Olympic Center organizes a basketball league match, and all the workers visit the Center to watch it. The autonomic system predicts the traffic burst for this time period and reserves radio resources for different services. For example, the BSs in the southern side of the center are switched to a RAT with low coverage but high throughput, as this location is reserved for the main-field fans. Different handsets are powered up by the users to receive the match news, to make phone calls, and so on when more than 5,000 fans enter the center. For people located in the back rows, the autonomic decision-making mechanism in the terminals specifies that it should be reconfigured to the DVB mode, since those users have difficulty seeing the game clearly. If by chance one BS is incautiously broken, the self-healing process starts. The system sends network and resource information to the users connected previously to the broken BS through the common cognitive pilot channel (CPC) [7], which helps users access other APs or BSs. If the QoS level falls below a defined threshold, the system automatically starts to negotiate with regulatory entities and other networks for additional spectrum band.

On the same day, the annual Beijing marathon is being held; it runs through the city and finishes in the Olympic Center. With autonomic devices, users can receive broadcasts via different services. The mobile broadcasting system is a mesh type that uses DVB-H for popular channels and 3G for on-demand niche channels, and alters dynamically to ensure optimal assurance of QoS. The content modulators receive information from operators and directly from devices to transcode the content to be optimized for the transmitting networks and receiving devices. A major television station transmits a signal that covers the entire race via DVB-H. The people within the stadium watch on-demand video broadcasts of certain camera viewpoints filmed by small independent TV crews. When the favorite runner drops from the lead, people switch to the on-demand channel of operator O1, causing congestion in the 3G network. O1 tries to exploit the underused GSM network, which it operates, to temporarily transfer the spectrum capacity and processing power to the 3G network. But since the congestion continues, O1 negotiates with the spectrum manager of O4, a competitor, to rent some spectrum capacity. Because the cost is too high, it negotiates with O2 to locally switch the on-demand signal to the Wi-Fi network. People outside the stadium switch to this camera point as the race progresses. The 3G networks become congested at many locations; broadcasting with DVB-H will be more efficient.

3.9   Conclusions

Multistandard base stations are expected to be key players in the future deployment of cognitive, heterogeneous wireless systems. With multistandard BSs, the network operator will be able to dynamically adapt the spectrum occupation strategy and the choice of the suitable RATs depending on the current context. The user terminals, on the other hand, have to adapt their respective radio resource usage strategies depending on a time-variant network-side configuration. An approach involving self-management and self-organization would take the specificities of heavily heterogeneous systems (which are based on collaborative and autonomous distributed decision-making principles) into account. While collaborative distributed decision-making requires some interactions between the network and its edges and/or the network and user terminals, any autonomous approach is supposed to enact its decisions unilaterally. In this context, two target areas can be identified: the edges of the network, which adapt autonomously to the time-variant needs of the system from a network perspective, and the terminals (e.g., operating in a multioperator domain), which are able to select link providers in a noncollaborative way and exchange context information mutually with neighboring entities. Autonomous cognitive radio schemes are a key enabler for distributed decision making and are, therefore, in the scope of the future research.

References

[1] IST-2005-027714 Project E2R II, http://www.e2r.motlabs.com

[2] Dillinger, M., K. Madami, and N. Alonistioti (eds.), *Software Defined Radio: Architectures, Systems, and Functions*, Wiley & Sons, April 2003.

[3] Haykin, S., "Cognitive Radio: Brain-Empowered Wireless Communications," *IEEE J-SAC*, Vol. 23, No. 2, February 2005.

[4] IBM Research, *Autonomic Computing*, http://www.research.ibm.com/autonomic/.

[5] Samaan, N., and A. Karmouch, "An Automated Policy-Based Management Framework for Wired/Wireless Differentiated Communications Systems," *IEEE J-SAC*, Vol. 23, No. 12, December 2005.

[6] ICT project E3R, https://ict-e3.eu/.

[7] ICT project E2R II, "Title." Deliverable D2.1, IST-2005-027714.

[8] Murch, R., *Autonomic Computing*, Prentice Hall, March 2004.

[9] Kephart, J., and D. Chess, "The Vision of Autonomic Computing," *Computer*, Vol. 36, No. 1, January 2003, pp. 41–50.

[10] Patouni, E. et al., "Exploiting Protocol Reconfiguration for Autonomic Communication Environments," *Proc. IST 2006*, Mykonos, Greece, June 2006.

[11] IBM Corporation, "An Architectural Blueprint for Autonomic Computing" (white paper), April 2003, http://www-01.ibm.com/software/tivoli/autonomic/.

[12] IBM Corporation, *Autonomic Computing Toolkit: Developer's Guide*, 2nd ed., August 2004, http://www-106.ibm.com/developerworks/autonomic/books/fpy0mst.htm

[13] Appavoo, J., et al. "Enabling Autonomic Behaviour in System Software with Soft Swapping," *IBM Systems Journal*, Vol. 42, No. 1, 2003.

[14] Fujii, K., and T. Suda, "Semantics-Based Dynamic Service Composition," *IEEE J-SAC*, Vol. 23, No. 12, December 2005.

[15] Mullany, F. et al. "Self-Deployment, Self-Configuration: Critical Future Paradigms for Wireless Access Networks," *Proc. 1st International Workshop on Autonomic Communications (WAC 2004)*, Berlin, Germany, October 2004.

[16] Gu, X. et al, "Towards Self-Optimizing Protocol Stack for Autonomic Communications: Initial Experience," *Proc. 2nd International Workshop for Autonomic Communications (WAC 2006)*, Athens, Greece, October 2005.

[17] Diao, Y. et al, "Self-Managing Systems: A Control Theory Foundation," *Proc. 12ᵗʰ IEEE International Conference and Workshops on the Engineering of Computer Based Systems (ECBS 2005)*, Greenbelt, Massachusetts, U.S.A., April 2005.

[18] Fei, W. and L. Fan-Zhang, "The Design of an Autonomic Computing Model and the Algo-
      rithm for Decision Making," *Proc. IEEE International Conference on Granular Comput-
      ing*, Beijing, China, July 2005.

[19] Vidales, P. et al. "Autonomic System for Mobility Support in 4G Networks," *IEEE J-SAC*,
      Vol. 23, No. 12, December 2005.

[20] Microsoft Passport, http://www.passport.net.

[21] The Liberty Alliance Project, http://projectliberty.org

[22] 3GPP, TS 29.240, "3GPP Generic User Profile (GUP); Stage 3; Network (Release 6),"
      V6.1.0, June 2005.

[23] ETSI EG 202 249 V1.1.1., "Universal Communications Identifier (UCI), Guidelines on the
      Usability of UCI-Based Systems," November 2003.

[24] "XRI and XDI Explained," http://www.xns.org/xri-and-xdi-explained.html

[25] Klyne, G. et al., "Composite Capability/Preference Profile (CC/PP): Structure and Vocabu-
      laries   1.0,"   *W3C   Recommendation*,   January   2004,   http://www.w3.org/
      TR/CCPP-struct-vocab

[26] OMA User Agent Profile (UAProf) Specifications, http://www.openmobilealliance.org/
      Technical/release_program/uap_v2_0.aspx

[27] Distributed   Management   Task   Force,   CIM   Network   Model   White   Paper,
      http://www.dmtf.org/standards/published_documents/DSP0152.pdf.

[28] TeleManagement Forum, SID, Overview, http://tmforum.org/browse.aspx?catid=2008.

[29] Gruber, T., "A Translation Approach to Portable Ontology Specifications," *Knowledge
      Acquisition*, Vol. 5, No. 2, June 1993, pp. 199–220.

[30] Floridi, L., *The Blackwell Guide to the Philosophy of Computing and Information*,
      Blackwell, August 2003, pp. 155–166.

[31] IST project E2R, "State-of-the-Art and Outlooks for Reconfiguration and Download Pro-
      cedures, Network Support Functions, and Protocol Architectures and Flexible Service Pro-
      vision, Enabling Platforms," Deliverable D3.1, June 2004.

[32] Segev, A., and A. Gal, "Putting Things in Context: A Topological Approach to Mapping
      Context and Ontologies," *Proc. AAAI Workshop on Contexts and Ontologies*, Pittsburgh,
      Pennsylvania, U.S.A., July 2005.

[33] Genersereth, M., and R. Fikes, "Knowledge Interchange Format, Version 3.0," Computer
      Science Department, Stanford University, Technical Report Logic-92-1, March 1992.

[34] Bechhofer, S. et al., "OWL Web Ontology Language Reference," W3C recommendation,
      February 2004, http://www.w3.org/TR/owl-ref/

[35] Resource Description Framework (RDF), http://www.w3.org/RDF/

[36] The DARPA Agent Mark-up Language, http://www.daml.org/index.html

[37] Strang, T., and C. Linnhoff-Popien, "A Context Modeling Survey," *Proc. Workshop on
      Advanced Context Modeling, Reasoning and Management 2004*, Nottingham, U.K., Sep-
      tember 2004.

[38] Chun, B. et al., "Design Considerations for Information Planes," *Proc. Workshop on Real,
      Large, Distributed Systems (WORLDS 2004)*, San Francisco, California, U.S.A., Decem-
      ber 2004.

[39] Keeney, J., et al., "Ontology-based Semantics for Composable Autnomic Elements," *Proc.
      Workshop on AI in Autonomic Communications*, Edinburgh, U.K., July 2005.

[40] Fleck, J., "An Overview of the NGOSS Architecture," TeleManagement Forum white
      paper, May 2003.

[41] Noll, J. et al, "Roaming of Advanced Telecom Services through Semantic Annotation,"
      Proc. IEEE ICPS 2006, Lyon, France, June 2006.

[42] Flury, T. et al., "OWL-based Location Ontology for Context-Aware Services," *Proc.
      Workshop in Artificial Intelligence in Mobile Systems (AIMS 2004)*, Nottingham, U.K.,
      September 2004.

[43] Lewis, D., et al., "Semantic-based Policy Engineering for Autonomic Systems," *Proc. 1st IFIP WG6.6 International Workshop on Autonomic Communication Principles*, Berlin, Germany, October 2004.

[44] Laukken, M., "Semantic Web Technologies in Context-Aware Systems," seminar, University of Helsinki, Finland, March 2004.

[45] Open Mobile Alliance, "OMA Device Management Protocol," http://member.openmobilealliance.org/ftp/Public_documents/DM/

[46] Open Mobile Alliance, "Device Management Architecture," http://member.openmobilealliance.org/ftp/Public_documents/DM/

[47] Open Mobile Alliance, "FUMO Architecure," http://member.openmobilealliance.org/ftp/Public_documents/DM/

[48] Open Mobile Alliance, "Software Component Management Object Requirements," http://member.openmobilealliance.org/ftp/Public_documents/DM/

[49] Open Mobile Alliance, "DM Diagnostics and Monitoring Requirements," http://member.openmobilealliance.org/ftp/Public_documents/DM/

[50] Open Mobile Alliance, "Device Management Scheduling Requirements," http://member.openmobilealliance.org/ftp/Public_documents/DM/

[51] Telecommunication Management Network (TMN), http://www.itu.int/itudoc/itu-t/com4/tmn/tmn0402.html.

[52] Telecommunication Information Networking Architecture (TINA), http://www.tinac.com

[53] TeleManagement Forum (TMF), http://www.tmforum.org

[54] ETSI TISPAN Service-Oriented Architecture (SOA), http://www.etsi.com/tispan/tispan.htm

[55] Nakina Systems Multi-Vendor Element Management Solution, http://www.nakinasystems.com/

[56] Cisco Active Network Abstraction (ANA), http://www.cisco.com/en/US/products/ps6776/index.html

[57] Open Communications Architecture Forum (OCAF), http://www.itu.int/ITU-T/ocaf

[58] ITU-T SG13 NGN group, http://www.itu.int/ITU-T/studygroups/com13/index.asp

[59] IST-2004 027965, "Definition of Context Filtering Mechanisms and Policy Framework," ORACLE Deliverable D4.2, May 2007, http://www.ist-oracle.org/.

[60] Lobo, J. et al., "A Policy Description Language," *Proceedings of AAAI*, Orlando, Florida, U.S.A., July 1999.

[61] Damianou, N. et al., "The Ponder Policy Specification Language," *Proceedings of the Policy 2001, Workshop Policies for Distributed Systems and Networks*, Springer-Verlag, 2001.

[62] DARPA XG Working Group, "DARPA XG Policy Language Framework, Request for Comments," prepared by BBN Technologies, Cambridge Massachusetts, U.S.A., 2004.

[63] Bray, T. et al., "Extensible Markup Language (XML) 1.0, 4th Edition—Origin and Goals," *WWW Consortium*, September 2006.

[64] Moessner, K. et al, "Draft OR Policy Framework," Deliverable D4.1, ORACLE, 2006.

[65] E2R Deliverable D1.1, "Scenarios and Requirements, State of the Art," June 2004.

CHAPTER 4

System Capabilities

This chapter discusses the capabilities of end-to-end reconfigurable systems that need to be supported by the system architectures. Several functional requirements are outlined and explained using use-case diagrams. The key thematic areas that are addressed are policy management, cognitive service provision and discovery, emergency services, context interpretation, self-configuring protocols, mass upgrades of mobile terminals, handover, formation of network compartments and base station reconfiguration, traffic load prediction and balancing, network resource management, and RAT discovery and selection. The main contributor to this project is the IST FP6 Project E2R [1, 2].

This chapter is organized as follows. Section 4.1 gives an introduction to the topic. Section 4.2 identifies the main actors and functional requirements for the policy management procedure proposed for reconfigurable systems. Section 4.3 describes the important actors and the functional requirements for cognitive service provision and discovery and analyzes several aspects related to it. Section 4.4 extends the basic concepts of Section 4.3 for the specific case of cognitive service provision in an emergency scenario. Section 4.5 describes the main actors and functional requirements for the context interpretation procedure. Section 4.6 describes the requirements for the protocols that support self-configuration. The mass upgrade of mobile terminals is described in section 4.7. Section 4.8 describes the main actors and functional requirements for support of handover. Section 4.9 describes the main actors and functional requirements for the formation of network compartments and for base station reconfiguration. Section 4.10 describes the main actors and functional requirements for the traffic load prediction and balancing procedure. Section 4.11 describes the main actors and functionalities involved in the network resource management procedure. Section 4.12 describes the main actors and functionalities involved in the RAT discovery and selection procedure. Section 4.13 concludes the chapter.

4.1 Introduction

The following is a list of capabilities that represent a class of common characteristics or features end-to-end systems need to provide to manage reconfigurable terminals [1]:

- *Service level agreement:* This agreement permits the involved parties to establish the minimum performance criteria for service provision and the actions to be taken if the services do not meet the criteria.

- *Equipment reconfiguration:* This means that the equipment should be able to change its configuration and operating parameters (e.g., frequency, modulation, transmitted power etc.) autonomously using software (i.e., without any external maintenance).

- *Security:* The equipment reconfiguration performed during an upgrade of one or more elements of the system should be done in a secure way.

- *No radio interference:* Protection is needed against the possible interference from badly reconfigured equipment.

- *Download:* Mechanisms that allow the equipment to identify a needed software module and download it to reconfigure to a new configuration in a new wireless network, should be available.

- *Reconfiguration management:* The process for reconfiguration could be initiated either by the equipment or by the network. A virtual identity, like a reconfiguration manager, could be used to control the end-to-end reconfiguration management actions.

- *Service adaptation:* The active services should be able to adapt to changes in the network status (e.g., congestion), modifications of the equipment (e.g., reconfiguration), or alteration of the access network used by an equipment (e.g., vertical handover). Equipment reconfiguration may also take place in an attempt to avoid major disruption on the executed service.

- *Vertical handover:* Equipments should be able to move through different access networks without losing their active connections. The handover could be either network or equipment-initiated.

- *Service provision:* These are basic telecommunication services and also the value-added services offered by operators or independent VAPs.

- *System monitoring:* The equipment and the network should be able to monitor the current state of the system operation, such as the traffic, used spectrum, available technologies, and so forth to estimate available resources and to facilitate the best use of existing and available resources.

- *Dynamic resource management:* To make the best use of any resource, the network operator assigns that resource dynamically to the different tasks performed.

- *Spectrum transfer:* The spectrum owners can transfer their spectrum to other parties according to commercial agreements such as resale or lease. To make new spectrum available, change an existing technology to a different band, or change the technology assigned currently to a certain band, the regulator may decide to perform a reallocation of the spectrum assigned to communication systems.

4.2 Policy Management

Figure 4.1 shows the use-case diagram that depicts the relation of policy management with autonomic decision making along with the concerned actors.

The concerned actors are as follows [2]:

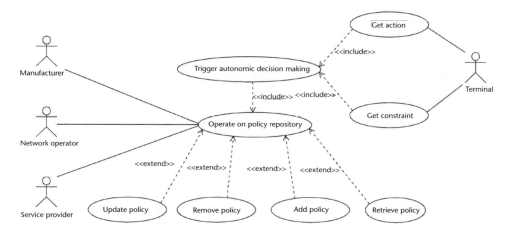

Figure 4.1 Use-case diagram for policy management [2].

- The device manufacturer;
- The network operator;
- The service provider;
- The terminal [1].

The first three actors can perform operations on the policy repository and can *retrieve*, *add*, *update* or *remove* policies. The terminal device can *get an action* or *get a constraint*, which triggers autonomic decision making.

The use-cases that have been defined in Figure 4.1 are as follows:

- Operate on repository;
- Retrieve policy;
- Add policy;
- Remove policy;
- Update policy;
- Get action;
- Get constraint;
- Trigger autonomic decision making.

The *operate on repository* policy can be broken down into *retrieve*, *add*, *remove*, and *update* policy use-cases.

Both, *get action* and *get constraint* policies trigger autonomic decision making which includes the "Operate on repository" use-case. These use-cases also trigger the definition of a reconfiguration action or the generation of a set of limitations, which are performed by the autonomic decision making mechanisms.

Table 4.1 summarizes the capabilities derived from Policy Management use-cases.

Table 4.1   Capabilities Derived from Policy Management Use-Cases [2]

Capability	Definition
Retrieve Policy	Specifies the mechanisms for policy retrieval from the policy repository according to the given criteria.
Add Policy	Inserts a new policy into the policy repository.
Remove Policy	Deletes an existing policy from the repository according to specific criteria.
Update Policy	Modifies the properties of a given policy in the repository.
Get Action	Specifies a concrete reconfiguration action.
Get Constraint	Specifies the limitation or a set of limitations.

4.3   Cognitive Service Provision and Discovery

Cognitive service provision and discovery is explained here in three main aspects—namely, involved actors and their relationships; corresponding use-cases; and the overall architecture, which is enhanced both by the actors and the use-cases. The diagrams provided for the actors and use-cases should be perceived as complementary to the overall use-case diagram.

Some definitions have been given here to differentiate and clarify the semantics of the two distinct notions: the application and the service.

An *application* is a type of software that provides services to the users via service capability features [3]. A *service* is the user experience provided by one or more applications or by an aggregation of a number of service capability features [4]. A service is basically the realization of the aggregation of the effects generated by the use of service capabilities features by a number of applications.

The main services that have been considered are as follows [5]:

- Advanced application services;
- Conversational services;
- Fixed/mobile converged services;
- Group communication services;
- Integrated services ;
- Media-streaming applications;
- Moving networks, ad hoc networks, personal networks, and personal area networks;
- Nonreal-time, interactive applications;
- Real-time, interactive applications;
- Ubiquitous services.

For example, IP networks are meant to provide a seamless user experience for all services within and across the various access systems.

Figure 4.2 shows the various actors of the cognitive service provision and discovery subsystem. The right part shows the actors that can be considered service providers and the left part identifies important roles that trigger certain events in the overall procedure.

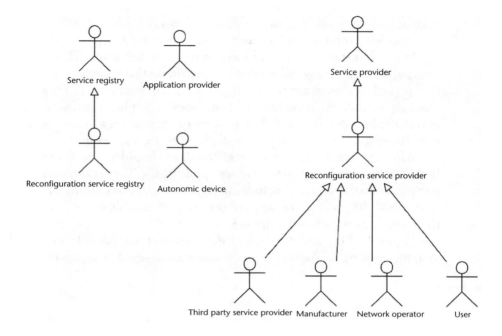

Figure 4.2   Actors involved in cognitive service provision and discovery [2].

The following actors are involved in cognitive service provision:

• Terminal (autonomic);
• Application provider;
• Service registry;
• Reconfiguration service registry;
• Service provider;
• Reconfiguration service provider;
• Third-party service provider;
• Manufacturer;
• Network operator;
• User.

Service providers directly or indirectly provide services to consumers or to the other suppliers. The reconfiguration service provider extends this notion by adding certain capabilities. Most of the network-related actors can be perceived as the reconfiguration service providers or simple service providers. Also, the user in ad hoc networks can become a service provider who supplies services realized by applications situated in the corresponding device. The reconfiguration service registry, a subclass of service registry, supports reconfiguration service registrations. A terminal with autonomic features and the application provider are actors who trigger important events.

In Figure 4.3, the use-cases cover the main cognitive service provision issues, such as service provision, service discovery, service adaptation, reconfiguration service provision, and reconfiguration service discovery.

In the first step, the registration of a service takes place. This registration has an internal part in the overall procedure, providing the necessary descriptive information regarding the nature and the requirements of a service. The registration of a service may include the registration of one or more applications, accompanied by their corresponding descriptions. The process of registering a reconfiguration service is equivalent to this; hence, it can be depicted using a generalization relationship.

After the service has been registered, it should be published; this may also include the publishing of one or more applications. The service-related action, under some circumstances, may include corresponding actions taking place on an application, while the same action applied on a reconfigurable service inherits and extends the former, as shown in Figure 4.3.

Figure 4.4, 4.5, and 4.6 depict three use-cases that provide an overall view of the system, focusing on the way the use-cases are triggered by actors or other use-cases.

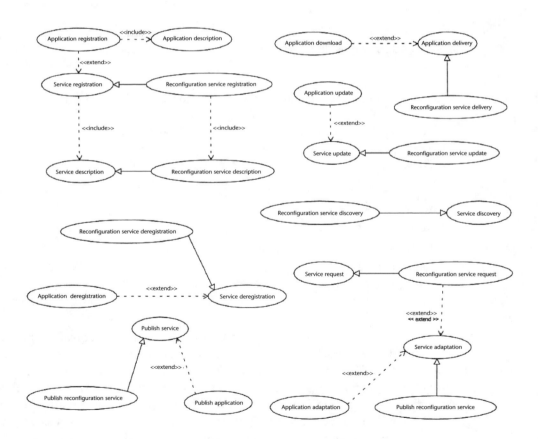

Figure 4.3 Use-cases for cognitive service provision and discovery, and the relations among them [2].

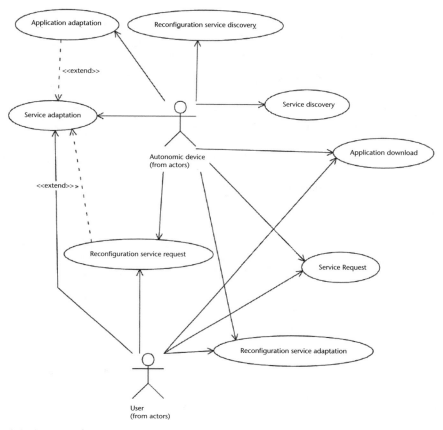

Figure 4.4 Use case for cognitive service provision and discovery (part 1) [2].

Table 4.2 summarizes the capabilities derived from the cognitive service provision and discovery use case [2].

4.4 Emergency Services

This section describes the main actors and functional requirements for the emergency service procedure and enhances specific aspects related to emergency situations. For example, when a natural disaster strikes, the area may be inaccessible, as some deployed access points may be completely devastated. A smart step would be to reconfigure the remaining access points in an autonomous and intelligent way such that even at a short notice, the system can become operational once again, offering vital services to users.

The use-cases defined are the following:

* Network monitoring;
* Traffic load prediction;
* Self-healing;
* Autonomic network dimensioning.

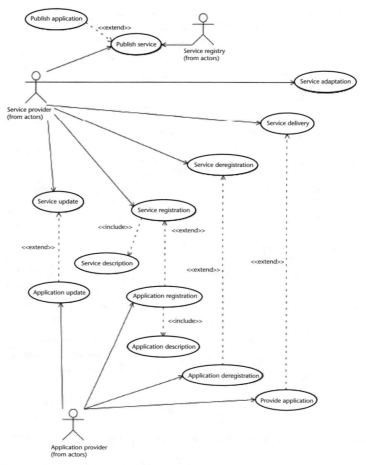

Figure 4.5   Use case for cognitive service provision and discovery (part 2) [2].

Figure 4.7 shows the use-case diagram for the emergency services scenario. The network operator who is the involved actor in this case is responsible for guaranteeing the secure and proper operation of both the network and equipment.

The network operator realizes the network monitoring and traffic load prediction procedures and, based on the acquired information, may trigger the self-healing and autonomic network dimensioning procedures to meet the requirements posed by the emergency situation.

Table 4.3 summarizes the capabilities related to emergency services, taking as an example the case of a service area where a disaster has taken place and there is a need to reconfigure the remaining access points in order to support their resulting traffic increase.

4.5   Context Interpretation

The *context information* can be classified into raw context information and abstract context information. Raw context information is obtained directly by the process of

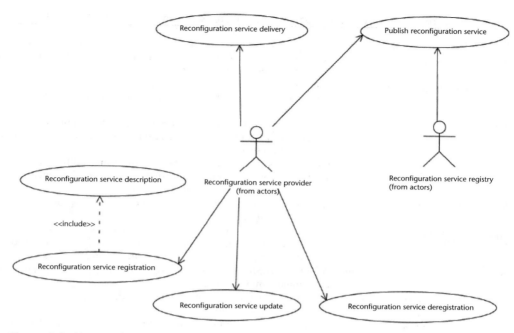

Figure 4.6   Use case for cognitive service provision and discovery (part 3) [2].

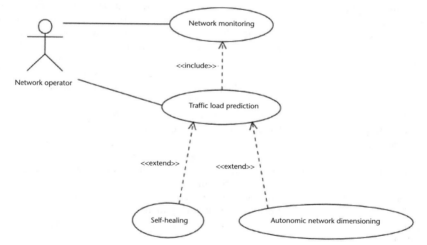

Figure 4.7   Use-case for emergency services [2].

monitoring the state of the world; it could be too voluminous, and it may not be meaningful to higher level adaptation algorithms. It has to be condensed into more abstract context information. *Context abstraction* can be understood as the process of obtaining abstract context information from the raw context information. This process is analogous to that in biological systems, where raw context information is produced by the perceptual transducers and abstract context information is produced by the neural pathways connecting the transducer to the brain. For example,

Table 4.2   Capabilities Derived from Cognitive Service Provision and Discover Use-Cases [2]

Capability	Definition
Service request	Issues a request for a specific service based on the contextual information and requirements at a certain time.
Reconfiguration service request	Issues a request for a specific reconfiguration service based on the user preferences, the contextual information, and the device requirements at a certain time.
Service discovery	Encapsulates all the necessary steps for querying and browsing for any available services that meet the requirements posed by context information and the issuer.
Reconfiguration service discovery	Encapsulates all the necessary steps for querying and browsing for any available reconfiguration services that meet the requirements posed by context information and the issuer.
Service adaptation	Adapts a service based on the requirements posed by the environment in which it operates.
Application adaptation	Adapts an application based on the requirements posed by service it realizes. Under certain circumstances, this event can be accomplished by exploiting component-based software engineering principles.
Reconfiguration service adaptation	Adapts a reconfiguration service based on the requirements posed by the environment in which it operates.
Application download	Embraces all aspects of software downloading in which the manipulated content is an application.
Publish application	Publishes application listings and defines how each particular application is available.
Publish service	Publishes service listings and defines how a particular service is to be published.
Service registration	Registers a service in a service registry for that service to be published.
Service description	Includes all the steps that provide the necessary descriptive information regarding the nature and the requirements of a service.
Service deregistration	Makes the service generally unavailable.
Service delivery	Provisions service through a service provider to one or more clients. Considering the type of service, this process includes the functions that take place at the provider's side and the streaming or transmission of the desired information.
Service update	Updates service registration information.
Application update	Updates application registration information.
Application registration	Involves the initial registration of a new application in an application registry.
Application description	Defines the main characteristics of the service via the application providers during the initial registration.
Provide application	Provisions applications through application providers, including the functions for downloading an application.
Application deregistration	Deletes the application from the application registry.
Reconfiguration service delivery	Provisions reconfiguration service through a reconfiguration service provider. The process includes the functions that take place at the provider's side and the streaming of the desired information, depending on the type of reconfiguration service.
Reconfiguration service registration	Adds the reconfiguration service to the reconfiguration service registry.
Reconfiguration service description	Describes the reconfiguration service characteristics and important information after the initial registration.
Publish reconfiguration service	Publishes reconfiguration service listings and defines how a particular reconfiguration service is available.
Reconfiguration service update	Modifies of the reconfiguration service's registration information.
Reconfiguration service deregistration	Deletes a particular reconfiguration service record from the corresponding registry.

Table 4.2 (continued)

Publish reconfiguration service	Publishes reconfiguration service listings and defines how a particular reconfiguration service is available.
Reconfiguration service update	Modifies of the reconfiguration service's registration information.
Reconfiguration service deregistration	Deletes a particular reconfiguration service record from the corresponding registry.

Table 4.3   Capabilities Derived from Emergency Services Use-Case [2]

Capability	Definition
Traffic load prediction	Predicts future traffic demand across different traffic types so the network can respond autonomously to the traffic load prediction, enabling resource provisioning/allocation to take place in time to accommodate vital user needs.
Self-healing	Diagnoses irregular operation, reactivity in case of malfunctioning, and the capability of automatic remedy.
Network monitoring	Constantly, or upon demand, monitors the network status and network usage evolution via statistics collection and analysis.
Autonomic network dimensioning	Realizes network dimensioning per node and interface without human intervention and analysis (based on network monitoring and traffic prediction).

in humans, the human features are the abstract context information and not the raw context information that is an input to the visual and auditory associative context.

The context abstraction extracts the features from the raw context data; these features are not directly usable by applications. For this, the additional dimensions of semantics and the relevance have to be taken into consideration; this is done through the process of *context interpretation*.

Figure 4.8 shows the use-case diagram for the context interpretation procedure; the following actors are involved:

- A context sensor that monitors and collects the context information of various kinds;
- A context manager that maintains the database of contexts both in the raw form as well as in the abstracted form (temporal aggregation/historical is included);
- An ontology provider that maintains the knowledge database required for semantic interpretation of contextual information;
- A decision-making module that is part of the autonomic adaptation loop, which decides the reconfiguration actions based on the interpreted context change.

Table 4.4 shows the capabilities related to context interpretation derived from the previously described emergency use-case.

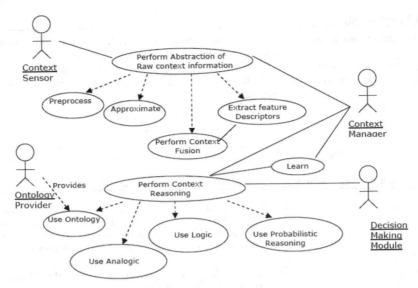

Figure 4.8 Use-case for context interpretation [2].

Table 4.4 Capabilities Derived from Emergency Services Use-Case [2]

Capability	Definition
Context preprocessing	Filters and reformats raw context information into a format acceptable to context management processes.
Context approximation	Trades functionality between quality of context (QoC) and the cost of acquiring for the context information.
Context fusion	Fuses context information of different types to provide a more abstract context description.
Context feature extraction	Performs "pattern recognition" on the raw context data for extracting higher level features of context information.
Ontology usage	Converts context info (raw/abstract) into knowledge using ontological information.
Reasoning	Reasons by combining context with knowledge (logical, analogical, and statistical).

4.6 Self-Configuring Protocols

The self-configuring protocols can be understood to be component-based protocols that incorporate advanced capabilities, enabling the semantics-based dynamic binding and on-the-fly replacement of protocol components through enhanced device management mechanisms [6]. The following actors are identified as key in the process:

- Terminal;
- Network operator;
- Platform module developer, which develops software modules to be installed and executed on reconfigurable network elements and/or user devices [1].

Figure 4.9 shows the use-case diagram for self-configuring protocols.

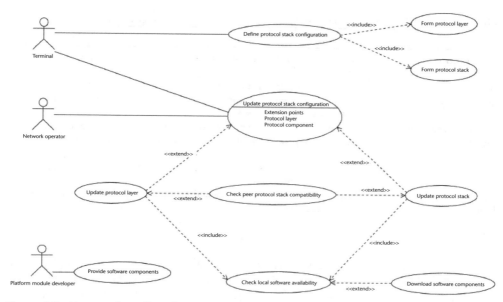

Figure 4.9   Use-case for self-configuring protocols [2].

During the boot-up process, the device terminal specifies the protocol stack configuration by specifying the protocol layers that are to be used and also the protocol components within these layers. Necessary mechanisms are needed to select, install, and activate different protocol versions or protocol stack instances to support the context-aware dynamic adaptation of mobile devices based on location, tariff information, QoS requirements, and network-level operational indicators. The update of protocol stack configuration is required and maybe triggered either by the device terminal or the network operator. The availability of the protocol components required for the selected configuration also needs to be checked, and if those components do not exist in the system, then their downloading and installation from a software repository should be triggered. The software repository has different versions of protocol(s) or protocol components that is provided by the platform module developer. Finally, after the protocol layer/component has been updated, the compatibility of the peer protocol stack is checked and the procedure of the respective update maybe triggered.

Table 4.5 summarizes the capabilities of the self-configuring protocols.

4.7   Mass Upgrade of Mobile Terminals

Mass upgrade of the mobile terminal is an important use-case of reconfiguration. The important actors involved in this case are as follows:

- Manufacturer;
- Platform modules developer;
- Network operator;
- Terminal.

Table 4.5   Capabilities related to use of self-configuring protocols use-case [2].

Capability	Definition
Form protocol layer	Composes a full-fledged protocol layer service based on protocol component services.
Form protocol stack	Composes a protocol stack graph from different protocol layers.
Update protocol stack	Replaces an existing protocol layer or add a new protocol layer to update the protocol stack.
Update protocol layer	Adds or replaces protocol components within a layer to update it.
Download software component	Specifies the procedure of software transfer to the system.
Check peer protocol stack compatibility	Defines the mechanisms that check whether the configuration of the peer protocol stack is compatible.
Check local software availability	Provides the means to verify whether the necessary software components are locally available in the system.

The user is provided with the terminal, along with its firmware, by the manufacturer. During the end-to-end (i.e., in the case of transport layer), mass-upgrade is performed. The manufacturer participates in scheduling the mass-upgrade download through its server. Dynamic switching between one-to-many download modes is done when appropriate. The manufacturer monitors the one-to-many download to determine whether a switch to another one-to-many download method is required. This process includes counting the number of receivers of the download. Figure 4.10 depicts the use-case for the procedure.

The platform modules developer provides the OS to the end user's terminal and is responsible for the terminal's OS upgrades. During an end-to-end mass upgrade of the OS, the platforms module developer participates through its server in the sched-

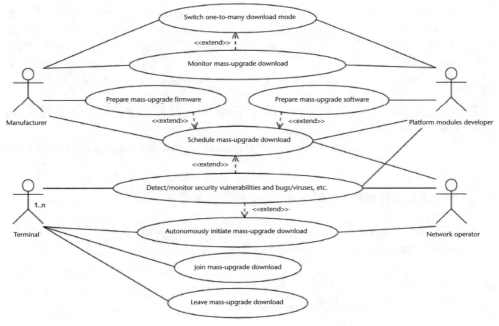

Figure 4.10   Use-case for mass-upgrade of mobile terminals [2].

uling of the mass upgrade download and also in the dynamic switching to the one-to-many download mode as needed. It also monitors the one-to-many download to determine if and when a switch to a different one-to-many download is needed. If there are hacking incidents, the platform modules developer can detect or be informed about the hacking event and thus be aware that a mass upgrade is required.

The network operator whose network the terminal is using may monitor attempts to compromise security of the system to assess whether a mass upgrade is needed. The operator also participates in scheduling the mass upgrade download to minimize the effect of the upgrade on competing traffic. Otherwise, it might have a role in autonomously initiating a mass upgrade in conjunction with a group of terminals and monitors, making decisions based on various aspects of lower layers of a mass upgrade download.

The user's terminal to which the mass upgrade is destined should join or leave the upgrade download; it might also participate in autonomously initiating a mass upgrade in conjunction with a group of terminals and/or the network operator. The terminal may also monitor the bugs and attempts to hack into its OS to determine whether an upgrade is needed. It may also send bug/virus reports to the platform modules developer so that the platform modules developer can assess whether a mass upgrade is required.

Table 4.6 summarizes the capabilities derived from the mass upgrade of mobile terminals use-case.

4.8   Handover

A vertical handover can take place between the RATs of one operator or between those of different operators. When handover occurs between the RATs of different operators, the mechanisms that enable the transfer of the context of the applications/services for the seamless operation of third-party applications/services need to be in place.

Table 4.6   Capabilities Derived from Mass Upgrade of Mobile Terminals Use Case [2]

Capabilities	Definition
Schedule/initiate mass upgrade download	Handles the scheduling/initiation of the mass upgrade download, which may be assisted by autonomic mechanisms.
Switch one-to-many download mode	Dynamically adapts the one-to-many download mode, dependent on the measured parameters such as the proportion of terminals in the system that are performing the mass upgrade download. Adaptation may occur by switching between many unicast and reliable multicast modes or vice versa, or from reliable multicast to/from reliable network broadcast.
Join mass upgrade download	Joins the mass upgrade download.
Leave mass upgrade download	Leaves the mass upgrade download.
Detect/monitor hacking attempts/system vulnerabilities	Detects or monitors attempts to gain unauthorized access to the system and/or the terminals' operating systems. The monitoring can be carried out by the terminals or software providers.
Monitor characteristics of mass upgrade download	Monitors aspects such as the proportion or number of terminals in the system receiving the mass upgrade to determine whether the dynamic switch is necessary. Most often, this is done by the software provider

The decision about the handover should be based on contextual information, such as users' paying capabilities, users' QoS demands, system capabilities, and system load; additionally, the transfer of services between cooperative RATs during the reconfiguration process must be seamless to the user. For this, the handover process should be completed in real time without delays due to service interruption; also packet loss should be minimized.

To reach a compromise among various profile parameters, the seamless service and autonomous behavior of the different handover algorithms may be available to download and execute.

Figure 4.11 gives the use-case diagram and shows the incidence of mobility management through a location update process that can be triggered by either the service provider or network operator. The latter would be a logical choice; however, the former could also be the case if the target RAT is offered by an operator who holds no agreements with the current operator. The central role of the operator at the time of transferring context, reserving resources, and routing the datastream is also shown in Figure 4.11. The latter stage may include buffering and multihoming in dependency of the constraints imposed by the scenario in place (i.e., multiple RATs, multiple operators).

The involved actors are the following:

- Service provider;
- Network operator;
- Terminal.

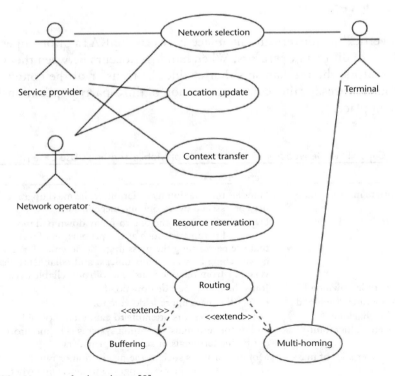

Figure 4.11 Use-case for handover [2].

Table 4.7 summarizes the system capabilities derived from the handover use-case.

4.9   Formation of Network Compartments and Base Station Reconfiguration

To accommodate a maximum of users and their specific needs, network compartments are created—that is, the area is divided into compartments, and for each compartment there is a wireless network solution provided for use (e.g., a 802.11 hotspot may be deployed in a small zone where high throughput is required or at an event in which specific VIP areas are given better access that the other areas). This maybe required during occasions such as exhibitions, concerts, or sport events, where thousands of people gather in a relatively reduced space and each user has different needs for using the wireless networks.

The base stations have to be configured to create the network clusters, suiting the kind of use to each compartment and accommodating a maximum of users while minimizing radio interference and optimizing the spectrum allocation.

The following actors are involved:

- *Network operator:* This creates a clustering scheme in either a fixed (e.g., telecommunication exhibition) or dynamic (e.g., high load for the basketball game) way.
- *Spectrum manager:* This decides the spectrum allocation, balances the load, and minimizes the interferences between the different clusters. The spectrum manager and the network operator can be the same entity.
- *Base station (BS):*
- *Terminal:* This gives information about the required QoS.

Table 4.7   Capabilities Derived from Handover Use-Case [2]

Capabilities	Definition
Location update	Manages of the user's mobility among different radio network subsystems that offer different RATs. Though it is typically controlled by the network operator, it can be controlled by the service provider.
Context transfer	Presents the user context information (i.e., PDP), describing the services and quality to the new radio network subsystem to maintain service seamlessness. Again, it is typically controlled by the network operator; however, it can possibly be controlled by the service provider.
Routing	Manages the changes of the user-plane datapath during the handover process, aiming at the optimization of delivery information from source to end point. It is typically controlled by the network operator.
Buffering	Executes the storage of user-plane data during initiation of a new connection in target network. It is typically controlled by the network operator.
Multihoming	Involves management of two connections (i.e., a soft-handover) that belong to different RATs. It is typically controlled by the network operator but often assisted by the terminal.
Resource reservation	Negotiates and books resources in the target network that will maintain the same service and quality for the user. It is typically controlled by the network operator.

Figure 4.12 shows the use-case for the formation of network compartments and the base station reconfiguration procedure.

Table 4.8 summarizes the capabilities derived from the handover use-case.

4.10   Traffic Load Prediction and Balancing

Traffic load prediction and balancing is very important for successful network management and support of the decision process for handover (see Section 4.8). The context of load control may include the traffic load balance and overload control over overlapped RAT networks. The traffic load imbalance could lock up network resources in an underutilized network and also cause blocking problems in a heavily used or overloaded network. Balancing the traffic load among overlapped networks

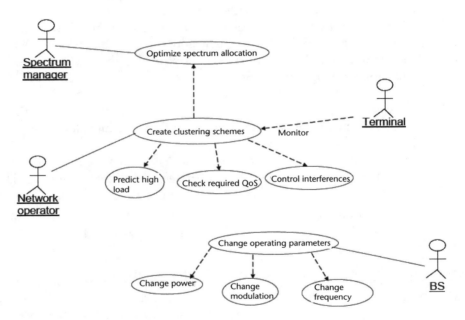

Figure 4.12   Use-case for formation of network compartments and base station reconfiguration [2].

Table 4.8   Capabilities Derived from Formation of Network Compartments and Base Station Reconfiguration Use-Case [2].

Capabilities	Definition
Change operating parameters	Changes the BS operating parameters, such as modulation, frequency, and power.
Control interference	Monitors the possible interfrequency interferences to optimize spectrum allocation and transmission power in each network compartment.
Predict load	Predicts when and/or where a specific cluster should be created, based on the traffic load.
Check needed QoS	Checks the QoS requirement by a group of users in the same cluster.
Optimize spectrum allocation	Optimizes the spectrum allocation and negotiates spectrum for each cluster.

can alleviate the problem of blocking and create headroom for traffic growth, especially if traffic load prediction can be applied.

The important actors are the following:

- *The OAM entity*, located within the RMP, can trigger the load-balancing function for necessary load-balancing action in coordination with the RAT-specific OAM. The load prediction function can gather corresponding information from each RAT-specific OAM to perform traffic load prediction.
- The *resource and spectrum manager* is responsible for the load-balancing decision, which in turn can be used for admission control and handover control, for resource and spectrum management in coordination with existing radio resource management (RRM) in the respective RAT-specific network. The purpose of overload control and load balancing requires the reconfiguration of the available spectrum, which is also a task of the resource and spectrum manager.

Figure 4.13 shows the use-case for the traffic load prediction and balancing.

Table 4.9 summarizes the capabilities derived from the traffic load prediction and balancing use-case.

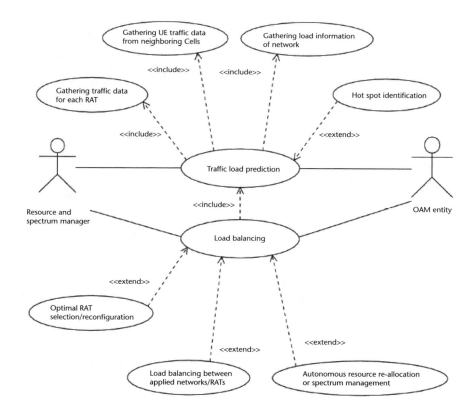

Figure 4.13 Use-case for traffic load prediction and balancing [2].

Table 4.9   Capabilities Derived from Traffic Load Prediction and Balancing Use-Case [2]

Capability	Definition
Gathering traffic data	Acquires the current traffic load for a specific RAT.
Gathering UE traffic	Acquires the predicted incoming or outgoing traffic load from or to a certain area for a specific RAT.
Gathering load information from network	Acquires the predicted network load related to incoming or outgoing services, intended mass upgrade of mobile terminals, upcoming events etc.
Hot-spot identification	Identifies hot-spot area, based on load statistic and load prediction, together with location information.
Optimal RAT selection/ reconfiguration	Selects and reconfigures an optimal RAT for a specific area based on current and predicted load, network resource utility, UE capability, interferences, and so forth.
Load balancing	Reassigns or adjusts the load between different RAT networks in an overlapped area based on current or predicted load and policies.
Resource reallocation and spectrum management	Reallocates and provides or releases resources among different RATs.

4.11   Network Resource Management

Network resource management includes also the procedures described in the previous sections but here, the main decision-makers are identified as follows:

- The *terminal* provides the context information.
- The *network operator* chooses the reconfiguration strategy of its resources. For example, the operator may decide to switch the RAT of an access from WiMAX to Wi-Fi if a large number of users are in proximity to the access point.

Figure 4.14 shows the use-case for reconfigurability during network resource management. The following specific use-cases have been defined: *network configuration, user context recovery, RAT reconfiguration, scan of network status*, and *user context update*.

The network configuration reprograms the different network entities, such as base stations. The user context recovery observes the current user status and context; the RAT reconfiguration executes the reprogramming of network entities;, the scan network status analyzes the current state of the network; and the user context update contains a constantly updated database comprising information on the user context.

The user terminal resource management is performed in a distributed way. The users obtain the policies from the network, limiting their choices of resource usage strategies. These policies, when available in the terminals, allow them to act autonomously and adapt at any time to a challenging context.

Table 4.10 summarizes the capabilities derived from the network resource management use-case.

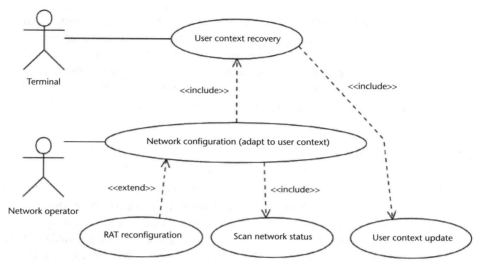

Figure 4.14   Use-case for network resource management [2].

Table 4.10   Capabilities Derived from the Network Resource Management Use-Case [2]

Capability	Definition
Network configuration	Specifies the mechanisms for the reconfiguration of network equipment.
User context recovery	Recovers the latest information of the user context.
RAT reconfiguration	Executes the network device reconfiguration.
Scan of network status	Recovers information on the latest status of the network.
User context update	Provides a dynamically updated database containing user context–related information.

4.12   RAT Discovery and Selection

Several options for wireless connectivity are available in the B3G reconfigurable environment. In this context, the elements of a reconfigurable system dynamically adapt to the changing conditions in the environment—for example, the introduction of an additional alternate RAT in the composite radio infrastructure.

A reconfigurable element monitors the performance of the current configuration as well as the overall status of the environment. The reconfigurable equipment can identify failures or performance degradation, or it may discover new alternate configurations. To efficiently handle the encountered environmental conditions, the reconfigurable element considers a set of feasible remediation actions and selects the most appropriate reconfiguration solution, taking into account the corresponding performance, cost, user and application requirements, network element, and equipment characteristics. Both the mobile terminal and the network operator are important actors in this process.

The reconfigurable elements require some management functionality, enabling the discovery of the compatible reconfigurations so as to appropriately handle dynamic changes in the environment, such as performance degradations.

Figure 4.15 shows the use-case for RAT discovery and selection.

Table 4.11 summarizes the capabilities needed by the system components as derived from the RAT discovery and selection use-case.

4.13   Conclusions

This chapter described the high-level system requirements related to reconfiguration capabilities. In total, 85 high-level system requirements were identified within the project E2R; these were grouped into 12 system capabilities in support of the reconfigurable system architecture development. The identified requirements for the system capabilities can provide a clear insight into essential architectural issues, which would pave the way for more detailed representations of the system architectural aspects and their impact. This is one of the important pillars for future research work in reconfigurability. The requirements are an important achievement made by the project E2R; they have a structural impact on the reconfigurable system architecture. Although this chapter provided a mature list of capabilities and system requirements, further refinements can be expected from future research work in this area.

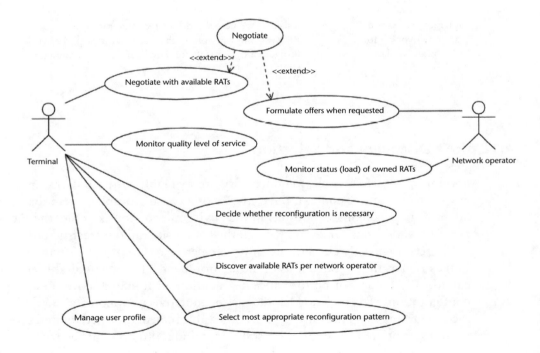

Figure 4.15   Use-case for RAT discovery and selection [2].

Table 4.11 Capabilities Derived from the RAT Discovery and Selection Use-Case [2].

Capability	Definition
Context acquisition	Acquires context information such as user and application requirements, network element and equipment characteristics, and business policies in order to assess the performance of the current configuration and the overall status of the environment. It corresponds to the self-healing and self-knowledge aspects of self-management.
Reconfigurable action selection	Makes a decision as to the most appropriate reconfiguration action and/or selection of the most appropriate RAT, taking into account the performance and the cost of the potential solutions, user and application requirements, network element, and equipment characteristics. This capability corresponds to the self-healing and self-optimization aspects of self-management.
Reconfigurable action implementation	Implements the selected reconfigurable actions. This capability corresponds to the self-healing and self-configuration aspects of self-management.

References

[1]   ICT project E2R, "E2R Reconfigurability Management System-Level Architecture," Deliverable D1.4, December 2005, http://e2r2.motlabs.com/

[2]   ICT project E2R II "Evolution of Reconfiguration Management and Control System Architecture," Deliverable D2.1, August 2006, IST-2005-027714.

[3]   3GPP TS 22.127, "Service Requirements for the Open Services Access (OSA): Stage 1 (Release 7)," v7.1.0, March 2006.

[4]   3GPP TS 22.105, "Services and Service Capabilities (Release 8)," v8.0.0, June 2006.

[5]   3GPP TS 22.258, "Service Requirements for the All-IP Network (AIPN); Stage 1, (Release 8),"v.8.0.0, March 2006.

[6]   Patouni, E., and N. Alonistioti, "Towards Self-Configuring Protocols for Reconfigurable Systems," *Proc. 13th International Conference on Telecommunications (ICT 2006)*, Madeira, Portugal, May 2006.

CHAPTER 5

Principles and Analysis of Reconfigurable Architectures

This chapter describes the realization a reconfigurable physical layer. Full system reconfigurability would depend on how the different reconfigurable components are integrated and controlled in a reconfigurable network. The design and development of an architectural framework for reconfigurable devices, base stations, and supporting system functions is required to offer an expanded set of operational choices to the users, applications, service providers, operators, and regulators in the context of heterogeneous mobile radio systems.

A number of EU-funded projects conducted as part of the IST FP6 have performed research and development work in this direction. The most prominent projects in the area were the IST projects E2R and E2R II [1]. Most of the material included in this chapter is based on their research results.

This chapter is organized as follows. Section 5.1 gives an introduction to the basics of reconfigurable design and development. Further, it outlines the main challenges related to such design. Section 5.2 describes in details the various reconfigurable elements. Both analog and digital architectural elements are described. In particular, it discusses the digital baseband section, the reconfigurable elements, and the system requirements. Section 5.3 gives the principles of the reconfigurable architectural design. A logical partitioning of the RE front end and the baseband processing section, along with the interface, is presented. Section 5.4 presents an analysis of the designed physical layer architecture. Section 5.5 concludes the chapter.

5.1 Introduction

Reconfigurability can bring the full benefit of the valuable diversity within the radio ecospace, which is composed of a wide range of systems, such as cellular, wireless local area, broadcast, and next generation. In the scope of NGNs, where it is envisioned that each layer in the service and transport stratums will have its own user, control, and management planes [2], conflicts or inefficiencies may occur because the layers operate independently. Reconfigurable architectures can prevent this by enabling cooperation between the layered networks, considering both the interlayer and intralayer aspects. Figure 5.1 shows a scenario in which different layers belonging to disparate domains cooperate efficiently [3].

Figure 5.2 shows intra- and interlayer interactions in next generation communication systems.

Related to the motivation for developing reconfigurable PHY-layer architectures, it is worth mentioning that another trend requiring such developments is the

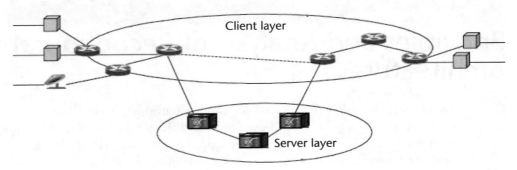

Figure 5.1 Cooperation between layered networks.

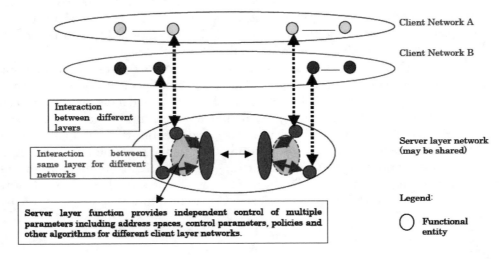

Figure 5.2 Intra- and interlayer interactions.

migration towards an all-IP network (AIPN). An AIPN is "a network based on IP technology where various access technologies can be connected. The AIPN provides common capabilities independently from access technologies for all aspects (including mobility, security, service provisioning, charging and QoS), which enable the provision of services to users and connectivity to other external networks" [4]. Figure 5.3 shows an overview of AIPN. A view on the support of ubiquitous services is shown in Figure 5.4. As a required capability expansion, relationships with an existing system are a must. The focus of AIPN is the packet-switched (PS) domain, including the IP Multimedia System (IMS). AIPN does not consider evolution from the circuit-switched (CS) domain. The CS infrastructure will be in use in real networks well after the rollout of AIPN, but it is expected to be slowly replaced by the AIPN infrastructure once AIPN is introduced. In addition, although currently there is no detailed specification for accommodation of the other access technologies, it is expanded to include different access technologies within an AIPN.

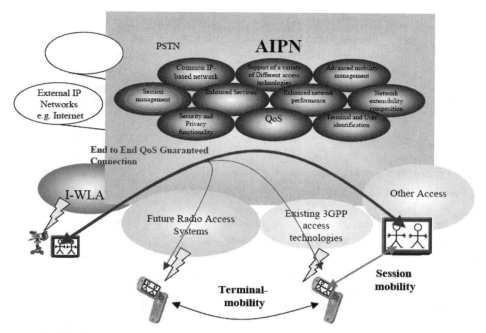

Figure 5.3 Visual representation and key aspects of the AIPN.

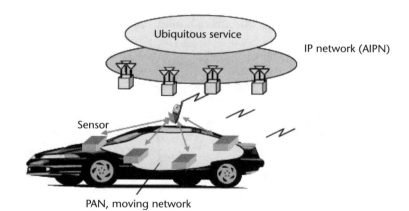

Figure 5.4 Support of ubiquitous services.

Reconfigurability in the PHY layer will be very important in a scenario of a moving network, as shown in Figure 5.4. The AIPN plays also an important role here.

5.1.1 Multistandard Base Stations

The problem of today's telecommunication equipment is the relatively rigid arrangement to one standard and one set of functionality. Thus, the introduction of new functionality or a new standard is cost-intensive for both the manufacturer and the operator. One solution to this problem is the introduction of multistandard base stations [5]. This kind of base station includes a couple of different standards and sufficient processing power for future extension and standards. Thus, the operator can easily update the base station via a software update. In addition, the operator can switch between actual standards without exchanging the existing telecommunication hardware. For the operation and maintenance of this complex telecommunication hardware, a set of new management functions is needed.

A multistandard base-band platform must fulfill the requirements of the applications to be supported (the type of radio standards and operating modes) and has to support dynamically varying traffic mixes. In relation to conventional solutions, clear advantages can be obtained regarding more efficient use of the existing hardware (HW) resources and the attainable QoS. A high granularity of resources must be allocated for the downlink and the uplink signal processing. This requires architectural enhancements and intelligent and dynamic HW resources management. A number of digital signal processors (DSPs) and HW accelerator blocks are used for high-bitrate and high-volume data processing as required for bit-stream oriented functions. These elements are interconnected with a control system—the so-called general purpose processor (GPP)—the DSP, and the input/output interface by high-speed data and control buses in a flexible manner. A software library provides signal processing modules for the radio functionality to be downloaded to the processing elements as required by the actual configuration, as shown in Figure 5.5.

A suitable operational software (SW) provides the functions to operate the HW elements, (e.g., procedures for task management, scheduling, and resources and load management).

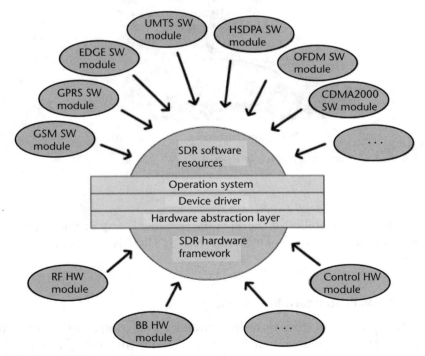

Figure 5.5 Overall SDR baseband software structure.

To be compliant with the overall requirements for a multistandard base station, a corresponding architecture for the SW installed at the system has to be realized. Furthermore, it is a strong requirement to reduce the costs for SW development and integration.

5.1.2 Programmable Reconfigurable Radio

A reconfigurable radio should have unrestricted programmability of functions and services across all the layers in the OSI-model of the mobile communication system. Figure 5.6 shows the concept of an appropriate HW architecture for the PHY layer [6]. It is an ideal software radio, in which the analog-to-digital converters (ADCs) and digital-to-analog converters (DACs) allow all radio to transmit, to receive, to generate signals, and to perform modulation, demodulation, timing, control, coding, and decoding functions via the software running on top of the hardware. This architecture offers complete programmability and requires no or little component selection.

The drawbacks of the ideal software radio are that the ADC would be exposed to numerous signals other than the desired signal and would require a huge dynamic range at radio frequencies, which would affect the system sensitivity and selectivity.

Figure 5.7 proposes a more practical architecture. In this architecture, the various elements are designed to be configurable and/or programmable.

A device is considered *programmable* when it is able to execute an instruction set, whereas it is *reprogrammable* when there is a certain capability provided to

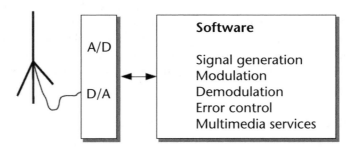

Figure 5.6 Ideal software radio.

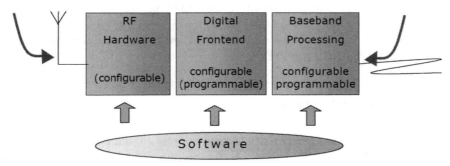

Figure 5.7 Realistic software radio.

change, for example, the program memory. A *configurable* device is one that can be configured only once, whereas a *reconfigurable* device can be configured frequently.

For an efficient reconfigurable architecture, the possibilities for reconfiguration in the analog front end should be minimized to reduce the silicon area and cost. The digital front end provides additional flexibility to adapt to the filtering and sampling functions. In particular, a receive ADC with high linearity and high dynamic range, tuneable and selective RF components such as power amplifiers (PAs), low-noise amplifiers (LNAs), and filters can be very efficient. An important feature is that channelization is performed in the digital domain, and a resampling capability is foreseen, which makes it possible to process any standard having modulation bandwidth within a given range and, hence, a suitable data rate is realized at the interface. In principle, no restrictions are imposed on the capabilities; therefore, the architecture can support the implementation of multiple standards and allow flexible spectrum management.

The concepts of partitioning and the architectural concept of the reconfigurable PHY layer are applicable to reconfiguration at the base station and the mobile device. The PHY layer can be partitioned into main modules, such as configuration control, operational software environment, and hardware resources.

5.1.3   Requirements for a Reconfigurable PHY Layer

Since reconfigurability affects all layers individually, it is important to examine the effect of reconfigurability on the PHY layer alone. Different PHY layer scenarios must be examined in order to determine all aspects of this impact.

From the PHY layer point of view, high-level scenarios can be too abstract to cover the basic and fundamental principles and properties of the reconfigurable PHY layer. Therefore, the focus is on the available and future technologies to be developed in the digital baseband, RF front-end, and configuration control. The following requirements can be summarized in relation to reconfigurable design in the PHY layer:

- Reconfiguration capabilities of the PHY layer will experience certain limitations because of the manufacturer-owned implementations.
- From the PHY layer point of view, reconfiguration is most likely independent from the initiator (i.e., the user, service, or network).
- From a business and user point of view, there might be differences in requirements related to a reconfigurable PHY layer.

Section 5.2 describes the most important issues related to the elements requiring reconfigurability in the PHY layer.

5.2   Reconfigurable Elements

To keep the RF front-end flexible, and not just switch between parallel implemented standard dedicated blocks, a certain amount of variable analog components are helpful.

5.2.1   Analog Reconfigurable Elements

The analog components include microelectromechanical systems (MEMS); variable gain amplifiers (VGAs) and gain trim amplifiers (GTAs); variable filters and composed components; and so forth.

MEMS switches overcome the problem of nonlinearity that exists in diode switches, since the control path is mechanically separated from the data path. It may also be used to configure functional blocks, such as variable filters, matching networks, and so on, and may be implemented in variable capacitors or coils. It is usually applied for band selection and change of standard. Though it is linear and capable of integration, it has a relatively high voltage requirement for operation [7].

The VGAs and GTAs are used to control the gain and adjust the TX/RX chain levels. But their dynamic range, limited BW, and noise figure are challenging issues. VGAs and GTAs are usually applied for change of standard, change of sensitivity, and power control.

One of the biggest problems in the design of reconfigurable equipment is the lack of tuneable filters. Tuneable filters are either realized by switching parallel structures or by mechanically changing elements. Another option could be the use of log-domain filters, but these have limitations, such as power consumption and noise generation. Variable filters could be applied for change of standard, change of bandwidth, and change of frequency band [8].

The single function components can be composed to new multifunction reconfigurable components—for example, IQ demodulators comprising LO-prescalers, switchable BB-filters, variable LNA, and switchable LNA matching network.

5.2.2   Digital Reconfigurable Elements

The digital reconfigurable components include dedicated signal processors, accelerators (ASICs), programmable logic (FPGA, EPLD, CPLD), the algorithm-/application-specific instruction processing accelerator (ASIP), the digital signal processor (DSP) and communication elements.

5.2.2.1   ASICs

The ASICs are able to process the datastream in a timely manner. These elements are also software reconfigurable so that they can support requirements of modular multistandard radio. ASICs are applicable for multiple modes and multiple channels. They consume less power and are easily reconfigurable. Their drawback is their long development cycle and limited flexibility.

5.2.2.2   Programmable Logic

Programmable logic elements are FPGA, EPLD, and CPLD. These elements can be used for the same functionality and applications as the dedicated processors discussed previously, but they have short development cycle and unlimited flexibility. As with ASICs, these are also applicable to multiple modes and multiple channels. These elements are comparatively more expensive and have a slower loading process. Further, the partial reconfiguration capability is limited.

5.2.2.3   Algorithm-/Application-Specific Instruction Processing Accelerators (ASIPs)

The ASIPs are a promising approach to soft-reconfigurable signal processing and typically are useful for a system involving a microprocessor, DSP, memory, I/O logic, and user logic. The ASIP enhances the flexibility after production by allowing a wide range of programmability for the algorithms that are executed, but it is too computation intensive to be solved by a microprocessor or DSP. The advantages of ASIPs are their flexibility, short development cycle, and partial reload capability related to bug fixing and enhancements. The limitation is that these elements have to be embedded in the complete processor environment.

5.2.2.4   Digital Signal Processors (DSPs)

DSPs avoid the overhead of dynamic instruction scheduling. DSPs supply a complex instruction set computing (CISC) instruction set, which does not consume a lot of memory and exploits the full degree of parallelism. Flexibility and a short development cycle are the DSPs' advantages, but they also have limitations, such as their cost and the complex reconfiguration process.

Usually the communication paths between the processing elements are fixed. The elements should be able to allow connecting or disconnecting to a bus or cross-connected serial connections, because in a reconfigurable system, a completely new matrix of processing elements may have to be built. A typical complex instruction in a DSP is the multiple-accumulate instruction, which combines multiplication and addition. As a result of combining several generic operations into complex instructions, DSPs are limited to performing a small set of basic algorithms efficiently. The most important algorithms for DSP are listed below:

- Component-wise addition and multiplication of vectors;
- Multiplication of vectors by a scalar, dot-product of two vectors;
- Matrix-vector multiplications, fast Fourier transformations, and so forth.

Key contributors for the efficiency of DSPs are address generation units (AGUs) and an extension to the program sequencer to support zero overhead loops (ZOLs). A ZOL-extended program sequencer is called program control unit (PCU).

5.2.2.5 Communication Elements

In dedicated baseband processing solutions, the communication paths between the processing elements are fixed. They consist of arbitrated bus systems, fast serial interconnections, or shared memory access. In a reconfigurable system, full reconfiguration could require building a completely new matrix of processing elements. Therefore, elements have to be provided that allow connecting or disconnecting to a bus or a serial interconnection.

5.3   Physical Layer–Related Scenarios and Requirements

Some of the scenarios developed by the IST project E2R II for capturing a range of physical layer operational instances include adaptation to environmental changes, mode switching, service addition, service enhancement, interRAT handover, bug fixing, and functionality enhancement.

These are described as follows:

- *Adaptation to environmental changes:* The reconfigurable terminals are exposed to various environmental changes, such as changing propagation conditions, changing signal strength, mobile and stationary environments, and changing network capacity. An adaptive digital baseband can be used to provide the user or an application with a defined QoS level. The adaptation should be local, to follow fast changes; however, network-initiated adaptation also can be considered. Adaptive algorithms, in general, are able to decrease the power consumption. The adaptation considered can be based on an algorithm selection, on an algorithms parameter selection, or on an algorithm precision. The relationship of the various functions inside the adaptation scenario and various actors is given in [1].

- *Mode switching:* A given RAT is capable of operating in several modes, all of which are defined in the standard. In a frequency band switch, the RAT remains in the same mode but seamlessly switches from one frequency to another. Standards (e.g., UMTS) support several possible throughputs by keying the number of transport channels, and the physical layer can manage the number of transport channels by reconfiguring itself to instantiate or suppress one of them. Switching from GPRS to EDGE can also be considered mode switching. Reconfiguration with the mode switching approach also allows standards such as IEEE 802.11a to adapt its throughput at runtime, depending on a various set of parameters. The switching could also take place between the FDD and TDD mode of the standards. It is important that switching from one operating mode to another be seamless, without interrupting the data stream. Additionally, the reconfiguring and switching should be tightly coupled and synchronized with the network side.

- *Service addition:* The users expect their handsets to connect to different networks with different RATs and offer them various services. The first step would be when a single device handles, in a static way, several standards. The equipment could be configured to support one RAT when the user asks for that service. But when the user does not need that service, the device can be reconfigured again to support another RAT to provide some other service required by the user. The next step would be to introduce flexibility, so that the user could to use more than one service at a time. For this, there should be enough resources—such as memory size, processing power, FPGA surface, and so forth—left to instantiate the new RAT with proper scheduling so that it does not interfere with the already running one. And once it has been instantiated, it should run without perturbing the other one.

- *Service enhancement:* The QoS has to achieve a certain bit error rate or access latency. Different coding schemes provide different quality on a service level. The service routing may run as software routines with dedicated hardware support.

- *InterRAT handover:* The interRAT handover should be seamless from the users' perspectives, without suffering any loss of data. An example of an intersystem handover (e.g., WCDMA to WLAN) is available in [1].
- *Bug fixing and functionality enhancement:* Bug fixing and functionality enhancements could lead to similar configuration controls at the generic resource interface. The bug fixing is mostly network driven, while in some cases the users' decisions could be asked before executing an enhancement. There are some differences in the installation of bug fixes and enhancements in the PHY layer between the device and the base station. There is not much difference in the reconfiguration process itself, but the timing of the procedure is very critical. The different actors and the relationship between their functions are available in full detail in the IST project E2R II Deliverable 4.3 [1].

The high-level system requirements for reconfiguration PHY-layer architectures are summarized as follows [1]:

- More than one RAT running at the same time;
- RAT monitoring functionality;
- Partial reconfiguration functionality;
- Runtime dynamic reconfiguration capability;
- Static reconfiguration capability;
- Immediate reconfiguration to support adaptation and configuration.

5.4 Physical Layer Architecture Principles

This section describes the partitioning of the physical layer and gives a high-level view of the architecture.

5.4.1 Partitioning Overview

The physical layer architecture of a reconfigurable modem inside a terminal or a base station/access point comprises three main modules: the configuration control module (CCM), the operational software module (OSM), and a module that includes the hardware resources, the real elements that guarantee the functionality of the physical layer, as shown in Figure 5.8.

These modules are described as follows:

- The *CCM module* supplies an abstract configuration interface to the signal processing system in a wireless terminal or base station/access point. The configuration procedure may include software download as well as the download of parameters into resources for implementing different functions, as well as the configuration of the communication fabric to implement the required data flow structures and associated bandwidths.
- The *hardware resources* may consist of instruction programmable elements— for example, GPPs, DSPs, and other configurable logic elements (e.g., FPGA),

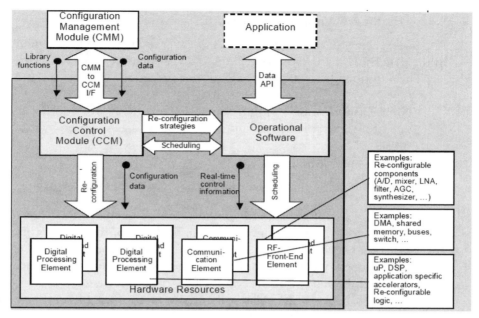

Figure 5.8 Basic partitioning of physical layer into three main modules.

as well as special parameterized accelerators (e.g., ASICS) and communication elements (e.g., switches, bus control logic, and multi-port memory). The hardware components of the RF front end, such as the oscillators, converters, filters, and mixers, behave as one configurable element. The single elements maybe converted into configuration execution modules (CEMs) that implement a certain level of configuration functionality in accordance with the resource interface provided by the CCM. Configuration allows these elements to be combined by functional modules for implementing a specified functionality, which may also guarantee the required data throughput.

- The *OSM module* provides for appropriate operational software support for the management of the different abstraction levels and temporal scheduling of hardware and software resources. It is challenging to manage all the different processes required to maintain the system reconfigurability while also guaranteeing proper functioning of the underlying data processing system. The execution platform of the OSM consists of GPP(s) connected to a system of DSP(s) and FPGA(s), ASIC(s), and a configurable RF front-end module, which are all connected by various communication elements. It may also incorporate a real-time operating system and some control interface to real-time mechanisms of other programmable processing elements (PPEs) while allocating and managing the processing resources for performing the scheduled tasks. While the CCM has to manage the data flows between various processing elements, the OSM has to provide the logical interfaces to configure and manage these resources. It has to support runtime reconfiguration. Additional services for development and execution, such as nonvolatile storage, file system, logging service, and diagnostics, are also provided.

5.4.2 High-Level View of Architecture

A high-level view of a reconfigurable PHY-layer architecture is shown in Figure 5.9. In the case shown, a GPP is controlling the HW platform.

The additionally included hardware elements are:

- Nonvolatile memory, comprising the basic software for booting, start-up, and operating, and for the configuration databases as required by the CCM;
- RAM;
- Interrupt control mechanism;
- Additional logic, such as clock, bus control, I/Os, and so forth.

The configuration execution modules are presented as a "farm" of DSPs, FPGAs, and accelerators to support the applicability of this architecture to devices and base stations, where, depending on the multiuser processing, a set of executing elements is necessary and a device can manage the processing load using a single element. In a multihoming system, several RF front ends have to be provided. Also, the data flows can be assumed to be separated into configuration/measurement data flows and processing data flows.

The configuration and measurement; control and user plane interfaces; and interface to higher layers are briefly described as follows:

- *Configuration and measurement:* The interface comprises the signals from the CCM side required to send configuration commands from the data to configuration execution modules, and also to receive the general capability and current measurement information. The physical implementation of a single interface will depend on the bandwidth and latency requirements; therefore, an FPGA or DSP will need a higher configuration interface bandwidth than an accelerator, which is controlled by a few parameters.

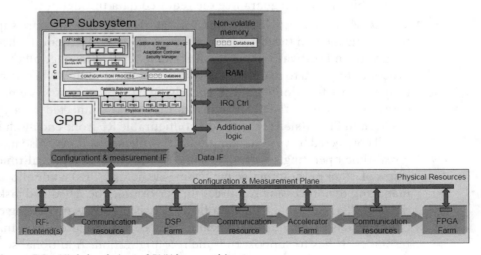

Figure 5.9 High-level view of PHY-layer architecture.

- *Control and user plane interfaces:* The connection between the processing data interfaces depends on the functionality it is being used for; thus, the communication channels between the executing elements may be subject to complete reconfiguration, for which a flexible switching mechanism will be needed. A switching plane may be implemented to control the data flow. In this case, all the required data flow and internal control flow are channelized by a configurable communication fabric realized by a network-on-chip (NoC), which is a type of configuration execution module as shown in Figure 5.10.

The interfaces to higher layers are required to define the borders of the PHY layers' responsibility and the changes that are necessary during the integration of a network controlled reconfiguration mechanism. Figure 5.11 shows the radio interface protocol architecture, in which the general functionality of the interfaces to layer 2 and layer 3 is guaranteed as required by the standard specification. Configuration protocols are exchanged via additional logical channels or in combination with the RRC; the communication from RRC to layer 1 has to be maintained and the RRC-I/F derived measurement data has to be stored for capability negotiation reasons.

To summarize, this section described the partitioning of the physical layer into three main modules. The high-level view for connecting the configurable executable elements to the configuration control module and an appropriate control of the processing data was presented. A configuration and measurement plane connected to the physical configuration interfaces of the hardware resources, presented as "farms" and "arrays" of configurable elements, was shown. These devices are arranged per the required functionality, and the data flow between them is controlled by heterogeneous communication elements.

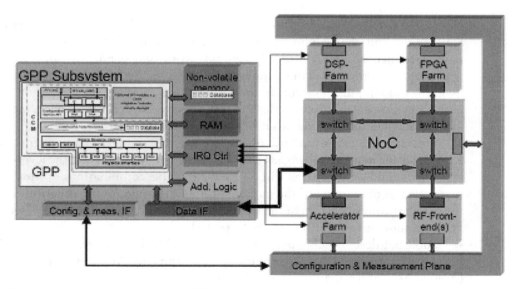

Figure 5.10 Implementation of switching plane using NoC.

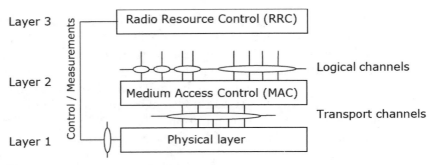

Figure 5.11   Radio interface protocol architecture.

5.5   Physical Layer Architecture Analysis

The physical layer is broadly classified into the radio frequency front end (RF-FE) and the digital baseband processing part. The communication architectures also play an important role in reconfigurable environment. These topics, along with the interface between RF-FE and baseband, are addressed in this section.

5.5.1   RF Front End Architecture

The RF-FE is a key component of the PHY layer. The RATs may share the same hardware, depending on the reconfiguration capabilities, leading towards a power- and cost-efficient solution system. This section focuses on the key components of the RF-FE architecture; the different transmitter and receiver architectures are discussed, as well as the architectural differences between a base station and user equipment.

An RF front end, which is dedicated to one standard, comprises a transmit and receive path that are operated either simultaneously or in time multiplex mode. For SDR equipment that supports multiple standards, multiple transmit and receive paths are needed, which ideally should work independently and should be configurable to any one of the standards required. Figure 5.12 contains a basic block diagram showing this concept. In the case of a base station, the receive or transmit paths may process multiple carrier radio signals, whereas for the device, one path is usually devoted to radio signals modulated on a single carrier.

This sort of multistandard operation will create new requirements for the RF-FE, posing a huge challenge. The configuration of the RF-FE at the level of parallel transmit and receive paths will define the possible combinations of simultaneous or sequential operations of the paths. For a base station, more attention is paid to the simultaneous operation of the resources, but for a device, a mixture of sequential and simultaneous operation may be expected.

Table 5.1 describes the functions of the main blocks in the transceiver. Here, the RX/TX path has been divided into the analog front end (AFE), the converter section (ADC/DAC), and a digital front end (DFE).

Configuration is mainly done at the baseband level, because topological changes are possible in the analog part as well as in the DFE. In the RF part, undesired para-

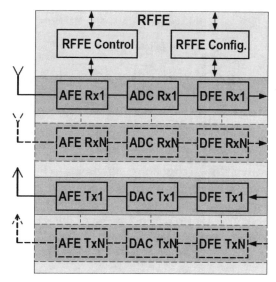

Figure 5.12 Basic block diagram of RF-FE architecture for SDR device.

Table 5.1 Important Blocks of a Transceiver and Their Functions

Block Name	Functions on RX Side	Functions on TX Side
AFE	Signal amplification and filtering; signal frequency conversion from RF to ADC input frequency.	Signal amplification, attenuation and filtering; signal frequency conversion.
ADC / DAC	Analog to digital signal conversion, either at baseband or intermediate frequency (IF)	Digital to analog signal conversion, definition of analog signal range.
DFE	Signal post filtering, gain control, sample rate reduction, radio signal gain control, frequency conversion, sample rate conversion, correction of analog impairments, data buffering.	Upsampling and interpolation, frequency conversion, modulation pulse shaping, radio signal gain control, data buffering.

sitic elements and paths are introduced with changes in circuit topology, due to which such changes will be difficult. Thus, with a selected architecture, configuration is achieved by parameterization of the hardware.

The architectures implemented in the reconfigurable equipment depend on its suitability for multistandard use. A brief description of the currently used architectures in mobile and base station equipment are given here, along with an evolutionary path toward the multistandard implementation.

Figure 5.13 shows an ideal SDR receiver architecture, which imposes enormous requirements on the selectivity and linearity of the RF filter and the ADC. The dynamic range has to cover the entire radio signal dynamics of around 100 dB (or more in the case of multiple radio channels and cellular applications) and show the corresponding linearity. The accurate sampling of the signal of the RF signal requires a sampling clock jitter in the subpico range which is very difficult to achieve.

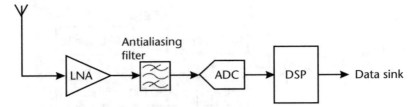

Figure 5.13   Ideal SDR receiver architecture.

A direct conversion architecture, as shown in Figure 5.14 [9], is the most promising architecture, since it has a lower number of building blocks at the RF than its heterodyne counterpart; this makes it suitable for high-level integration. Some of the other advantages are that there is no image of the RF signal, the baseband processing can make use of mainstream technology, and so forth. It has certain disadvantages, however, such as LO self-mixing, large varying DC offset, flicker noise, baseband filter linearity, and noise issues.

Figure 5.15 shows the architecture of the heterodyne receiver. Unlike the direct conversion receiver, the amplification and filtering are distributed over the stages, due to which the requirements for the individual are comparatively relaxed. However, the drawback is the use of external IF filters, which limits the signal dynamic range for the baseband analog processing, thus making the receiver less flexible for multistandard use, since multiple filters maybe needed to cover the various standards. Also, the IF stages may not be very effective, or different parallel stages with external filters may be needed to accommodate a multistandard operation. On the

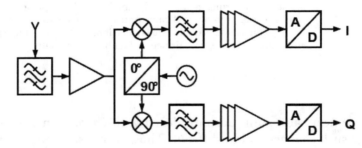

Figure 5.14   Direct conversion architecture for receivers.

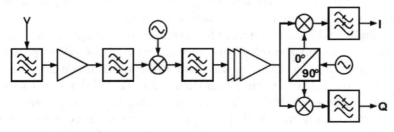

Figure 5.15   Heterodyne architecture for receivers.

positive side, the largest gain can be realized at the IF; it is free from flicker noise, offset, and AM detection.

Figure 5.16 shows a Hartley receiver architecture, which has the advantages of the direct conversion architecture and also overcomes the DC offset problem. Some sort of phasing techniques are required to suppress the image signals resulting from signal down-conversion to an intermediate frequency. A 90° phase shifter is used for this purpose. The reader is referred for further details and a description of a very low IF architecture to the IST project E2R II Deliverable 4.3 [1].

Another receiver architecture is the IF-sampling architecture, which has hard ADC requirements in terms of sampling jitter and dynamic range, and hence may not be considered for reconfigurability.

Now the transmitter architectures will be discussed briefly. There is a direct up-conversion architecture, shown in Figure 5.17, in which the transmitter carrier frequency is equal to the local oscillator frequency. But this architecture suffers from a problem termed as *injection pulling* or *injection locking*, in which the power amplifier corrupts the transmitted LO spectrum. Another inherent problem with this architecture is that the LO always lies in the transmit band, which causes high requirements on the LO-RF isolation. It also suffers from issues related to I/Q phase mismatches.

In a heterodyne transmitter architecture, as shown in Figure 5.18, the up-conversion process is done in more than one step. It circumvents the problem of LO pulling, and the I- and Q-matching is also better, since quadrature modulation is performed at the lower frequencies. But the need for an external IF filter increases the cost. This kind of architecture is not suitable where high integration is required.

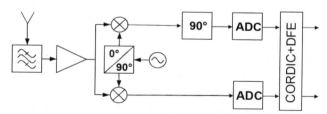

Figure 5.16   Hartley architecture for receivers.

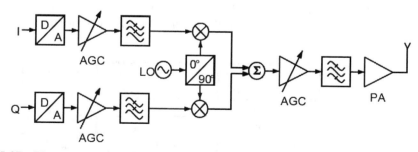

Figure 5.17   Direct up-conversion architecture for transmitters.

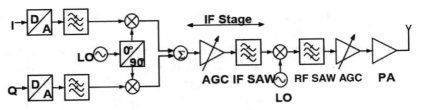

Figure 5.18 Heterodyne architecture for transmitters.

Frequency planning may also become difficult if more than one transmission band has to be supported.

The modulation domain is shifted from the rectangular system to polar coordinates in the polar modulation architecture, as shown in Figure 5.19. The most important feature is its low power consumption and, partially, the count reduction. The PLL generates the phase modulation, and the amplitude modulation is applied directly by modulating the already-phase-modulated signal at the PA. This results in a higher efficiency than conventional TX architectures, as a switched-mode PA is used for nonconstant envelope signals. Usually, digital synchronization techniques are applied. The challenges include the modulation mechanisms within the PA and the PA characterization.

The important functional blocks of the RF-FE are described below as follows:

- *ADC:* Depending on the application, a set of parameters, such as the required SNR for baseband coding, the peak to average signal ratio of the modulation scheme, and the channel noise present in the ADC input determine sthe required dynamic range of the converters. The huge requirements for a GSM/W-CDMA application with out-of-band filtering are given in Figure 5.20. These requirements are not met by the state-of-the-art converters [10, 11].

  Power consumption is a major issue of concern. A very low-power IF sampled ADC for dual mode GSM/W-CDMA applications was described by Burger [12] and a few other implementations are available from Veldhoven [13]. For terminals, the $\Delta\Sigma$-based converters are a very promising approach, as they offer a natural trade-off between resolution and speed at a low power consumption (in the range of milliwatts). Oversampling allows filter specifications of the baseband filter to be relaxed and also offers different choices for postfiltering. For the base stations, the pipelined converters [10, 11] and the $\Delta\Sigma$-based converters [14] offer similar levels of performance.

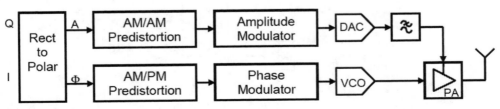

Figure 5.19 Polar modulation architecture for transmitters.

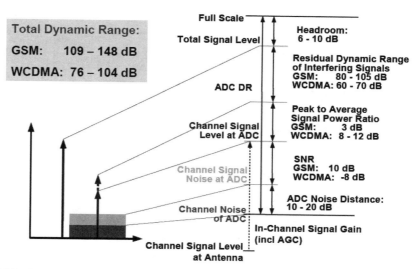

Figure 5.20   Dynamic range requirements for GSM/W-CDMA RF band converter.

- *DFE:* RF-CMOS technology is a very promising approach; however, the tight requirements in terms of power consumption and the anticipated delay compared to other standard CMOS technology for digital baseband applications, necessitates the use of efficient filter-structures within DFE. The cascaded-integrator-comb (CIC) filters and wave-digital filters (WDFs) are used for signal filtering and decimation/interpolation; they can be implemented with shift and add logic to avoid the use of multipliers. Taking into account the high sampling rates before and after the ADCs/DACs, the suitability of these structures for decimation/interpolation in terms of die space and power consumption is more pronounced. CORDIC algorithms can be used to limit the needed chip space frequency correction functions. Automatic gain control can be partitioned into analog gain control (AGC) and digital gain control (DGC), since the latter can be efficiently used for small and highly precise gain steps within the TX-DFE that are difficult to realize in the analog domain.

- *Frequency Synthesizer:* This is important due to the required frequency agility of SDR-capable terminals. Some of the requirements are as follows:

  - Multiband/-system capable;
  - Multiband VCOs;
  - One reference clock or baseband A/D clock and RF PLL reference;
  - Low jitter/phase noise;
  - Highest integration level;
  - Low chip area and power consumption.

A fractional N-synthesizer with noise shaping could be considered for this purpose and further investigated. Multiband VCOs play an important role in covering the large frequency bands on the VCO side.

The IST Project E2R II proposed an RF-FE constellation with two receive paths and one transmit path, shown in Figure 5.21. This constellation enables two inde-

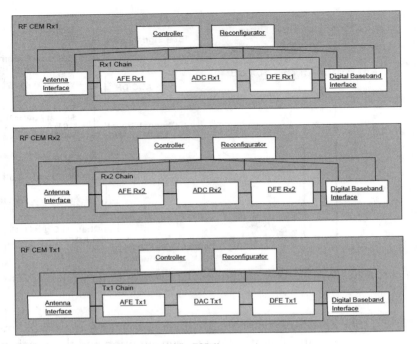

Figure 5.21 RF-FE constellation proposed in E2R II.

pendent RATs to be operated simultaneously; the second receive path is mainly for monitoring purposes. The controller and reconfigurator contain digital logic to perform the conventional control tasks and the reconfiguration tasks, respectively. A number of scenarios can be considered with Figure 5.16, such as adaptation to environmental changes, mode switching, service addition, service enhancement, interRAT handover, and bug fixing and spectrum management. More details on this are available in IST project E2R II Deliverable 4.3[1].

The most important parameters of the proposed receiver are summarized in Table 5.2.

The direct conversion architecture suffers from flicker noise and DC offset problems for the narrowband systems. Systems such as this, with low adjacent channel selectivity requirements, could use the V-LIF receiver structure. If the dynamic range and selectivity requirements are pushed to the digital domain, this would boost the reconfiguration capabilities of the RF IC and lead to a future-proof architecture, as semiconductor developments in the field of digital signal processing outperform the development speed of technologies that are purely based on analog design. It also has other advantages, such as stability, reconfigurability, power consumption, and cost. An approach based on the above reasoning is shown in Figure 5.22.

Table 5.2 Important Parameters for the Working Assumption Receiver in E2R II

Parameter	Range
Signal Bandwidth	100 KHz – 10 MHz
RF Frequency Range	400 MHz – 6 GHz
Sample Rate at Interface	2x–4x standards symbol rate

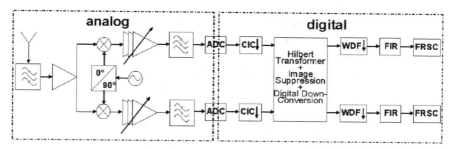

Figure 5.22   Direct conversion/V-LIF receiver for reconfigurable systems.

Some special aspects related to the user equipment are as follows:

- *RF impairment cancellation:* Due to the large tolerances of integrated resistors, capacitances and inductances, the respective circuits exhibit large performance spreads. The online calibration functionality locally implemented in the DFE can limit the impairments caused by these tolerances. RF impairments can be canceled by using digital correction blocks within the DFE, such as I/Q amplitude and phase imbalance, in-channel amplitude and phase distortion, nonlinear amplification effects, and so forth But this kind of cancellation requires prior knowledge of the magnitude and stability over time of the signal impairment. Hence, extraction of the impairment from the RF signal is an important prerequisite for RF impairment cancellation.
- *Size and weight:* The size of the device is a very important aspect. Small, lightweight, portable multiband radios are preferable. Hence, RF sections and antennas need to be combined and cannot be stacked upon each other.
- *Power consumption:* This is a very critical issue, and there is need for optimization during a standby mode of state and operational power consumption using the latest silicon technologies.

Some special aspects related to the base station are summarized as follows:

- *Power consumption:* Architectures and new technologies for increasing PA efficiency are needed. ADCs and DACs with high sampling rates and high input bandwidths are required to be able to move the point of digitization as near as possible to the antenna. Converters in technologies, such as BiCMOS, maybe used for this purpose. Also, architectures (including the DFEs) that make use of undersampling techniques and deliver a rather oversampled signal toward an increased processing gain have been state-of-the-art for a relatively long time in BSs.
- *Multiband capability:* The need for a wide frequency range is linked with the linearity features of the modulator and demodulator components. The devices usually available for prototypes cover a frequency range from 0.4 GHz to 2.8 GHz. New PA technologies are also emerging; these support multiband capabilities because of higher internal transistor impedances, and because the requirements of external matching networks are much lower, which leads to reachable bandwidths of up to 3 GHz.

- *Multicarrier applications:* In contrast to 2G, the decreased dynamic range requirements of 3G air interface standards and the lack of a frequency hopping requirement lead to the application of multicarrier transceiver architectures in BSs which are supported by:
    - Evolving converter technologies, which enable IF-sampling at considerable IF-frequencies with wide bandwidths;
    - Converter-related progress in DFE architectures;
    - Enhanced PA linearization technologies, enabling efficient and intermodulation-free multicarrier applications.

    The realization of a flexible multicarrier RF-FE for BSs which can be configured either for GSM1800 or for UMTS operation is given by Koenig and Walter [15].

- *Remote front-ends:* A remote RF-FE comprises a digital connection to the central BS that can be solved by an electrical or optical wire. In contrast to a terminal RF-FE, these devices are completely self-contained (power supply, embedded control, single/multiple carrier, receiver diversity, control signal multiplexed into data, etc.). The success of remote RF-FEs is strongly linked to the development of reconfigurable multiband front ends [16].

5.5.2 Digital Baseband Architecture

A digital baseband modem consists of components that modulate and demodulate signals in a digital baseband, (i.e., it processes signals after down-conversion and before up-conversion and provides an interface to the protocol stack or user interface). Parts of the protocol stack may also be done in the digital module when a short response time is required. The major functions are summarized as follows:

- Modulation;
- Error correction/control;
- Filtering;
- Equalization;
- Interleaving;
- FFT;
- Special functions: synchronization, rake receiver, and automatic gain control.

Partitioning is very important for digital baseband architecture. The component may be implemented either in hardware or in software.

Hardware accelerators have low energy consumption and low die size to execute a specific function, which is beneficial. Digital signal processors allow reuse of the hardware resources and programming flexibility. It is envisioned that these components could be configured at runtime and enable the reuse of hardware resources. Communication standards that require few modes are usually implemented in hardware, whereas standards with many modes would use software implementation.

Another criterion for partitioning is the time-to-market. A pure software implementation on multiple DSPs and microprocessors is faster than a new hardware design, and the manufacturing and testing of new hardware devices can be avoided.

Such an approach is also cheaper for smaller volumes, compared to designing a new SoC. This is very beneficial for base station development.

There are several architectural concepts existing for digital baseband processing. However, for every new standard, the companies must invest in new architectures, which are programmable, configurable, and robust. Those that fulfill the requirements of the standards have a small die size and package size and a low bill of materials. Low power is very important, especially for device equipment. Instead of developing new platforms for every change required, companies are now focusing on building platforms, in which it is possible to incorporate the changes so that a complete redesign may not be necessary. For SDR systems, multiple standards can be implemented into digital baseband modems, in which the hardware can be reused extensively to maintain a reasonable die size and power consumption. Figure 5.23 shows an example of a single digital baseband architecture platform on which multiple standards have been implemented. Software programmability allows the extension of such a system through software download or reconfiguration.

5.5.2.1 Configuration Execution Modules

The configuration execution modules are described here with a focus on the configuration capabilities and on the flexible communication connections.

The three types of processors that are implemented in the baseband module are the GPPs, the DSPs, and the ASIPs:

- The *GPP* (general purpose processor) is used for control tasks and for applications in which die size and energy consumption are not as important as programmability. Branch prediction, speculation, and caching may improve the GPP's performance targeted to software with unpredictable runtime behavior. The operating systems exist for advanced task scheduling and process management. Hardware accelerators may be used to enhance the performance of the GPP, and software reconfiguration is also easily employed.

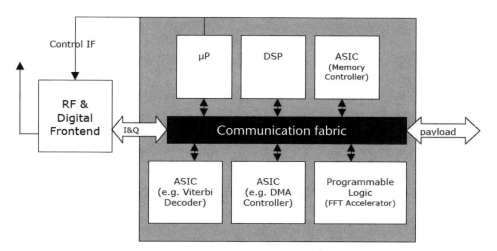

Figure 5.23 Digital baseband architecture supporting multiple standards.

However, due to high power consumption and a large die size, such systems are not suitable for use in terminals but are feasible for use in base stations.

- The *DSP* (digital signal processor) is used to enable fast signal processing with a lack of programmability. It has a low overhead compared to microprocessors.
- The *ASIP* (algorithm-/application-specific instruction processing accelerator), like the DSP, is used to enable fast signal processing with a lack of programmability. These components are designed for optimized computation of specific applications. For digital baseband processing, it has specially tailored paths for used algorithms. More work on automatically generated compilers for ASIP development tools is required.

5.5.2.2   Configurable Devices

A pure gate-level configuration can be done in programmable logic devices (PLDs). It can be done at gate- and basic block–level in the common FPGAs. Currently, FPGAs with large gate counts, such as 10 million gates, can be found, with an internal clock operating at 500 MHz. High-end FPGAs offer 179K of logic cells, with more than 350 embedded multipliers. A high-end FPGA has a configuration memory, which can be rewritten during operation. Their capability for in-system reprogrammability makes them a good choice for SDR applications in the cases when reconfigurable systems use variable-length data paths to provide different system performances. The major drawbacks of such systems are their high component cost and high power consumption; dedicated accelerators and IP cores may be used to reduce the unnecessary overhead for reconfigurability. The challenge would then be to suitably partition the digital baseband functionality and identify the functional modules. In systems with multiple cores and accelerators, a trade-off must be sought between the efficient use of silicon area and the lack of flexibility for every application. Interconnect and routing delays require very careful HDL design. DSP block elements are used to prohibit high latencies and area consumption of gate-level configured complex DSP operations. Functional level configuration is done in application-specific accelerators. An important point to note here is that FPGAs have an area times power product, which is 1,000 times more than that for a custom ASIC, and hence coarse-grain elements need to be added to these architectures. Fine-grain configurability will allow the DSPs and FPGAs to handle an infinite number of configurations and it will be challenging to manage this degree of freedom without constraints.

5.5.2.3   Architectural Concepts

Digital baseband processing comprises communications configuration execution modules (CEMs), programmable devices (processors), and configurable devices (FPGAs, accelerators). The configurable devices also include the nonconfigurable devices, such as ASICs. The modules constituting the digital baseband architecture are called CEMs; each one can execute one or more functions of the functional chain or signal processing algorithms chain. Each processor is a single CEM. The processors can execute almost all signal processing algorithms and can run multiple signal processing algorithms simultaneously. Metric information is used to select the processor best suited for an application; also, it is needed for checking the timing constraints. Software download is also a vital issue for processors. Processors may be configured in three ways: by enabling or disabling implemented functions and changing the parameters, by loading new functions on the DSP through the device driver, or by running unified source code on an implemented virtual machine. In case of FPGAs, the bit-stream synthesis and download is the main task, and reconfigurability can be divided into three general classes: complete only reconfigurable, partly reconfigurable, and self-reconfigurable. Configuration data management is an important task in this case. The process of mapping the functions onto the CEMs is shown in Figure 5.24. Depending on the granularity level, the three reconfiguration cases are classified as follows:

- *Static analysis:* The hard-coded configurations delivered by the manufacturer define the mapping in a fixed template.
- *Semidynamic analysis:* The manufacturer guides most of the reconfiguration, but the terminal can decide between some options.
- *Dynamic analysis:* Of the three cases, this is the most flexible but also the most complex. The device can decide by itself, and this is more relevant for future-proofing or for base stations.

The multimode operation of digital baseband depends on the selection of the reconfiguration architecture and its reconfiguration. After the various operating modes have been identified, their corresponding functional baseband processing blocks can be designed and implemented in such a way that they can be configured by parameterization for operation in a specific mode.

The baseband transmitter and receiver structure (with parameterizable modules) is shown in Figures 5.25 and 5.26, respectively [17]. The problem with this

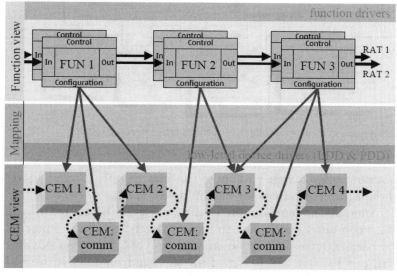

Figure 5.24   A simple example of functions mapping to CEM for reconfiguration.

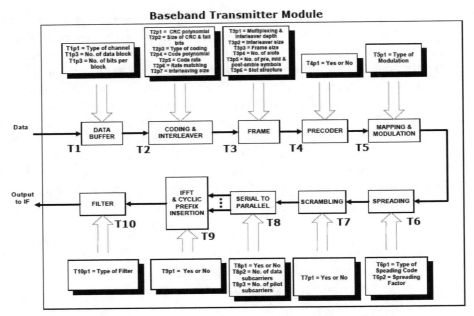

Figure 5.25   Baseband transmitter architecture with parameterizable modules.

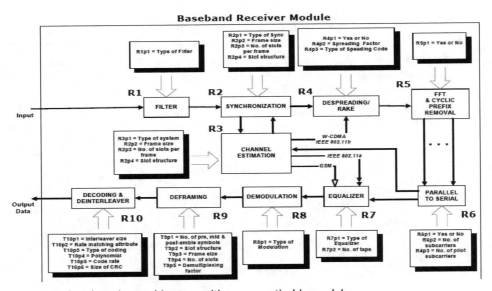

Figure 5.26   Baseband receiver architecture with parameertizable modules.

structure is that a huge amount of runtime memory is required to store and retrieve all the parameter values for the multiple systems.

Another implementation is the parameter-based reconfigurable transceiver, which combines the implementation transceivers for a certain class of air interfaces and parameterizes its functional modules [18]. The other possible baseband architecture is to have a superset of all baseband modules to form a transmitter and receiver chain. This architecture is shown in Figure 5.27 and Figure 5.28. It is challenging to partition the modules into different functions and to specify the inter-

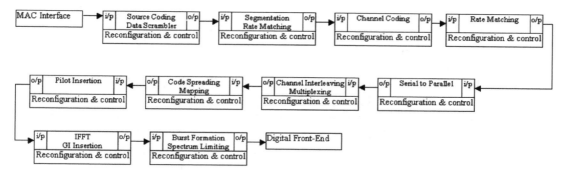

Figure 5.27   Reconfigurable baseband transmitter.

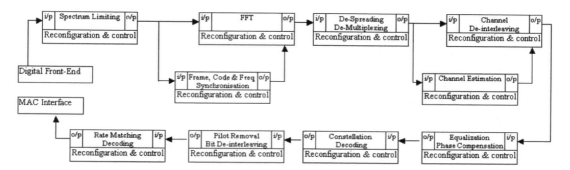

Figure 5.28   Reconfigurable baseband receiver.

faces. The same transceiver architecture, with parallel structures processing individual transport streams/channels, may be implemented for the configuration of systems with multiple transport streams/channels. The hardware platform on which such a system would be implemented should provide sufficient performance for the execution of the transceiver chain.

5.5.3   Communication Architectures

Effective communication across the CEMs is a very important issue. The communication resource or reconfigurable communication fabric and its execution module (communication configuration execution module, or C-CEM) is a hardware resource dedicated to communicating the data. The data may be application data, configuration control data, control measurement data, baseband data, or IF data. Each one has its own attributes [1].

The data transfer performed should be done at an acceptable latency level. Along with the reconfiguration process, the characteristics of the data tend to change, which would require the C-CEM to be adaptable to different data widths and formats. Dedicated hardware resources to provide the needed data services are compulsory. Some of the possible realizations include NoC, reconfigurable switch matrix, or bus architectures.

Proprietary bus interfaces are used to provide a connection between baseband processing elements. This technique has a high cost and a low degree of flexibility. In simple bus architectures, all the devices share the same communication paths, which results in an increased response time and low system throughput—especially when the system becomes busy. Power constraints should also be considered. It would be possible to tune each subsystem to achieve latency and frequency goals by using a segmented bus approach, in which bridges are used between segments. But a segmentation bridge between higher and lower frequency devices adds to the latency between segments and some up-front portioning work.

When multiple parallel buses are used, point-to-point connections between masters and peripherals are created, which allow simultaneous transfers without the problem of matching peripheral clocks to bus clocks. In switched bus architectures, the challenge would be how to make the point-to-point connections (e.g., the advanced microprocessor bus architecture [19]) . Switched buses are more versatile than multiple parallel buses, but are limited to a low number of bus layers when implementing point-to-point connections. Simultaneous transfers on the same clock cycle are possible with a cross-point switch. Some of the new contenders for the disruptive technology that will shift high-performance from a shred-bus to switched-bus architecture are HyperTransport and RapidIO, which impact chip-to-chip communication, board design, and system topology. Serial and parallel architectures are the main options for the multiprocessor design. A serial approach strictly partitions processing across processors; it is better suited for more deterministic processing, in which each data set requires no more than a known limit of resources. The parallel processing approach places multiple processors in equal relationship to each other, unlike the serial approach; the data can travel along a variety of paths, depending on the current state of the processor usage. Coherency is a big challenge for such interconnected technologies. The decision related to the reconfigurable communication interface is based on a cycle of optimization and reoptimization in terms of flexibility, cost, power, and space consumption.

The shared memory architectures have to be able to transfer data from one processor node to another and maintain the data coherency. Scalable, reconfigurable high-bandwidth communication resources between processing elements are needed. A cross-bar communication system is scalable, programmable, and has high-bandwidth streaming communication, which is ideal for applications such as digital cross-connects, add-drop multiplexers, video, broadcast and ATM switches. The switch structure shown in Figure 5.29 results in considerable area savings.

The problem with the current bus structures is that they limit the number of communication partners; also, long global wires are undesirable due to their low performance, high power consumption, and noise phenomenon [20]. Clock skew is also a problem. But a network may allow an asynchronous communication between the different system parts, and the dedicated parts are locally clocked. The amount of overhead is reduced, which also makes the implementation simpler. More details on this plus the comparison of the existing NoCs and overview of the state-of-the-art is available in the IST project E2R II Deliverable 4.3 [1].

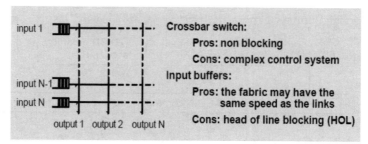

Figure 5.29   Crossbar switch architecture with significant area savings.

5.5.4   RF Front End to Digital Baseband Interface

The functionality of the front end includes channelization, filtering, digitization, sample-rate conversion, and RF impairment cancellation. This subsection discusses the RF-FE interface at the terminal as well as the base station.

From a terminal point of view, the SDR IF requires higher bandwidth and reconfiguration requirements. The electromagnetic interference (EMI) between pins of the data-IF and the highly sensitive RF input pins is a significant problem, as is the low power dissipation. The currently standardized high-speed serial interfaces are not suitable for an SDR-compliant digital RF-BB data IF.

The project E2R II proposed the low-power LVDS approach and a current-mode single-ended signaling approach. Both of these methods need a common reference clock generated on the RF chip. In LVDS, differential signaling is used, with the advantage that the electromagnetic field of the two wires should ideally destructively interfere. The drawback is that an additional second pin is needed for each serial link, but the number of pins in this case will be limited, since small packages are used. In the second approach, a current mode is used to limit the IF EMI.

Figure 5.30 shows the mapping of the architecture proposed by the project E2R II onto the OMAP 2 platform architecture. The requirements taken into account are the size, the configuration download, the power consumption, the processing power, third-party downloads, and the comparison to commercial architectures. A typical OMAP 2 platform consists of a RISC processor, DSP processor, accelerators, and the interconnect bus to connect all these devices. Different applications can run independently on the individual devices in the platform. Due to the open nature, third-party vendors can develop applications that utilize the processing power of the individual devices in the architecture.

From the point of view of a base station, the EMI is not the major issue that it is in the terminals. When bus-interfaces are used, the necessary data, such as user plane data, control, and management, as well as synchronization signals, must be completely separated, but these are multiplexed with fast serial connections onto a unique interface, depending on the direction. For example, the receive direction would include framing structure sync signal, multiple data, digital gain information, and RRC acknowledgement signals, depending on the requests to the transmit interface. The transmit direction may include framing sync signal, clock reference, multiple data, and RRC requests. The control and measurement plane should also be multiplexed into the serial data streams.

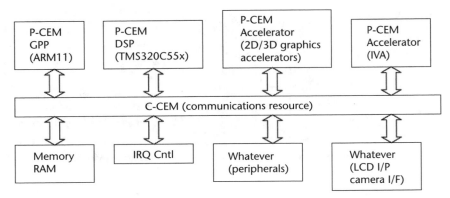

Figure 5.30 Mapping of architecture proposed by E2R II onto OMAP 2 platform architecture.

Figure 5.31 shows the overview of the open base station architecture initiative (OBSAI) [21]. There are three reference points: RP1, RP2, and RP3. RP1 interchanges control, performance, status, alarm and provisioning data between the control and clock bus and other BTS blocks. RP2 interchanges user data between transport block and the baseband block. RP3 interchanges formatted air interface user and fast control data between the baseband block and the RF block. It is divided into four functional modules, namely, transport module, processing module, radio module, and control module. More details on this are available in the IST project E2R II Deliverable 4.3 [1].

5.5.5 Transmitter Architecture for Opportunistic Radio

A novel digital radio transmitter for opportunistic radio was developed in IST FP6 ORACLE [22] based on the digital modulator concept. The goal was to design the

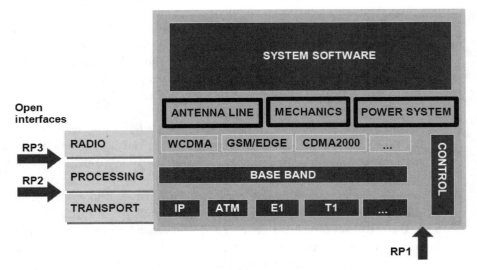

Figure 5.31 Overview of OBSAI base station architecture.

architecture that is the most suitable for establishing an OR link, both in licensed and unlicensed bands.

Many multistandard terminals are made of a set of RF silicon devices which is optimized to operate in a particular band of interest. The drawback to this approach is that it is very inefficient and costly. Traditional analog solutions are very sensitive to temperature, frequency, and process variations, and make use of inaccurate external components whose values cannot be modified. But the OR RFICs need to be flexible and reconfigurable. An all-digital circuit based on CMOS technology offers a high degree of reconfigurability and flexibility to enable the circuit performance to fit with the tight specifications of the modern wireless communication systems.

Figure 5.32 shows the architecture type as a function of power level and channel bandwidth.

Closed-loop transmit architectures, such as the polar loop, are a good choice for narrowband signals; however, they will struggle to provide sufficient bandwidth for W-CDMA or OFDM signals due to the circuit time delays, which translates into phase margin reduction and could also lead to system instability. The direct modulation transmit architectures are suitable for signals of all channel bandwidths; however, they struggle to provide good power efficiency. The limitations of feed-forward architectures is the requirement of two PAs, heavy calibration, costly and cumbersome circuitry, and thus they are usually considered for implementation in base stations. The transmitter in IST FP6 ORACLE was based on direct modulation architecture. The overall architecture of the OR terminal front end is shown in Figure 5.33.

The aim was to design a single hardware arrangement that could be reconfigured to adapt to different situations. Some interesting points to be noted are as fol-

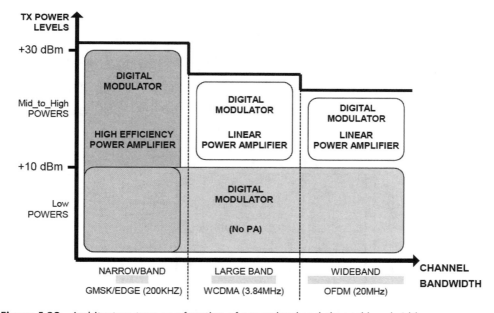

Figure 5.32 Architecture type as a function of power level and channel bandwidth.

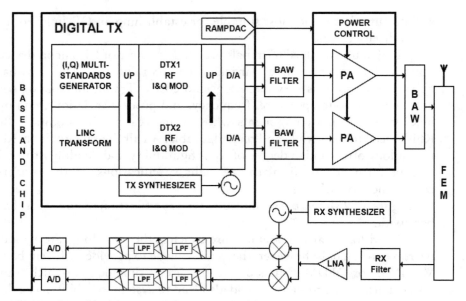

Figure 5.33 Overall architecture of the OR terminal front end.

lows. To be able to opportunistically establish 4G communications, the transmitter architecture carried out frequency modulation with an all-digital circuit to provide the high degree of reconfigurability and flexibility needed for communication in EDGE; WCDMA or OFDM modes. The proposed architecture, shown in Figure 5.34, is convenient for all types of modulation with constant or variable envelope and narrow- or wide-channel bandwidth. The digital transmitter chain for low output power combined several technological advantages, such as highly adaptive and reactive transmitter circuits for OR systems for exploiting temporal and spectral opportunities, enlarging the digital perimeter from base-band frequencies up to radio frequencies that allow higher integration, lower power consumption, higher reconfigurability, insensitivity to temperature, frequency, production variations, and so forth.

The digital predistortion technique has been implemented to provide the flexibility for controlling the trade off between power efficiency and the level of

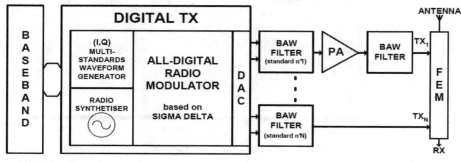

Figure 5.34 Proposed digital transmitter architecture for OR terminals.

nonlinearity of an amplifier for different targeted QoS. Techniques for correcting the memory effect were also considered and introduced in the simulation model. Another approach was to transform the traditional analog linear architecture into a digital LINC architecture, which brought high linearity, good filtering performances, and improved power efficiency at higher power levels, but was found to be not suitable for wideband signals. Multiband antennas were investigated, and the need for new approaches to improve the performance of small antennas to fulfill the multiwideband operation requirements of future wireless mobile communications terminals was identified. Key elements of the antenna design evolution were studied, and agile antennas were introduced as a promising radiating structures for OR. Special emphasis was laid on notch antennas, since these are a radiating structure presenting interesting perspectives in terms of miniaturization and integration inside future mobile terminals. Experimental results were obtained with static prototypes that validated the feasibility of agile notch antenna properties by using a simple capacitive loading technique. Improvements have to be envisaged before implementing tunable varactors on the notch antenna structure to study distortions and losses introduced by active components on antenna performances. The advantage of an all-digital PLL is that all the loop filter coefficients are fully programmable, and it is possible to changes them dynamically, which greatly improves the speed of convergence while programming to a new RF frequency. But realizing frequency synthesis with only digital components is a new technology and requires further research work.

The key results of the IST FP6 project ORACLE in this regard are summarized as follows. For more details, please refer to Deliverable 3.3 [22]. The signal delta modulator is an important element of the digital circuit that fixes the performances in terms of achievable transmit band and SNR, and it was seen that fourth-order modulators with an oversampling ratio of 12 or more are good compromise values. Baseband modulators offer more transition bandwidth compared to band-pass modulators; however, they are limited in terms of transmit frequency due to the high clock frequency. But the property of reconfigurability allows the selection of system parameters that best fit the transmission requirements. System-level simulations were done in MATLAB and found to be in line with the results in literature. The digital architecture concept has been verified in hardware after designing a dedicated radio board to filter and amplify digital signals at gigabit rates, and high-quality spectrum and EVM were measured. The device under test was limited to QPSK and 16-QAM constellations. Figure 5.35 and 5.36 shows the results of the EVM measurements; these are good results, since the W-CDMA signals should exhibit less than 17.5% EVM.

The proposed solution, based on complementary association of the digital modulator with the high-rejection BAW-CRF filters, was very compact, highly adaptive,and responsive, which is needed for systems that are required to exploit temporal and spectral opening in the context of opportunistic use. However, there are some challenging issues to be solved, such as the maximum speed at which the signal delta modulator can operate, the optimization of the system power efficiency, the noise effect associated with finite resolution, and so on.

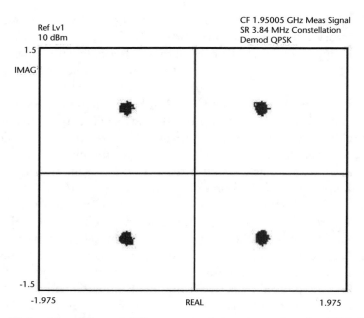

Figure 5.35   RF signal I and Q constellation measured at vector analyzer for QPSK, EVM$_{PEAK}$ = 7.03%, EVM$_{RMS}$ = 2.86 %.

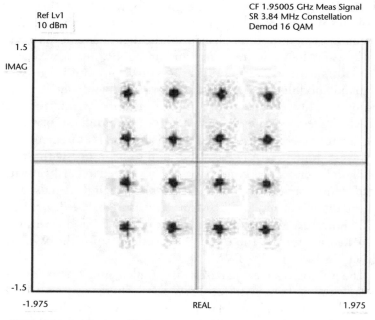

Figure 5.36   RF signal I and Q constellation measured at vector analyzer for QPSK, EVM$_{PEAK}$ = 6.39%, EVM$_{RMS}$ = 2.45 %.

To conclude, it was seen that radio transmitter technologies in line with the OR requirements are appearing at research level but are not mature enough to be put into production.

5.6 Conclusions

This chapter described the research effort toward the design of key elements of a reconfigurable architecture envisioned for the support of heterogeneous mobile systems. It tackled the main challenges in the configuration technologies for the physical layer of the reconfigurable equipment (e.g., devices, base stations).

Wireless systems are becoming more and more heterogeneous in terms of standards and used frequency bands. In the case of the base station and the radio network, the reusability of existing sites is particularly important when it comes to introducing a new radio standard, because it has significant benefits for operators in terms of savings in both capital expenditure (CAPEX) and operating expenses (OPEX).

Multistandard multiband solutions are essential to enable equipment suppliers to reduce the number of development cycles, thereby reducing the research and development costs and shortening the time-to-market. Fewer variants mean much lower maintenance, manufacturing, and module improvement (cost-reduction) efforts. However, all these advantages have to compensate for the possibly slightly higher cost of SDR solutions compared with dedicated solutions.

In "beyond 3G" systems, new air interfaces (e.g., WiMAX for hot-spots and hot-zones, also known as WINNER) and an extension of the multistandard radio resource management procedures are making SDR solutions ever more relevant for base station platforms.

For the operator, SDR offers greater flexibility in the deployment of different standards and multiple frequency bands. For example, it can be used to rapidly introduce hot-spots and hot-zones that reuse existing sites.

References

[1] IST project E2R II, "Functional Physical Layer Architecture," WP 4 Deliverable 4.3, January 2005, http://e2r2.motlabs.com/

[2] ITU-T Recommendation Y.2011, "General Principles and General Reference Model for Next Generation Networks," October 2004.

[3] IST project E2R, "E2R Reconfigurability Management System-Level Architecture," Deliverable D1.4, January 2006.

[4] 3GPP TR 22.978 v1.0.0 "All-IP Network (AIPN) Feasibility Study" http://www.3gpp.org/ftp/Specs/archive/22_series/22.978.

[5] IST Project E2R, "Function Split and Procedures for Flexible Partitioning of Network Entities," Deliverable D3.4, October 2005.

[6] Mitola, J., "Software Radios: Survey, Critical Evaluation and Further Directions," *IEEE Communications Magazine*, May 1995.

[7] Siemens, "3G Wireless Standards for Cellular Mobile Services," white paper, Munich, Germany, 2002.

[8] 3 GPP TS 25.101 V6.3.0 (2003-12), User Equipment Radio Transmission and Reception (Release 6).

[9] Springer, A. et al., "RF System Concepts for Highly Integrated RFICs for WCDMA Mobile Radio Terminals," *IEEE Transactions on Microwave Theory and Techniques*, Vol. MTT-50, No. 1, January 2002.

[10] Liu, H. et al., "A 15b 20 MS/s CMOS Pipelined ADC with Digital Background Calibration," *IEEE International Solid-State Conference, Digest of Technical Papers*, U.S.A., February 2004.

[11] Siragusa, E. et al., "A Digitally Enhanced 1.8V 15b 40 MS/s CMOS Pipelined ADC," *IEEE International Solid-State Conference, Digest of Technical Papers*, U.S.A., February 2004.

[12] Burger, T., *On the Optimum Design of Operational Transconductance Amplifiers with Application to Delta-Sigma Modulators*, Series in Microelectronics, Vol. 133, Hartung-Gorre Verlag, 2002.

[13] Veldhoven, V., "A Tri-Mode Continuous-Time Delta-Sigma Modulator with Switched Capacitor Feedback DAC for a GSM-EDGE/CDMA2000/UMTS Receiver," *IEEE International Solid-State Conference, Digest of Technical Papers*, U.S.A., February 2003.

[14] Balmelli, P. and Q. Huang, "A 25 MS/s 14b 200 mW Delta-Sigma Modulator in 0.18μm CMOS," *IEEE International Solid-State Conference, Digest of Technical Papers*, U.S.A., February 2004.

[15] Koenig, W. and S. Walter, "Following the Software-Defined-Radio-Idea in Design Concept of Base Stations: Possibilities and Limitations," *Annales des telecommunications*, no. 7–8, Paris 2002.

[16] Koenig, W. and S. Walter, "Architecture of a Multi-Band Front-End for a Medium Range Base Station," *European Microwave Week*, Munich, 2003.

[17] Bourse, D., ed., "Reconfigurable SDR Equipment and Supporting Networks," Wireless World Research Forum, Working Group 3, white paper, http://wg6.ww-rf.org/images/pdfs/WWRF_WG3_SDR_WhitePaper1_v2.0.pdf.

[18] Liang, Y. et al., "A Universal Transceiver for Broadband Wireless Communications," *PMIRC E2R Workshop*, Barcelona, Spain, August 2004.

[19] http://www.arm.com/products/solutions/AMBAHomePage.html.

[20] Jantsch, A. and H. Tenhunen, *Networks on Chip*, Kluwer Academic Publishers, 2003.

[21] www.obsai.org/latest/OBSAIfirstspecslaunchtechfacts100903.pdf

[22] IST project ORACLE, "Advanced Transmitter Architecture for OR Terminals," Deliverable 3.3, July 2008, IST-2004 027956.

CHAPTER 6

Reconfigurable Radio Equipment and Its Management

Radio reconfigurable equipment is an inherent part of the seamless communication system. Radio devices with enhanced reconfiguration capabilities are required to ensure that customers (i.e., end users or network operators) have flexible, modular, and evolvable connectivity solutions. This chapter focuses on processes and mechanisms that enable such reconfigurable radio equipment to exist.

One of the major FP6 IST research projects that was active in the area was the project E2R. This project developed a three-tier reference architecture based on reconfiguration management and comprising *configuration management modules (CMMs)*, *reconfiguration control through configuration control modules* (CCMs), and *reconfigurable elements through configurable execution modules* (CEMs). The reconfiguration management could cover all the means inside the equipment, enabling it to contribute to make appropriate decisions concerning the RAT reconfiguration to be applied.

Other contributors to this area (on a more specific, but related, communication area) were FP6 projects, such as the IST project MAGNET Beyond and the IST project RESOLUTION [1]. These projects designed and implemented radio reconfigurable equipment to support personal network (PN) communications and the use of localization technologies in wireless sensor networks, respectively. The achievements and details of these implementations are described in detail in the book of this series entitled *Ad Hoc Networks*.

This chapter is further organized as follows. Section 6.1 gives an introduction to the topic. Section 6.2 describes the reconfigurable management solutions from the network and equipment viewpoint, as proposed within the IST project E2R. Section 6.3 proposes solutions for handling reconfigurability control. Section 6.4 describes implementation and other issues related to the reconfiguration elements. Section 6.5 concludes the chapter.

6.1 Introduction

Reconfigurable radio equipment (RRE) will be a decisive segment of the future end-to-end reconfigurable systems. On one hand, the heterogeneity of environments, standards, and devices are making the number of combinations explosively larger; on the other hand, traditional radio devices with a reduced and predefined set of available modes are evolving into largely reconfigurable devices or access points.

The architectural and technological impacts can be studied through the description of unified points of convergence across the main subsystems of reconfigurable equipment, addressing the available RAT processing chains and also the management and control modules that take the reconfiguration from one chain to another. APIs and data structures applicable at any relevant point of the architecture should be defined as candidates to further unification through standardization efforts.

The robustness constraints imposed by highly constrained embedded equipments, the severe and diverse security requirements, and the stringent reliability constraints imposed by customers can be investigated through context-aware security mechanisms, secure over-the-air downloads, and adaptive security frameworks.

The integration of RRE with the rest of the communication architecture is shown in Figure 6.1. The main characteristic of such equipment is that it supports distributed communication architecture.

Reconfigurable connectivity can be understood to be the capability of equipment to provide radio connections with reconfiguration possibilities from one radio standard to another. The extent of the reconfiguration is highly variable, as are the solutions that have been proposed. Flexible devices are expected to play an important role in answering the reconfigurable connectivity challenges [2]. The research that has been conducted related to the architecture deals with general perspectives as well as technological issues.

Reconfigurability has been addressed for mobile equipment, such as a user terminal, a laptop, or a PDA. Reconfigurability in the network addresses flexible network equipment, such as a smart base station, a reconfigurable access point, and so forth. The focus ranges from physical or media access issues to functionality at the high level on the layers of the active connectivity stack.

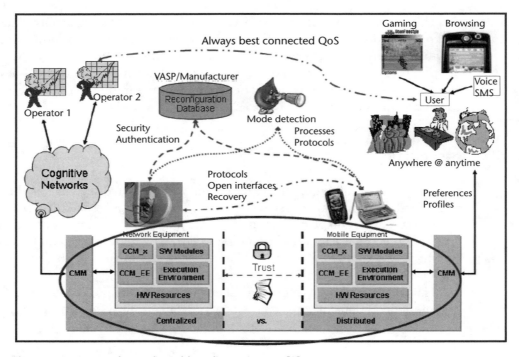

Figure 6.1   Scope of reconfigurable radio equipment [2].

In addition to the internal equipment architecture, network elements are also required. In this respect, a general supporting architecture of a cognitive network has been investigated. The topics are linked to remote device management facilities and reconfiguration control entities, which the core network needs to present to management of the reconfigurable elements appropriately.

The term *equipment* can indistinctively refer to user equipment (e.g., flexible mobile phone, etc.) or network equipment (e.g., flexible base station or flexible core network devices, etc.). It not only refer sto user terminals but also encompasses base stations and/or access points. The term *radio equipment* specifically refers to equipment that is capable of providing radio connectivity.

RRE has its internal architecture divided into a three-tier model, as shown in Figure 6.2.

Reconfiguration control, as one of the tiers of this architecture, can be investigated through approaches that focus on the overall simulation and verification of the architecture, study of the configuration control requested for base stations, presentation of the FDL usage as a way to provide functional-level inputs to reconfiguration control, and analysis of the way advanced results on spatial scheduling can be applied to increase efficiency of reconfigurable equipment implementations.

The research on reconfigurable elements described in this chapter focuses on the configuration execution module hardware abstraction layer (CEM-HAL) and CEM implementation for a SAMIRA DSP target, choices of implementation for adaptive execution environments, and results obtained concerning internal software architecture of embedded real-time processors, based on an abstraction layer, the definition of the microframework, and the transceiver APIs.

6.2   Reconfigurable Management

The reconfiguration management enables the reconfigurable equipment to make appropriate decisions concerning the RAT reconfiguration. The means range from

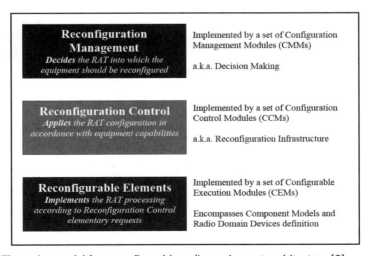

Figure 6.2   Three-tier model for reconfigurable radio equipment architecture [2].

very simple (e.g., a push-button on the front panel for the user to switch between one RAT to another) to more complex collaborative expert systems.

The contribution of the reconfigurable radio equipment (RRE) in defining the desired reconfiguration state varies from one RRE to another. A complex solution could be based on an autonomic decision-making engine coupled to the core network entities to collaborate for optimized decisions. The simplest of the cases could be the unilateral reconfiguration orders locally triggered by the end user.

The RRE obtains the necessary authorization, parameters, and files to perform reconfiguration from remote sources through the interactions between the RRE reconfiguration management modules and the remote modules located in the core network, such as a device management server. Appropriate protocols, such as the OMA device management standard, also support these interactions.

Reconfiguration management, therefore, is also known as decision-making architecture.

The reconfiguration management manages processes according to a specified semantics, protocols, and configuration data model, using functional level descriptions. Figure 6.3 shows the prime classes identified as part of the reconfiguration management proposed for the RRE [2].

6.2.1 Network Perspective

The topics covered in this section include adaptive security framework, policy-supported device management, and secure OTA download. First the problem is stated, and then the technical approach is explained.

6.2.1.1 Adaptive Security Framework

Generally, security is considered to be a static component in system design, and one security scheme is assumed to protect the system throughout its lifetime. The sys-

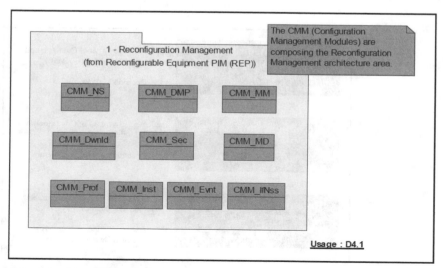

Figure 6.3 Overview of Reconfiguration Management [2].

tems using solutions based on one time assessment, however, may not be adequate when exposed to diverse operating environments. The resource-constrained mobile devices cannot deploy the best security all of the times, as this may not be feasible. Therefore, it is important to identify relevant security elements in the contextual space and develop suitable monitoring schemes to trigger the appropriate security adaptations. In a network setting, such information is distributed across various network identities, and an end-to-end aggregation of this information would be required for a complete view across the network. However, this would amount to overhead and it is important that the proposed security framework remain conscious of the trust structures interconnecting the different network elements (see Figure 6.1). A certain set of identified nodes in the network can be subjected to controlled aggregation.

An adaptive security framework provides a platform to realize the three aspects of reconfiguration, (i.e., monitoring, analyzing, and responding).

This framework shown in Figure 6.4 would help monitor the various security-related aspects of the contextual space using mechanisms to identify the gaps between existing and required levels of security, to perform the relative cost analysis of the security schemes for an optimal selection, and to expose the interface to the underlying functional units for dynamic reconfiguration.

One of the basic tasks of the adaptive security framework is to develop a security context from a sensing and monitoring point of view. The CMM_Prof module of the architecture in Figure 6.4 can be extended according to the architecture requirements for security. The identified aspects are monitored and the changes are reported as a new security context. The difference between the existing and the new context is calculated and, accordingly, appropriate security configurations are selected and recommended for provisioning. The CMM_DMP module includes the policies governing the decision making, and the CMM_Evnt module includes a priority scheme to prioritize the flow of critical events.

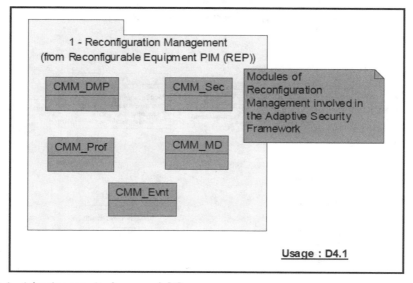

Figure 6.4   Adaptive security framework [2].

The transients should be avoided, while provisioning the new security configurations by the security manager and the new security contexts should be updated as the current contexts. The CMM_Sec includes mechanisms for this purpose.

The functionality distribution across various CMM modules is highlighted in Figure 6.5. As shown, CMM_PROF and CMM_MD play a role in monitory and sensory efforts, whereas the analyzer involves CMM_DMP, CMM_EVT, and CMM_SEC. The interface to the relevant CCM modules is responsible for provisioning the new configurations.

The existing CCM architecture can possibly be extended by introducing new operations and interfaces. This is shown in Figure 6.6. Full details about the functionalities in Figure 6.6 are available in [2].

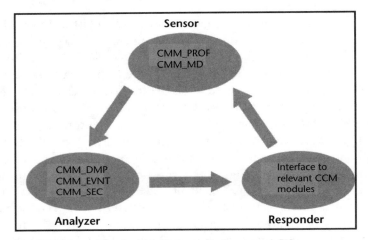

Figure 6.5   Roles of CMM modules in adaptive security framework [2].

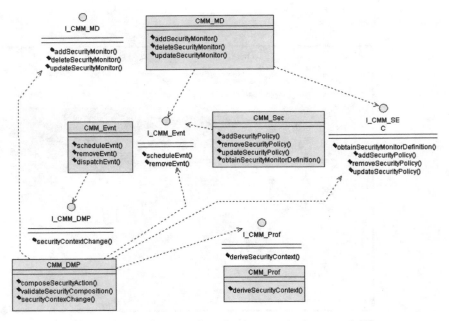

Figure 6.6   Extension of CCM architecture for adaptive security framework [2].

A monitoring mechanism for security points with respect to end-to-end security for user equipment is provided by a *monitoring network association*.

This mechanism can trigger dynamic policy creation and enforcement. It is also able to incorporate all the individual monitoring schemes and aggregate different types of sensory data obtained from sources across the network into the proposed framework to make it more effective.

Figure 6.7 shows the 5-tuple of source and destination addresses, port numbers, and protocols (also known as the network association), which is used to define the network security policies.

Dynamic port allocation is used by several applications, such as file transfer protocol (FTP), which assigns the port dynamically for the data connections; real time protocol (RTP)–based streaming applications that negotiate ports at runtime; and so on. The security requirements vary depending on the network association; for example, a control connection of FTP may require only authentication, whereas the data connection may require encryption. This fine-grained network security adaptation can be achieved by the so-called Berkeley socket system [2]. The sockets opened by various applications on the device are monitored to enforce high-level security policies on the protocol components as sockets are the most prevalent interface to the communication protocol. This kind of mechanism has been discussed in [3].The network layer protocol suite, IPSec, is also subject to reconfiguration [4]. The security policy determines the kind of security services that are offered to an IP flow and the security association specifies how they are to be processed. This is stored in different databases, called *Security Policy Database (SPD)* and *Security Association Database (SAD)*, respectively. Thus, for every new network association, the high-level security, which could either be user- or application-specified, is identified based on the tuples components and the corresponding IPSec specific security policies are created and added to the SPD and SAD.

The adaptive security framework proposed here was developed as an event-based system. First, the purpose of the event list is to identify where the com-

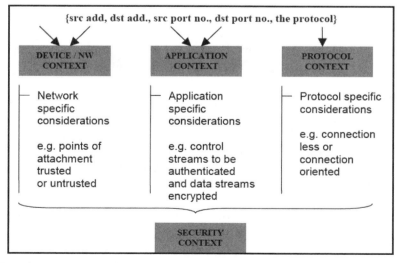

Figure 6.7 Relevant security information from the monitoring network association [2].

ponents can schedule events according to their interest. The event is serviced by an associated event listener for an appropriate action when dispatched. But the details of the scheduling algorithms are open for further investigation, along with the idea of a compact representation of the security context.

6.2.1.2 Policy-Supported Device Management (DM for Server)

An automated provisioning system is needed to launch new operator-relevant services that can handle end-user personalization, remote equipment configuration, new device detection, and device capability management. Device management helps to track, manage, and tailor the configuration settings for a variety of equipments without being dependent on the underlying platform. This approach would not only help the base stations but also enable the end-users to view and update configuration information in their own mobile equipment. There has been an evolution in device management from device provisioning to remote diagnostics and troubleshooting tasks [2].

Fine-grained access control to the remote device is needed to support the installation of third-party software components. Suitable defined policies can be used to solve issues related to the needs and rights of a variety of stakeholders (manufacturer, operator, regulator, user, etc.). The stakeholders can act in a domain relevant to their corresponding influence area, but they have no access to private user data, because private user data comes under a policy that belongs only to the user domain. For instance, when the e-mail service provider has to change the e-mail clients' settings, the new settings are wrapped into a policy with a certificate to authenticate the e-mail service provider. The DM server and the DM client only help to transfer the information; after the DM client receives the policy, the certificate is verified and confirmed only if the stakeholder is entitled to do this operation. Figure 6.8 shows the scope of the policy-supported device management.

The work done is in accordance with the OMA device management specification. The two key concepts involved are explained as follows:

- *DM client:* This is a device containing a device management client agent, which sends its modifications to the server (in some cases, the server can also initiate the synchronization) and receives responses from the OMA DM server. Usually, the OMA DM client is a mobile phone, PC, or PDA.
- *DM server:* This is a device containing a device management server agent and a sync engine, which sends the clients' modification to the server, usually after the OMA DM has started the synchronization. It is responsible for processing the sync analysis after receiving the client modifications, and it initiates synchronization if unsolicited commands from the server to the client are supported on the transport protocol level. Usually, the OMA DM server is a server device or a PC.

Figure 6.9 is an example of synchronization between an OMA DM client (e.g., mobile phone) and an OMA DM server (e.g., PC server). The SyncML message containing the data modifications made in the client is sent to the server by the client. It is responsible for synchronizing the data within the SyncML messages with data

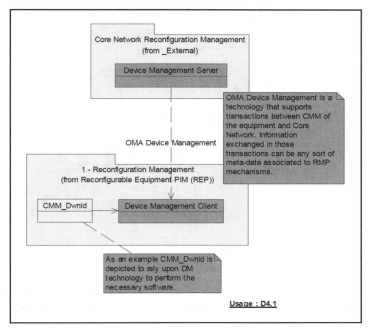

Figure 6.8   Scope of policy supported device management [2].

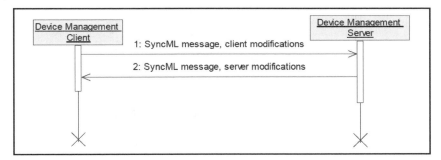

Figure 6.9   A high-level sequence of OMA DM process [2].

stored in the server, and then the server returns the modifications to the client. The use-cases covered are shown in Figure 6.10.

The important use-cases are described as follows:

- The *initial configuration provisioning* (as shown in Figure 6.11) is used to provide the initial access for the devices to the DM server and to restore the device manager information if the device has lost it. Provisioning is the process of changing a device from a "clean state" (i.e., no provisioning) to a device that can initiate a management session with the device management server. The http/https protocol is used for the initial provisioning of new parameters and settings to the device:
  · Local DM infrastructure discovers the device;

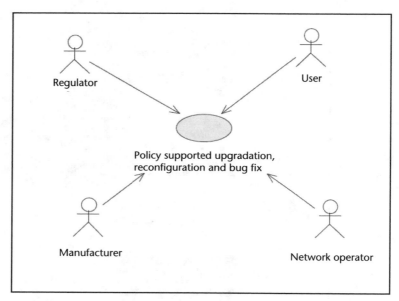

Figure 6.10   High-level use-case and important actors for policy supported DM [2].

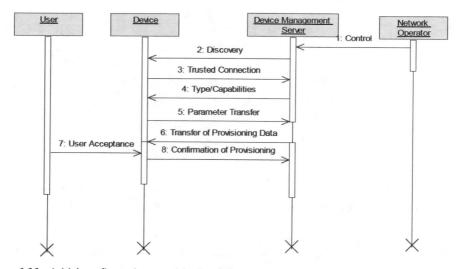

Figure 6.11   Initial configuration provisioning [2].

- Trusted relationship established through certification;
- Device is queried concerning the type and capabilities;
- Provisioning data is transferred to the device;
- Provisioning is confirmed.

    Figure 6.12 shows the main scenario for the provisioning use-case.

- *Bootstrap provisioning* (see Figure 6.13) is used for a case of wrong configuration. In that scenario, the device . might be highly misconfigured if the

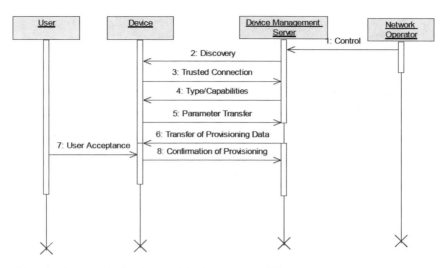

Figure 6.12   Main scenario for the provisioning use-case [2].

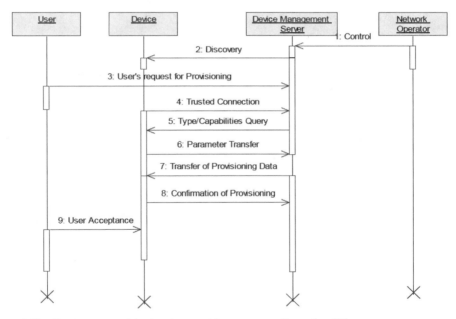

Figure 6.13   Bootstrap provisioning to override wrong configuration [2].

already-inserted configuration data is not overridden. The process is explained as follows:

· Discovery of the device;
· Request for provisioning the parameters by the user;
· Trusted relationship established between device and DM server;
· Management server queries the device for its capabilities;
· Device responds to the request;

- Appropriate configuration transferred by the DM server;
- Device provisioning confirmed.
- *Remote configuration* initiated by the DM server (see Figure 6.14) refers to the attempt to access the operator's infrastructure with the old parameter set, which could either trigger the transfer of the new parameter set or identify and automatically update all the affected devices regardless of the actual usage of the service. The procedure can be explained as follows:
  - Device with obsolete configuration settings is detected, (e.g., by addressing the wrong URL or IP address) or a query for the device settings is made actively (e.g., triggered by a customer care call);
  - Trusted relationship established between the device and the DM server;
  - Management server queries the device for the actual configuration;
  - Device responds to that query;
  - Transfer of effective configuration data to the device;
  - Update of the configuration data is confirmed.
- The *software upgrade, update, and installation* initiated by the management server/user (see Figure 6.15) is a procedure followed when the management server requests the software/hardware inventory of the device. It has the following steps:
  - The DM server issues a request to the device for an inventory of installed software;
  - The device issues a request to the user for authorization to send a response to DM server that contains an inventory of installed software and authorization to install upgrades;
  - The user confirms the request;
  - Device sends response to the DM server;
  - The DM server initiates software download, installation, and execution;

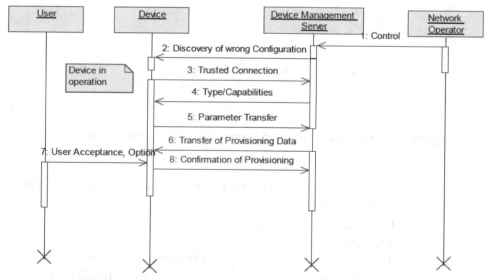

Figure 6.14   Remote configuration initiated by the DM server [2].

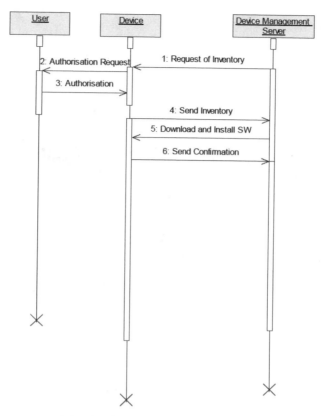

Figure 6.15   Software upgrade/update/installation initiated by management server/user [2].

- Device sends a confirmation back to the DM server.
- The *procedure for bug fixes* for an operational improvement includes the following steps:
  - The operator management authority installs the corrected software in the server;
  - The software is downloaded to the device using a standardized protocol, which includes device discovery, capability exchange, authentication, authorization, security, and other software download functions;
  - The user is notified about the software change.

Further in-depth investigations for the sequences presented above are required, but it is mandatory to fulfill them only for a case of software upgrade or software provisioning, when the procedure must comply with the OMA DM specifications.

6.2.1.3   Secure OTA Download

The reconfiguration parameters, user profiles, binary representation of hardware drivers, and so forth, as well as other transferred data, are kept private and protected by using encryption. Figure 6.16 shows the modules involved in a secure over-the-air download.

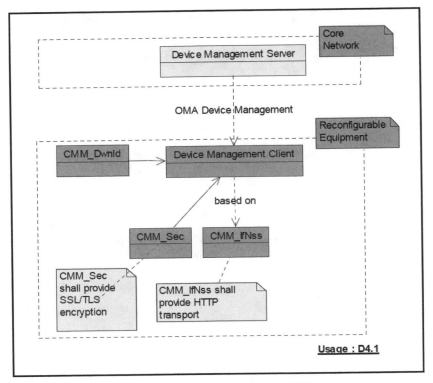

Figure 6.16   Modules involved in secure over-the-air download [2].

The figure demonstrates how the OMA DM is split between the server and the client. CMM_Inst, CMM_Dwnld, and CMM_Sec are important security modules for the OMA integration architecture. The CMM_Dwnld and CMM_lfNss are responsible for the download that manages and provides the basic HTTP-based transport layer respectively. The SSL/TLS security functionality is used to establish a secure channel between the server and the client. It is located within the CMM_Sec module.

HTTP is used for the PoC prototype implemented in combination with an additional encryption layer of SSL/TLS to protect the data traffic by providing privacy and data integrity. More information on the libraries embedded is available in [2].

6.2.1.4   Network Support for Negotiation Functionality

Reconfigurability enhances the terminal and network elements with the capability to dynamically select and adapt to the most appropriate technology and transparent, seamless communications in a robust way (i.e., safe, secure, and reliable). It is necessary that this selection be consistent with user preferences and equipment capabilities. Further, it should take into consideration the specific service area region conditions and time zones. It should not be restricted to technologies that are preinstalled on the device. Dynamic downloading, installation, and validation of software components needed for the reconfiguration and support of a potential RAT should be supported. The equipment should also be capable of negotiating offers with various available networks to select the most appropriate reconfiguration action.

Figure 6.17 shows the scope for the core network negotiation functionality.

For facilitating the negotiation to select the most appropriate reconfiguration, three main support entities are introduced as follows:

- The *reconfiguration control module (RCM)* emulates the network reconfiguration control module. Its main function is to respond to offer requests from CMM, and CMM_NS. The design is shown in Figure 6.18. More information on the interfaces and prototype is available in [2].
- The *cognitive pilot system (CPS) emulator* provides the support for the RAT discovery and the initial information for discovery of the available RATs of network operators in the area. It allows the network operator to advertize themselves with information about the RAT provided, the frequency range per RAT, and an RCM contact point, through which the terminal can communicate with the network side to initiate negotiations, download SW modules, etc. The design is shown in Figure 6.19.
- The *network monitoring and performance assessment (NMPA)* monitors the load per RAT of a certain operator and provides it to the RCM, where it is used to formulate pricing policies and offers. The design is shown in Figure 6.20. More information on the interfaces and prototype is available in [2].

Enhancements of the developed mechanisms to provide additional features, such as an efficient approach for cognitive networks, should be considered for future research. An important achievement would be to investigate how to integrate

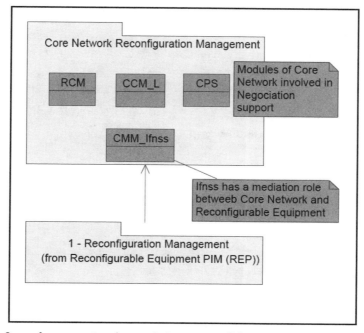

Figure 6.17   Scope for core network negotiation support [2].

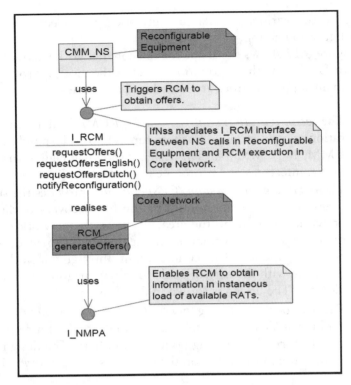

Figure 6.18 Design view of RCM [2].

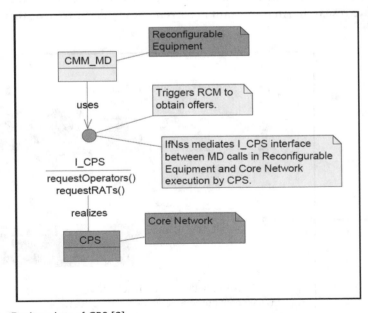

Figure 6.19 Design view of CPS [2].

and exploit the developed support entities in a composite radio environment comprising real network segments and user terminals.

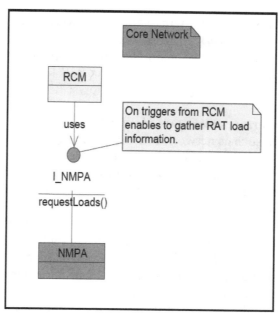

Figure 6.20   Design view of NMPA [2].

6.2.2   Equipment Perspective

Reconfigurability, from an equipment point of view, requires context-aware security mechanisms, policy-based device management, reconfiguration negotiation and selection, reconfigurable terminal QoS architecture, and reconfiguration management function for multistandard base stations. The following sections identify the problem for each topic, the technical approach for solving it, and the obtained results.

6.2.2.1   Context-Aware Security Mechanisms

To design a fully reconfigurable and modular telecommunication infrastructure that enables diversely connected devices to transparently access services in several contexts, (e.g., in different locations or at variable connectivity status), a guarantee is required for the protection of the resources contained in the mobile device.

The complexity increases when each networking environment has its own protection requirements, making it difficult to unify the corresponding security policies. The problem with a single policy is that it cannot by itself capture the entire range of security requirements of a federated system. The policies used by default are generally not adapted to the current environment, which makes it necessary for the equipment to support multiple security policies.

If the security mechanisms are made reconfigurable (i.e., the security services such as authentication or access control are adapted to the local security context), this would in turn enhance policy flexibility. Examples of applications of context-aware security mechanisms are the following: adapting key lengths depending on the geographic location of the user to comply with local regulations on cryptography and relaxing the strength of authentication depending on the ambient security level to increase network performance.

Figure 6.21 shows the main elements comprising the context-aware security architecture as proposed in the IST project E2R [2].

The components are explained as follows:

- The *security context provider (SCP)* collects low-level security-related information, such as network context or situational information from the environment (i.e., different sources, such as network or sensors), and aggregates the input into a higher-level description of the ambient security context. For implementing an SCP, dedicated interfaces in the CMM_Prof components are needed for profile management.

- The *security decision-making component (SDMC)* receives the description from the SCP and, based on that, decides whether to reconfigure the security infrastructure. Reconfiguration in that case can be used, for example, for relaxing the strength of the authentication or selecting the type of authorization model adapted to the current network. Once made, the decision is transmitted to the security mechanism that should be adapted (e.g., access control, authentication), which performs the reconfiguration operation by changing the needed component. Dedicated interfaces are required in the CMM_DMP component to implement the SDMC.

- The *CMM_Sec* guarantees a certain level of security to the elementary transactions involved in the context-aware decision loop. The reconfiguration process, which changes the authorization policies in ASM from a DMP trigger, would be secured transactions.

- The *adaptable security mechanism* relies on the CMM_Sec to ensure safety of the reconfiguration operation. It should be flexible enough to be reconfigured by tuning the security configuration parameters or to be replaced by another component offering similar security services. It may optionally notify the environment about the result of adaptation. An automatic loop with monitoring, decision, and action steps is set up for establishing a "self-protecting" system.

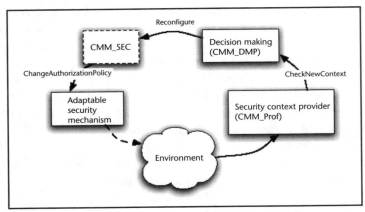

Figure 6.21   Components in context-aware security architecture [2].

Figure 6.22 is the UML diagram of the concepts described above.

The context-aware authentication adapts the security of the wireless protocol stack by plugging libraries for cryptography and authentication in and out according to the level of the security needed. If the ambient security level is sufficient, then the strength of authentication can be relaxed to increase the network performance. The framework's adaptability can be seen at two levels: the first is by fine-tuning some configuration parameters, such as sensitivity or key lengths of the authentication service, without changing the authentication algorithm or mechanism; and the second is by replacing the authentication mechanism with another. The desired degree of flexibility can be offered via a component-based design, similar to pluggable authentication modules (PAM) architecture, for the authentication framework. The coarse-grained architecture is shown in Figure 6.23.

The functions needed for the authentication framework have been described in detail in [2].

The context-aware access control applies an authorization policy within the infrastructure, depending on the context, and a wide range of authorization models can be enforced. A proposed architecture for context-aware access control is shown in Figure 6.24. It is entirely component-based, and each kernel is rectified at the software level by a component. The architecture has many similarities with the extensible access control markup language (XACML) authorization architecture, so the architecture can be integrated with XACML in terms of identification of components and in the policy specification language used.

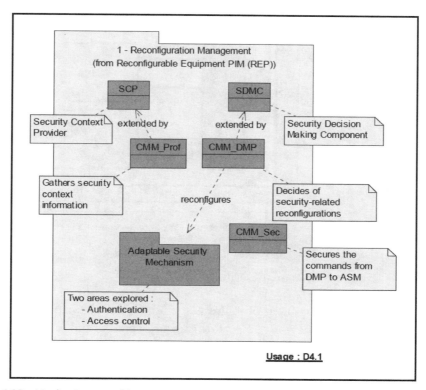

Figure 6.22   Mechanisms used in context-aware security [2].

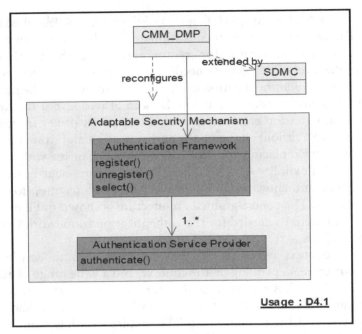

Figure 6.23 Adaptable authentication framework [2].

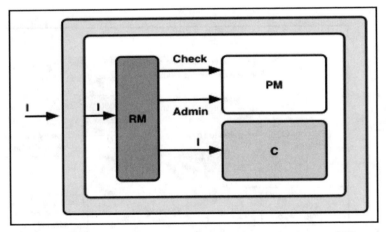

Figure 6.24 Component-based architecture for context-aware access control [2].

The components are described as follows:

- C is the component of interest encapsulated in the depicted framework.
- The *reference monitor* performs security checks and is automatically attached to the resource to be controlled.
- The *policy manager (PM)* encapsulates the policy decision making for a given authorization model.

• The *CMM_Sec* ensures the security of the reconfiguration process and also performs revocation operations, such as reinitializing the access matrix to host new permissions according to a new PM, interrupting the authorized threads beforehand, or not depending on the revocation policy.

The functions are discussed in detail in IST project E2R Deliverable D4.1 [2]. Precise modeling of the security context is needed to capture the sensitivity of the security infrastructure to the external environment of the equipment; much richer modeling can be achieved by using simple security ontology. Future work needs to be carried out in this direction.

6.2.2.2   Policy-Based Device Management (Client)

The DM client can be implemented in the device as the counterpart to the DM server at the network side. It provides the necessary functionality to manage the tremendous amount of different software components that are synchronized with the databases in the network in a secure manner.

The OMA DM, shown in Figure 6.25, is split between the server and the client.

The managed objects, such as the reconfiguration parameter, the binary representation of the hardware deriver, the complete protocol stack, and so forth, are entities that can be manipulated by the management functions of the OMA DM protocol. The protocol is agnostic about the contents and values of the management objects, and treats the mode values as opaque data in all cases. Objects are stored in a hierarchical key tree that can be dynamically changed.

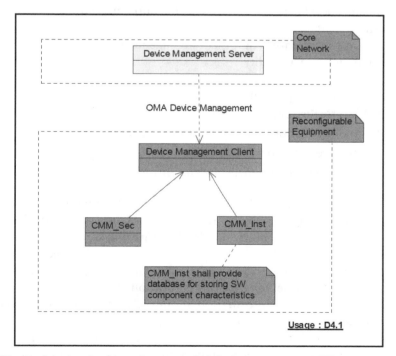

Figure 6.25   Modules involved in policy supported device management [2].

The requirements that apply are as follows:

- Clients implementing the OMA DM must support the OMA DM account management object, the DevInfo management object, and the DevDetail management object. Also, the OMA DM servers must support all the three management objects [5].
- The tree types of management objects are stored in a database [6].
- All relevant metrics and characteristics of the low-level software and underlying hardware platform are placed under DevDetail.
- Relevant CMM_x instances can access the database for query and storage of parameters. It should also trigger the CMM-embedded DM/DS client for upgrade and download of appropriate parameters and files.

Access to the internal parameters and low-level software components of the device will be regulated via the XACML policies, in cooperation with SELinux mechanisms, and the policies shall be dynamically generated.

The embedded libraries used in this context are described in detail in [2]. A feasible implementation of the footprint for a PoC has not been realized yet and is a good topic to be investigated.

6.2.2.3   Reconfiguration Negotiation and Selection

The capability for negotiating offers with the available networks and selecting the most appropriate reconfiguration action for the equipment is needed for efficient, reliable, and secure equipment operation in the context of reconfigurability. Below, the prototype platforms developed and implemented by the IST project E2R are presented.

The reconfigurable equipment management system (REMS) prototype provides the following functionality:

- Profile management;
- Monitoring and discovery;
- Reconfiguration negotiation and selection;
- Interface with control domain;
- Interface with network support entities.

The main modules addressed for REMS are shown in Figure 6.26.

In a typical high-level scenario for the operation of REMS, the CMM_MD scans the environment for the available RATs/networks continuously in the coverage area by using the CPS module. It also requests and collects the necessary profile information. After an analysis of the information (such as the RAT status and profile data), it decides whether the reconfiguration should be triggered. The CCM_RM cooperates to provide information about the status of the networks. If reconfiguration is needed, then the CMM_NS is triggered to request offers from the networks. This is usually done when a new technology is available that would offer better alternatives, or if the QoS of current services is degrading. The CMM_MD, CMM_Prof, and CCM_RM participate in completing the reconfiguration process.

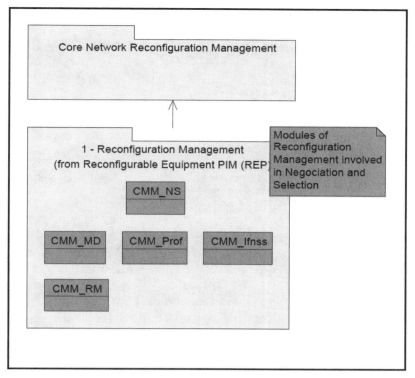

Figure 6.26   Scope for REMS prototype [2].

Figure 6.27 shows the classes and the relationships between the modules of the REMS prototype. The sequence of the interactions is shown in Figure 6.28.

The main functionality of the modules is briefly presented as follows, but more details about the prototyping and interfaces are given in IST project E2R Deliverable D4.1 [2].

- The CMM_Prof mainly maintains, manages, and provides the user and terminal profile information.
- The CMM_MD is responsible for discovering the available networks and monitoring their status.
- The CMM_NS is responsible for negotiating or exchanging offers with the available networks on specific services and for selecting the most appropriate one in terms of pricing, QoS level, and reconfiguration simplicity.
- The CMM_RM is responsible for emulating the network card of the terminal and the switch from one configuration to another.
- The CMM_IfNss is a mediator for the realization of the interaction between the CMM modules in the REMS and the network support modules.

6.2.2.4   Reconfigurable Terminal QoS Architecture

User mobility has brought about a rapid change in the network environment and in the services offered to a mobile user. Numerous proposals exist for the same service,

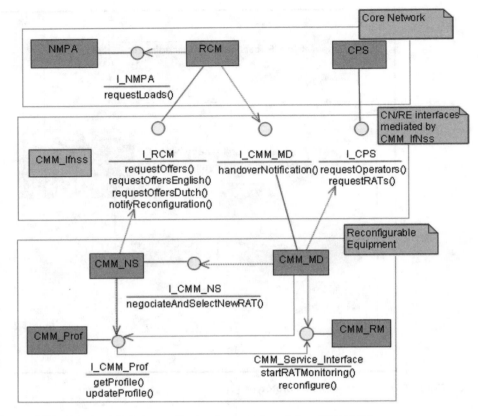

Figure 6.27   Relationship between the implemented modules of REMS [2].

which makes it a complex task for the mobile terminal operation. The IST project E2R II [2] designed architecture extensions for offered QoS from the network to the mobile users by using the methodology explained in Boukis et al. [7]. The architecture that can adapt between protocols offering the same network service using software reconfiguration is the one suitable for such operation. It should be modified by a secure method, dynamically, and be capable of eliminating the unnecessary alternations.

Figure 6.29 shows the scope for reconfigurable QoS signaling and the interaction of CMM_Inst with other CMM modules. The extra functionality is located inside CMM_Inst and interacts with other CMM modules. CMM_NS identifies the offered QoS protocol and informs CMM_Inst. The QoS is modified to operate according to the utilized QoS signaling protocol. CMM_Prof informs the QoS signaling component of the application's QoS requirements for the appropriate QoS signaling to take place. The interactions with the CMM modules are omitted from the description here.

The procedure for providing QoS can be divided into two subtasks:

- The network supports a mechanism to provide an enhanced service. This task is network dependent and the terminal in involved only in one framework.

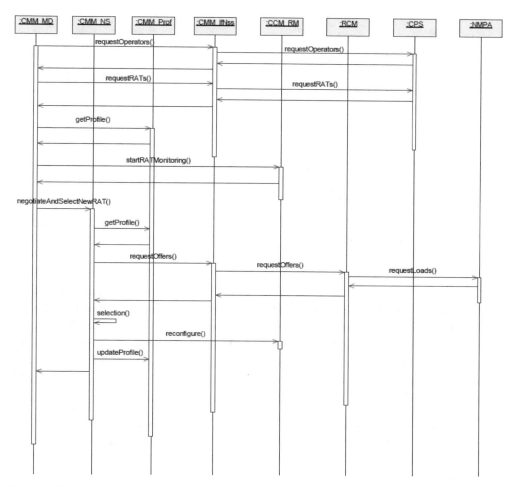

Figure 6.28   Sequence of interactions among REMS modules [2].

- A mechanism is provided for the user and the network to negotiate the per-ceived QoS. This task is more important from the terminal reconfiguration perspective.

Two major proposals have emerged for providing QoS in the network: differen-tiated services and integrated services.

In the differentiated services architecture [8], IP packets receive special service depending on the value carried in the type-of-service (TOS) field in the IP header. The framework does not specify how the privileged packets perceive the enhanced services; hence it varies from prioritized queuing to dedicated link bandwidth, or may be both. Users are not differentiated (i.e., packets from different users perceive an identical service, provided that the same value is carried in the TOS field). This allows scalable provision of QoS for networks of thousands of users, but it can also cause problems in situations where a single user produces excessive data traffic (i.e., the perceived service of all the users with the same value in the TOS field will be affected).

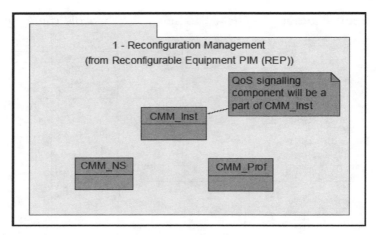

Figure 6.29   Scope for reconfigurable QoS signaling [2].

The integrated services architecture [9] adapts a different approach, in which the enhanced service is provided to distinct data flows. The packets belonging to a particular data flow attain identical values for the destination address, source address, protocol number, and the source and destination ports. The enhanced QoS is provided through two different mechanisms, namely, the controlled-load service [10] and the guaranteed service framework [11]. The former provides the client data flow with a QoS closely approximating the QoS that same flow would receive from unloaded network conditions and, in the latter, provides firm bounds on end-to-end packet delay. It does not suffer from the restrictions of the differentiated services approach, but its drawback is that it is not robust enough to maintain the forward information for every particular flow from every user in a network of a few thousand users.

The mechanisms for QoS negotiations are explained as follows. For a particular user to perceive an enhanced QoS, the QoS negotiation mechanism must communicate with the network routers and negotiate the desired QoS. The Resource Reservation Protocol (RSVP) [12] is one important proposal for this purpose. The reservations are made from the receiver application. The data transmitters send RSVP control messages to provide the network and receivers with an estimation of the transmitted traffic and inform the network. Traffic receivers submit RSVP control messages to make the appropriate reservations. Flows within the same application are treated separately, and the resources reserved for a user are used only for the incoming packets. Though it has been designed for integrated services [13], this approach has been extended to differentiated services [14], and extensions are provided also to work over IP tunnels. Examples of other signaling protocols are the Border Gateway Reservation protocol [15], YESSIR [16], and Mobile RSVP [17].

The internal structure of the QoS signaling components is briefly explained here. The mobile terminal is actively involved in signaling, and the mechanism for providing the QoS is network-specific and usually transparent to the mobile terminal. Though RSVP is a dominant proposal, its operation can vary as different protocols operate in the networks for QoS provision. Hence, a mobile user is confronted with the situation that different signaling protocols can be employed for QoS negotiation among the visited networks.

A QoS signaling component that is able to adapt to the utilized signaling protocol is needed to enable the mobile terminal to exploit the QoS service offered from the visited network. This component is shown in Figure 6.30.

The black circles represent the external sources or destinations of data. The rounded rectangles represent the functional entities that take the data as input and output this data after performing a task on it. The data flow is shown by arrows. The QoS control packets arrive from the network, and first any erroneous packets are identified. The erroneous packet is discarded and the remaining packets are sent for further processing. The next functional entity initiates a response according to the input stimulus. The detailed operation of this component is under research. The final construction entity constructs the protocol control messages.

The interfaces and interactions between the QoS signaling component and the CMM are explained in detail in IST project E2R Deliverable D4.1 [2].

6.2.2.5   Reconfiguration Management Function for Multistandard Base Stations

In the traditional Telecommunication Management Network (TMN) terminology, the MSBS reconfiguration management is an agent function for reconfiguration management at the network element level. The function comprises the following elements:

- RAT SW download;
- RAT SW activation;

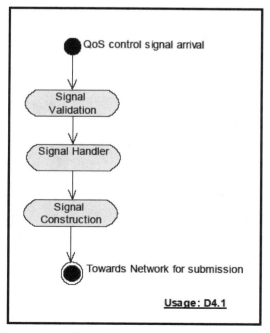

Figure 6.30   Activity diagram for QoS signaling protocols [2].

- RAT capability survey;
- Decision supporting measurement reporting;
- RAT change for a given UE.

Figure 6.31 shows the basic elements of this function.

The MSBS reconfiguration management function can be logically split into three parts:

- *SW-related management:* It is organized in a way that allows the manager to download or remove SW for a given RAT. For a download operation, the RAT name and the descriptor file name are relevant parameters. Removal is done on behalf of the RAT name and version. When a given RAT SW has been downloaded, its components can be distributed to file systems on the target boards that are established on behalf of the given RAT description. The manager can also create, retrieve, or modify SW profiles. A profile describes the group of RATs that can run in parallel in the MSBS, and it defines the concrete conditions and restrictions that the RATs must fulfill. When the *install()* method is applied, the concerned RATs are loaded into their target processors, initialized, and configured, and the data paths between the components are established (if required) and then finally started. The OMG's SW Radio Spec can be seen as a guideline for this process.

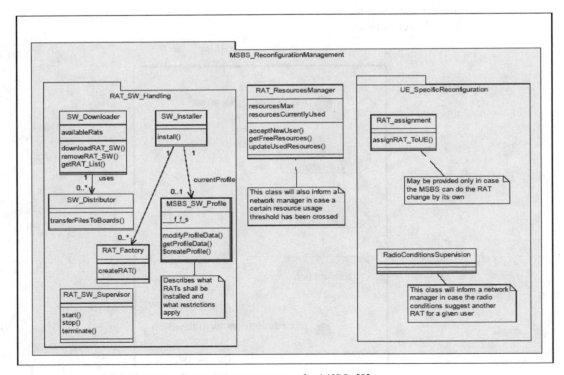

Figure 6.31   Overview for reconfiguration management for MSBSs [2].

- *Resource supervision:* This is done to prevent the overload of the MSBS for a given RAT. The resources spent by the users are monitored, and usually there is a limit and a current value for a given resource. The resources could relate to the RAT as a whole, or to a cell, or to a HW module being used. The underlying configuration control keeps the data up-to-date. A manager can also ask for free resources. The MSBS returns true if the necessary resources can be provided and false otherwise. In case of abnormal behavior, there is a method to free the resources. A manager may also be informed if the remaining resources become less than a certain predefined amount.

- *UE-specific part:* Considering a single user, the manager should be informed if the given radio conditions recommends to change the RAT of the user. A RAT switch for the user can be organized within an MSBS. This should be in line with the RAT assignment strategy as given by the corresponding standards, and is hence only an optional feature.

The PoC approach is explained in detail in IST project E2R Deliverable D4.1 [2].

6.3   Reconfiguration Control

Reconfiguration control, also known as reconfiguration infrastructure, covers all the means inside the equipment that enable it to take advantage of the reconfigurable elements of which it is composed.

The existing solutions are based on metadata descriptors of the target applications (the RAT configuration mode) and the hosting platform (a certain set of reconfigurable elements).Further enhanced features need to be investigated, taking into account the optimization strategies. The architecture and mechanisms of the configuration control module proposed here focus on physical layer processing and have been generalized to management of the whole processing chain in the current structure.

Physical descriptions of the appropriate configuration state that describe the hardware and software modules, the exact mapping of the software onto programmable hardware devices, the use of coprocessor or accelerators, and so on are used by reconfiguration control function. The physical configuration can be defined at the design time by the development teams or by using complex runtime mapping algorithms, taking as inputs the functional requests expressed by reconfiguration management and transforming them into applicable physical configurations.

To optimize the reconfiguration, physical configurations that answer to the functional configuration requests are applied without any principle obligation. These implementations can be determined at development time or can be selected at runtime by reconfiguration control for most of the advanced architectures.

This section focuses on the implementation of mechanisms to control the reconfigurable hardware, the CCM control module, and the links to the hardware abstraction layer (HAL). The control message flow between the different modules and the control signals inside the CCM submodules, interfaces, and their state machines, their final verifications and test cases are shown.

6.3.1　CCM Simulations and Verifications

CCM specification and verification finalize the actions that control the entity of the terminal by defining a coherent representation of the control interfaces of the CCM. A clear identification between the reconfiguration support and reconfiguration radio functionality is made.

The full set of operations and its specification's verifications of the CCM have been realized by SDL with a Telelogic simulation tool kit. The result from the verified system will produce the SDL specification of the complete system and a refinement of its interfaces using IDL. The simulation model is shown in Figure 6.32.

The CCM simulation model presents the process in which control mechanisms are developed in two parts: on one side, using formal definition language like SDL to specify the CCM and interfaces to reconfigurable part, and on the other, the simulation of the reconfigurable part in SystemC using real systems implementation baseband modules to verify the CCM part.

The demonstrator shown in Figure 6.33 contains a C++ code implementation of the CCM and links to the CMM and the CEM. The CEM representation includes SystemC three-radio chains—HDSPA, FDD, and TDD—with the possibility of interchanging the modules.

The efforts in the project E2R II were directed toward formal specification and implementation, which lead to code generation and translation to a Nokia 770 tablet that subsequently was connected and applied different test cases, together with real HW. It provided the possibility of using formal methods to verify the design and specification and of preparing for standardization submission in the format of SDL and IDL specifications [2].

6.3.2　Configuration Control for MSBSs

The configuration control module (CCM) services the requests according to the installation, configuration, and deletion of RATs coming from the related CMM and reports back the appropriate results. Figure 6.34 shows the main entities and the

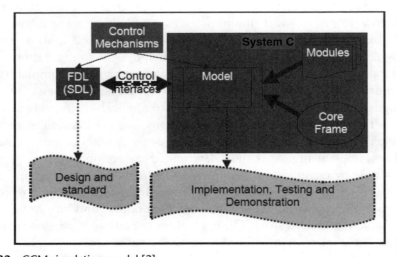

Figure 6.32　CCM simulation model [2].

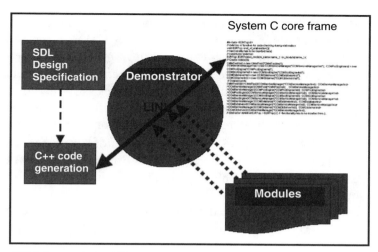

Figure 6.33   CCM demonstrator [2].

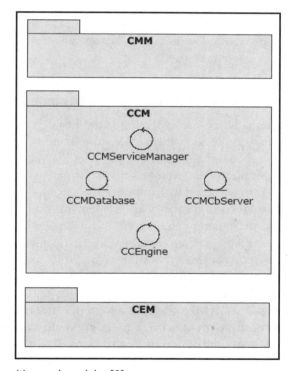

Figure 6.34   CCM position and modules [2].

position of the CCM between the CMM and the underlying configurable processing elements (CEMs), which could be P-CEMs (such as GPPs, DSPs, FPGAs, or ASICs) or C-CEMs (i.e., communication elements) or RF-CEMs (i.e., RF configuration execution modules).

The components of the CCM are explained as follows:

- The *CCM_ServiceManager* processes the configuration requests that come with the appropriate parameters from the CMM. Depending on the requested command, the primitives cause actions to the other elements of the CCM.
- The *CCM_Database* holds the set of functional RAT modules and the respective configuration parameters. It completes the "last good configuration" to enable autonomous recovery mechanisms to be performed after emergency events or after the detection of a "bad reconfiguration."
- The *CCM_CapabilityServer* stores the set of parameters describing the physical platform. The respective resources occupied by the RAT processes are registered in the CCMCbServer by the definition and continuous collection of metric parameters and, thus, at every point of time the number of available resources for a requested reconfiguration can be reported to the CMM.
- The *CCM_CCEngine* comprises the main processing elements used to download the RAT SW components from the CMM for the installation of the modules of the P-CEMs for the initial and further requested configuration and also the its removal.

The described functionality is implemented on the GPP part of a BB processing platform comprising GPP, DSPs, FPGAs, and HW accelerators. More details concerning the approach followed in E2R II are given in Deliverable 4.1 [2], along with the PoC system implementation.

6.3.3   Functional Description Language (FDL) Interpreter

FDL is a language based on XML that is used to describe the functional configurations for reconfigurable equipment. FDL equipment is interpreted by CCM and used to determine a set of signal processing modules (SPMs), which is a binary configuration of the target platform that meets the requirements of the functional description. The class diagram is shown in Figure 6.35.

The functional configurations are implementation-independent and capture the required signal processing behavior for each RAT as a dataflow model of the constituent signal-processes, their parameters, and constraints. FDL is made of up to two languages, one for describing the algorithms and the other for describing the process parameters and arguments. Both of these languages are defined by an XML schema. The FDL languages are candidates for standardization within the SDR community [18], and XML, a widely adopted, standard metalanguage that is effectively platform-independent is therefore a good choice.

The CCM reads the function descriptions and makes configuration decisions based on it. The parser implementation used is platform dependent but the interface to the parser functionality may be constant across different platforms and technologies.

Figure 6.36 shows how the FDL and SPMs are used in a scenario where an intelligent entity owned by the operator sends a new configuration request to a terminal so that a VOIP call can be set up using an alternate RAT.

Then the CCM requests a specific SPM to support the new configuration. The SPM is downloaded from the equipment manufacturer's database and its binary

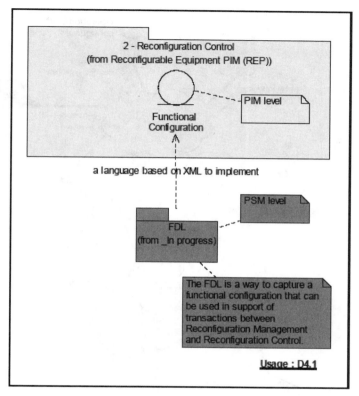

Figure 6.35   FDL class diagram [2].

part is installed on the relevant reconfigurable hardware (CEM). More details about the work are available in IST project E2R Deliverable 4.1 [2].

6.3.4   Spatial Scheduling

Figure 6.37 shows the position of spatial scheduling in the CCM module.

Figure 6.38 shows the mapping of a software realization of an application onto a heterogeneous hardware platform on a functional level.

During the equipment design time, when the application and HW platform are known, it will be possible to associate the application's modules with particular parts of the HW platform in an optimum way in terms of certain performance metrics, such as the processing time, transport time, power consumption, and so forth. But in reconfigurable systems, a situation may occur where an application unknown at the design time of an equipment will need to be deployed onto a HW platform consisting of several different processors. This situation would require an entity that can make the best possible integration of an application and its modules with the processing fabrics of the available heterogeneous HW platform.

Let us consider an application for cellular phones to draw their energy from a battery that has a limited amount of energy. The energy-conscious spatial scheduling can lead to drastic reductions of energy dissipation of the whole mobile system.

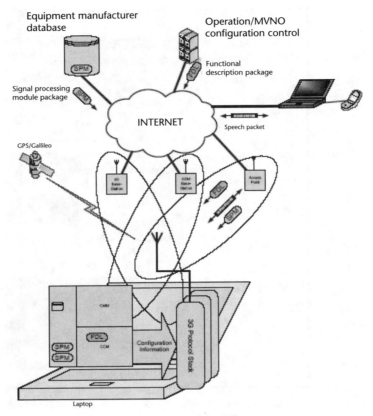

Figure 6.36 Use of FDL and signal processing modules [2].

The reconfigurable systems have a HW platform composed of a network of different architectural fabrics, such as ISA processor, DSP processor, FPGA, accelerators, and ASIC blocks. The instruction set architectures are often related to as soft programmable forms, and field programmable logic as hard programmable forms. This is why spatial scheduling is often called HW/SW partitioning.

The technical approach to this issue can be explained as follows. An optimal integrated schedule, an entity that integrates the application and the underlying HW platform in an optimum way, is composed of the following elements, as shown in Figure 6.39:

- *Input:* The system-level description language is used throughout the whole design process starting from algorithm design and evaluation to HW and SW implementation. The language can unify the representation of the system used to describe its functionality independent of the type of processors employed in the HW platform.
- *Concurrency analysis:* This finds and uses the inherent parallelism in various applications.
- *Dependency analysis:* This maintains the temporal scheduling.
- *Profiling:* This provides an estimated or cycle-accurate calculation of performance metrics.

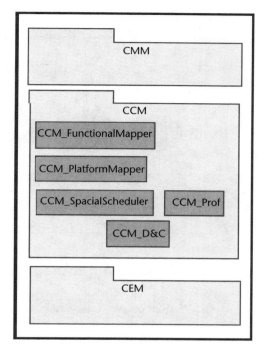

Figure 6.37   Position of spatial scheduling in CCM module [2].

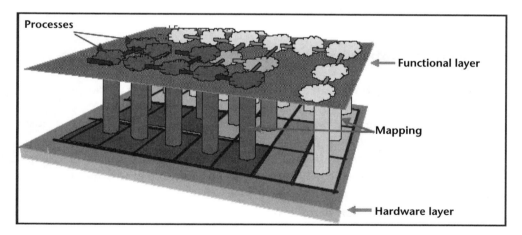

Figure 6.38   Mapping of functional layer onto the heterogeneous hardware platform [2].

- *Mapping:* The mapping of the spatial scheduler evaluates the different mapping solutions until it identifies the optimal one for the given working conditions.
- *Output:* This presents the integrated temporal and spatial schedulers.

With the current available technology, profiling should be performed off-line and provided to the optimal mapping module with a request for application deploy-

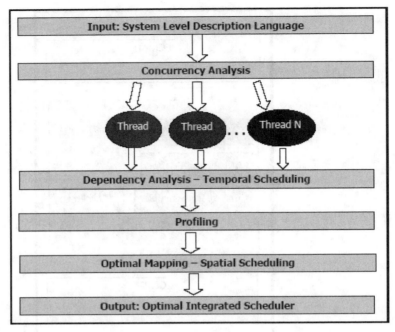

Figure 6.39   The functional components of optimal integrated scheduler [2].

ment. An assessment of the different methods available for mapping that have been investigated is presented in [2].

6.4   Reconfiguration Elements

The reconfiguration elements can be classified into two main categories, as follows:

- *Software programmable modules* are the processing units that host the RAT software modules (among which GPP, DSPs, FPGAs);
- *Parametrical modules* are the modules that reconfigure using parameters adjustments instead of software installation and execution; they can be further categorized into the following:
  - *Flexible hardware subsystems*: These are assembled with their driver software.
  - *Flexible software components*: These are autonomous software modules proposing by themselves a certain number of tunable parameters.

The reconfiguration control of the RREs takes into consideration the drivers that correspond to the reconfiguration of the processing entity. The choice for implementation is also variable, ranging from optimized implementations on embedded processors with no assumptions of the way to implement useful processing to component-based software architectures in which detailed architecture recommendation enables the development of useful modules with interesting properties, such as modularity, composability, reuse, and so forth.

This section describes the elements in the E2R configuration architecture, emphasizing CEMs, as shown in Figure 6.40. In general, most of the presented elements are contained in the terminals and also in the base stations. But their design and quantity depends on the implemented equipment.

6.4.1 CEM-HAL Implementation

CEM-HAL provides a common interface towards CCM. It also configures the CEMs and reports achievements according to the data passed within the configuration file. For some of the configuration tasks, CEM-HAL is independent of the hardware. This reduces the amount of data and the overhead needed by the configuration to a minimum.

The elements of CEM-HAL are shown in Figure 6.41. It consists of one service dispatcher and an arbitrary number of CEM logical device drivers (LDDs), each with its own database. The LDD can handle one or more physical device drivers (PDDs), each connected to one CEM.

The service dispatcher provides the configuration request initiated by the CCM-PCC engine or the CCM-SCC engine to the addressed LDD engine. It has two services for receiving these requests. One is the *primary service interface* (for PCC requests); the other handles the secondary service requests (for the SCC requests). Analysis of the requests is not done to avoid the need for a small database, an increase in latency, and also the required processing power. This component is very important, as it enables complete independence from the hardware. It provides a

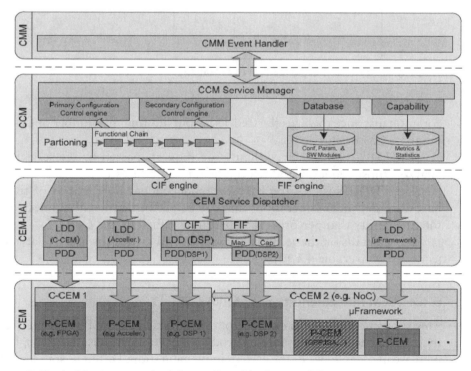

Figure 6.40 Architecture overview of reconfigurable elements [2].

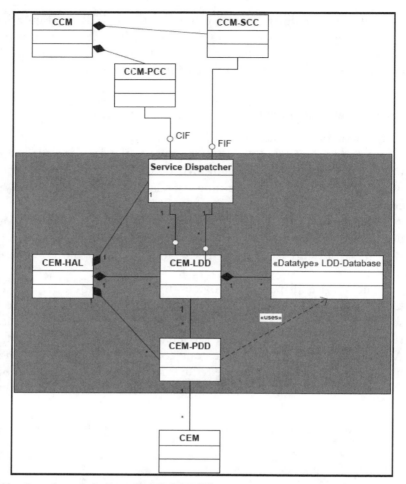

Figure 6.41   Overview of elements in CEM-HAL [2].

common interface towards the CCM without limiting the number of LDDs or showing the CCM design details.

The LDD receives the configuration requests passed by the service dispatcher,
analyzes those requests, and rejects them if not addressed.

The PDD translates the hardware-independent instructions of the LDD into
hardware-dependent operations that can be performed on the connected CEM. It is
the final entity that performs the operations called by the CCM.

More details on the technical approach and the results are available in IST project E2R Deliverable 4.1 [2].

6.4.2   CEM Implementation: SAMIRA DSP

This subsection covers an example of operations by the DSP needed to (re)configure
the CEM. The increasing rate of product development has also brought with it an
increase in complexity of the system implementation. Different standards have to be
implemented, and changes in the standards also need to be considered in the devel-

opment phase to support flexibility. At times, the implemented standards have to be reconfigured or a complete RAT has to be changed during the runtime of the system. Efficiency is also an important parameter, so devices should exhibit high performance and small battery consumption.

The SAMIRA DSP system is based on synchronous transfer architecture (STA) [19], which is an efficient and compiler-friendly processor. The DSP microarchitectural concept is based on basic modules shown in Figure 6.42.

The data-processing part of the processor is built up from basic modules. An interconnection of network formed by multiplexers connects the ports of same data type with each other. The processor instructions control the functionality of the basic modules and the input multiplexers. A synchronous network is formed, which consumes and produces data at each clock cycle. The STA offers a high degree of data reusability, which not only speeds up the computations but also reduces power consumption and registers file pressure. The generic processor architecture template shown in Figure 6.43 is used to support data (single instruction, multiple data, or SIMD) and instruction-level parallelism (very long instruction word, or VLIW), and thus offers a high degree of parallelism and efficiency. The SAMIRA processors are designed as follows:

- An *SIMD* yields high performance.
- An *STA*, a novel microarchitecture, enables the compiler to exploit the maximum instruction-level parallelism without expensive hardware overhead.
- A *tagged VLIW* (TVLIW) instruction reduces the memory consumption of instruction words.

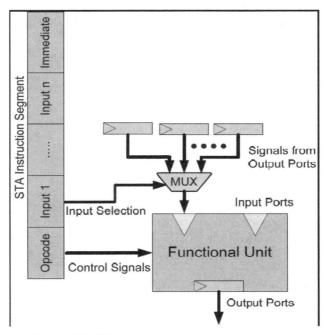

Figure 6.42 Basic modules in STA [2].

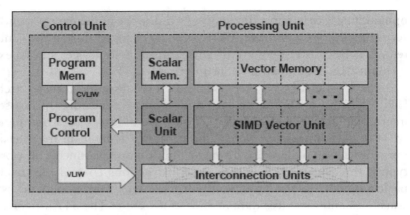

Figure 6.43   Generic architectural template used by SAMIRA [2].

- A *specialized memory architecture* enables efficient access to vectors and matrices.
- *Compressed instruction words* are used to avoid the overhead incurred in storing and fetching VLIWs directly from the memory.

The design methodology used here speeds up the design and implementation of the HW/SW solutions, and a template-based integrated automated design flow was developed. Processor cores with different characteristics, such as the size of the register file, memory capacity, interconnection network, functional units, data types, and amount of SIMD-vector parallelism, were generated by use of automatic generation of RTL, simulation models of the processor cores, and a matrix laboratory (MATLAB) compiler from a machine description. In this way, heterogeneous and highly reconfigurable low-power processors can be designed capable for use in mobile devices.

SAMIRA [2] is the proof of concept for the idea explained above. The SAMIRA DSP is a vector floating-point processor based on the STA microarchitecture designed for T-Mobile applications. It allows the efficient execution of the DSP algorithms developed using MATLAB. The technical details about the chip layout of the SAMIRA DSP, the architecture of the simulator used in integration with CEM, and the implementation details are given in IST project E2R Deliverable 4.1 [2].

6.4.3   Adaptive Execution Environment

An adaptive execution environment is the means for providing the basic mechanism required for dynamic reliable and secure change of equipment operation that can offer a consistent interface to the equipment reconfiguration manager for applying the needed reconfiguration actions. It consists of the following elements:

- A set of interconnected hardware components, which is a heterogeneous mix of different processing elements such as the GPPs, DSPs, and the FPGAs;
- The software abstractions, which access the hardware resources via a well-defined mechanism and interface (e.g., HAL and proxy software compo-

nents that represent the application-specific hardware functionality in the software domain). The system software can be considered a combination of an operating system (OS), a virtual machine (VM), middleware, component framework, firmware, and so forth, depending on the case.

The hardware Real-Time Research Platform (RTRP) used to validate the concept is shown in Figure 6.44. It has a compact PCI rack with several plug-in boards. The host board is a dual Pentium IV processor PC with Windows XP installed; four daughter modules are supported by carrier boards, such as processors, ADCs, DACs, TI DSPs, and FPGAs. The boards are linked together using FIFOs. A reconfigurable data transfer fabric between modules, called HEART, forms a ring that enables the nodes to be connecting dynamically with the guaranteed fixed bandwidths. The DSP module supports a 300 MHz TI C6203 DSP connected to the host system and other modules through the data transfer fabric. The FPGA module provides a Xilinx XC2VP7 Virtex-II Pro FPGA connected to the data transfer fabric. The FPGA is directly connected to the 100-MHz 32-bit wide HERON FIFO interface, which can potentially feed data to the FPGA at gigabit rates while simultaneously accepting data at similar rates.

Further details on the generic HAL interfaces for the RTRP and the block diagrams of the RTRP hardware modules are presented in IST project E2R Deliverable 4.1 [2].

6.4.4 SW Architecture for Embedded Real-Time Processors

Two main problems with the reconfigurable equipment can be reported as follows:

- *Blind RAT deployment:* This is the problem of deploying a RAT without knowing it. A complete RAT would require processing modules to be distributed among several processing CEMs and an air interface that includes interaction with the analogical world.
- *Multiple RAT deployment and management:* This requires the definition of a platform that can be flexible enough to support various RATs. The air inter-

Figure 6.44   RTRP used for PoC implementation in adaptive execution.

face, which is a key point of flexibility, has been more developed in software than in hardware.

An abstraction layer enabling RAT deployment and a flexible air interface will therefore be a must-have component of any reconfigurable platform.

Figure 6.45 shows the breakdown of the proposed CEM software whose two principal parts are the abstraction layer (i.e., it holds the complete resident software capable of hosting the RAT resource), and the P-CEM RAT resources (these make useful processing corresponding to the RATs under current operation).

As shown in Figure 6.45, the major interfaces for a reconfigurable equipment are as follows:

- ExecutableDevice and Resource Interface to access the P-CEM RAT resource, which requires the GPP software to support a Devices subsystem and a Resources Proxies subsystem to realize these interfaces;
- Abstraction layer connectivity used by the RAT resource located elsewhere to communicate with P-CEM RAT resource or radio facilities.

Figure 6.46 shows the proposed overall software architecture of the abstraction layer subsystem.

The microframework communicates with the CCM to deploy and manage a RAT within the CEM by realizing the transcription of the OMG SW radio specification (P^{2}SRC) ExecutableDevice, LoadableDevice, and ResourceMessages interfaces.

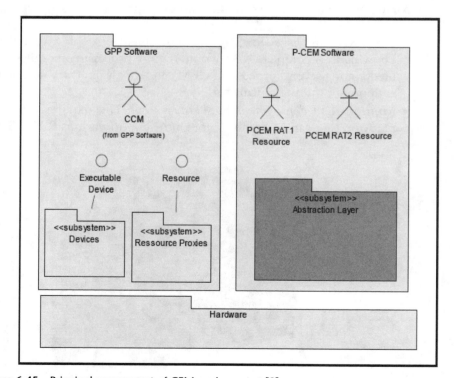

Figure 6.45 Principal component of CEM environment [2].

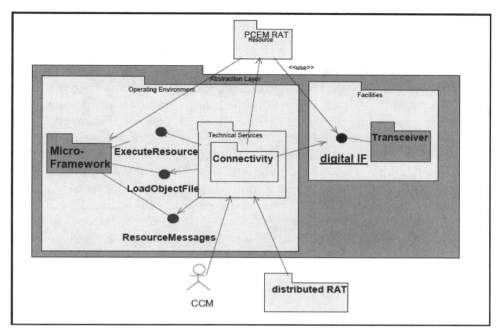

Figure 6.46 Proposed SW architecture for abstraction layer [2].

The transceiver communicates with the RAT resources and provides the way to access the antenna. A platform-independent model (PIM) defines the logical interface with the transceiver, which if standardized will allow multiple RAT access. This work will also serve as a basis for addressing the upper functionality definitions.

Further technical details on the deployment of the micro-framework and the transceiver is given in IST project E2R Deliverable 4.1 [2].

6.5 Conclusions

This chapter focused on key software technologies that enable the establishment of a coherent architectural vision for future flexible user and network equipment. Further, it proposed solutions for reconfigurable connectivity developed within the frames of the IST projects E2R and E2R II.

The reconfiguration control covers all the means inside equipment that are responsible for making decisions. It also spans the set of reconfigurable elements composing the flexible equipment. Reconfigurable elements refer to the actually reconfigured modules of the flexible equipment that perform the useful RAT, protocol, or user application processing.

This chapter covered achievements related to the improvement and development of the description of the general architecture. It presented work on some key subjects of the architecture areas. The presented results range from improvements in theoretical aspects to fully specified inputs for standardization at OMG or OMA level. A local proof-of-concept (PoC) was also performed to support the research work. The association of theoretical results, standardization outputs, and PoC support work is scientifically ambitious and sufficiently close to offer future industrial implementations.

Studies in reconfiguration management address the equipment aspects both at the user and network level. Further, the existing coupling to core network supporting concepts is also addressed by the reconfiguration management. The management architecture presented here was built around negotiation and selection. The following subjects support the decisions process: RAT load-reporting mechanisms, device management classmarks, resource monitoring, and the management of reconfigurable terminal QoS architectures. The OMA device management implementation provided an insight in the technology that supports transactions between the management modules and reconfigurable equipment. The improvement in robustness can be achieved by an adaptive security framework, context-aware security mechanisms, and secure over-the-air download technologies.

References

[1]   http://www2.ife.ee.ethz.ch/RESOLUTION/index.html

[2]   ICT project E2R, "Specification and Preliminary Design of Mechanisms and Processes Enabling End-to-End Dynamic Reconfiguration," Deliverable D4.1, August 2006, IST 2005-027714, http://e2r2.motlabs.com

[3]   Yin, H. and H. Wang, "Building an Application-Aware IPsec Policy System," *USENIC Security Symposium 2005*, Baltimore, MD, August 2005.

[4]   Security Architecture for the Internet Protocol, IETF Network Working Group, RFC2401.

[5]   OMA Device Management Tree and Description Serialization, Candidate version 1.2, 7th June 2005, OMA-TS-DM-TNDS-V1_2-20050607-C

[6]   OMA Device Management Standardized Objects, Candidate Version 1.2, 8th February 2006, OMA-TS-DM-StdObj-V1_2-20060208

[7]   Boukis, K. et al., "The Reconfigurable IP Mobility Component: Single Protocol Operation," *VTC*, Fall 2006.

[8]   Blake, S. et al., "An Architecture for Differentiated Services," RFC 2475.

[9]   Braden, R. et al., "Integrated Services in the Internet Architecture: an Overview," RFC 1633.

[10] Wroclawski, J., "Specification of the Controlled-Load Network Element Service," RFC 2211.

[11] Shenker, S. et al., "Specification of the Guaranteed Quality of Service," RFC 2212.

[12] Braden, R. et al., "Resource Reservation Protocol—Version 1 Functional Specifications," RFC 2005.

[13] Wroclawski, J., "The Use of RSVP with IETF Integrated Services," RFC 2210.

[14] Bernet, Y. et al., "A Framework for Integrated Services Operation over Diffserv Networks," RFC 2998.

[15] Pan, P. et al., "BGRP: Sink Tree Based Aggregation for Inter-Domain Reservation," *Journal of Communications and Networks*, Vol. 2, No. 2, June 2000, pp. 157–167

[16] Pan, P. and H. Schulzrinne, "YESSIR: A Simple Reservation Mechanism for the Internet," *Proceedings NOSSDAV 1998*.

[17] Talikdar et al., "MRSVP: A Reservation Protocol for an Integrated Services Packet Network with Mobile Hosts," Wireless Networks, Vol. 7, No. 1, pp 5–19, 2001.

[18]   www.sdrforum.org.

[19]   Cichon, G. and G. Fettweis, "MOUSE: A Shortcut from MatlabTM Source to SIMD DSP Assembly Code," *Proc. of Third International Workshop on Systems, Architectures, Modeling, and Simulation (SAMOS'03)*, pp 126–130, Samos, Greece, July 2003.

CHAPTER 7

Spectrum Management and Radio Resource Allocation

Earlier, it was believed that there was a problem of spectrum scarcity at frequencies that could be utilized for the purpose of wireless communications. But according to the studies conducted by the FCC's Spectrum Policy Task Force, there are vast temporal and geographical variations in the usage of allocated spectrum with the utilization, ranging from 15% to 85% [1]. Cognitive radio technology has been promoted as a prime technology to use these spectrum white spaces by implementing negotiated or opportunistic spectrum sharing.

Dynamic spectrum management techniques can be used to reduce interference and also to overcome the limitations of the existing static spectrum allocations, fixed radio functions, and limited network coordination. A cognitive radio is able to sense the environment and adapt accordingly. In this case, the radio sensitivity requirements and wideband frequency agility are some of the main challenges. The sensitivity maybe improved by enhancing the RF-FE sensitivity, exploiting the digital signal processing gain for specific primary user signal and network cooperation between users who share their spectrum sensing measurements.

The main drawback of RF and digital signal processing techniques that are used to detect primary users is that the performance of these techniques depends on the received signal strength. In these cases, cooperative sensing techniques can be used that alleviate the problem of detecting the primary users, since these techniques use the variability of signal strength at various locations. The mechanism of cooperation between the cognitive radios and the overhead involved are some of the main challenges to be addressed.

This chapter proposes solutions to spectrum management and the management of radio resources investigated and developed by the EU-funded IST Projects, ORACLE [2] and SURFACE [3]. This chapter is organized as follows. Section 7.1 gives an introduction to the topic of cognitive radio and opportunistic radio, on which the presented solutions are based. Section 7.2 presents spectrum sensing techniques and, in particular, discusses the newly emerging cooperative sensing technology. Section 7.3 proposes two approaches for cooperative sensing: centralized and distributed. Section 7.4 proposes algorithms for cooperative sensing. Section 7.5 discusses the issue of spectrum policies as a way to ensure better utilization of the available spectrum. Section 7.6 concludes the chapter.

7.1 Introduction

The cognitive radio (CR) is an intelligent wireless communication system that is aware of its surrounding environment. It uses the methodology of understanding by building to learn from the environment and adapt its internal states to statistical variations in the incoming RF stimuli by making corresponding changes in certain operating parameters in real time, with two important objectives: highly reliable communication, anytime, anywhere; and efficient utilization of the radio spectrum [4].

The term opportunistic radio (OR) may also be used; it addresses radio devices with the ability to perform opportunistic allocation of radio resources and the radio spectrum. They make use of appropriate sensing methods so they can modify the information about the radio resource space given the optimization criterion by radio transmission parameter adaptation. In this context, opportunity refers to temporarily vacant radio resources in terms of frequency, time, and space, which may be exploited to achieve certain transmission parameters. Another kind of opportunity for radio communication improvement is considering a given propagation environment or interference situation in which it might be beneficial to switch several transmission parameters to enhance the link quality [5].

The OR follows four main steps to interact with the external world: sensing, analysis, decision, and action, as shown in Figure 7.1 [6]. This idea was initially proposed by Mitola [6]. In the sensing stage, the OR gathers information about the internal state and surrounding environment in a continuous way. During analysis, the acquired data is processed and analyzed to provide the system with a higher-level synthetic representation of the context. Then the OR has to decide about the most proper action to the received external stimulus using the information in the embedded internal knowledge, the past experience, and the current context. Finally, there will be an active interaction with the external environment.

The OR system is designed to be aware of and sensitive to the changes in the surrounding environment. Spectrum opportunities have to be detected so that the OR may adapt accordingly. One of the means to do this would be to detect the primary users that are receiving the data within the communication range of the OR user. Practically, it would be difficult to obtain direct measurement of the channel between the primary receiver and transmitter for a cognitive radio. A practical approach would be to use the local observations of OR users for primary user detection. But this may turn out to be difficult, due to the varying conditions of the channel between the primary user and the OR user.

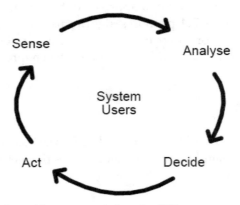

Figure 7.1   Basic steps followed by opportunistic radio (OR).

In general, the spectrum sensing methods can be classified into local and cooperative. The local methods use the local observations of the OR users, whereas cooperative sensing uses information from multiple ORs.

The analysis phase of the cycle illustrated in Figure 7.1 comprises orientation, planning, and learning. External intelligent sources are used in this phase.

7.2   Spectrum Sensing and Cooperative Sensing

Spectrum sensing can be classified broadly into three categories as follows:

- Transmitter detection;
- Cooperative detection;
- Interference-based detection.

The transmitter detection includes matched filter detection, energy detection, and cyclostationary feature detection, the latter being the most complex. The cooperative mechanisms for sensing are also becoming increasingly important.

The primary user detection can be classified into known and unknown parameters. The known parameters include the matched filter and cyclostationary feature detection, whereas the unknown parameter is further classified into known noise level and unknown noise level. The former includes energy detection, and the latter includes cooperative detection, blind detection, and interference-based detection.

These techniques have been described briefly as follows [7]:

- *Matched filter detection:* The structure of the primary signal is known to the secondary user. The optimal detector in stationary Gaussian noise is a matched-filter followed by a threshold test. It is a type of coherent detector, which means that synchronization should be achieved before detection to maximize the detector output SNR. Extra dedicated circuits will be needed in the secondary user devices to achieve synchronization with each type of primary user. At times, it is difficult to use the matched-filter detector due to lack of knowledge about the primary signals' structure, in which case a noncoherent detector scheme, like energy detector, will be more appropriate.
- *Cyclostationary detection:* The communication signals are modeled as cyclostationary signals where the parameters vary in time with single or multiple periodicities. Examples of cyclic spectrum density are shown in Figure 7.2 [7].
- *Energy detection:* The energy detector was first proposed by Urkowitz [8]. A typical block diagram is shown in Figure 7.3 [7]. The input band-pass filter's function is to select the center frequency and the bandwidth of interest (W). The received energy is measured by the squaring device. A detailed mathematical description of this block diagram is given in [7].

  For different values of false alarm probability, the graphs for minimum required SNR versus the time bandwidth product (TW) have been obtained as shown in Figure 7.4 [7]. The signals can be detected at SNR as low as desired, provided the detection interval is long enough and the noise power spectral

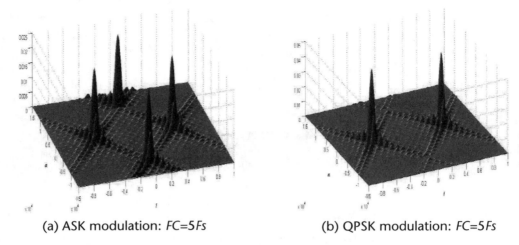

(a) ASK modulation: $FC=5Fs$ (b) QPSK modulation: $FC=5Fs$

Figure 7.2   Examples of cyclic spectrum densities, where Fc is carrier frequency and Fs is symbol frequency.

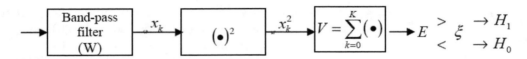

Figure 7.3   Block diagram of an energy detector.

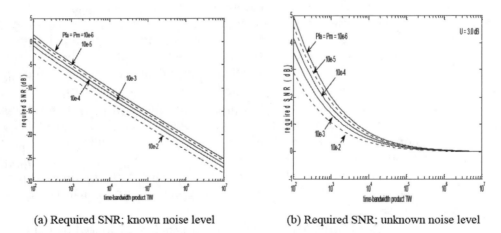

(a) Required SNR; known noise level   (b) Required SNR; unknown noise level

Figure 7.4   SNR versus TW for different values of false alarm probability.

density is known. But realistic limitations on the detectors' knowledge of the noise level can produce serious degradation in the detectors' performance.

Despite its simplicity, the main drawback of the energy detector is the susceptibility of its performance to the noise power uncertainty [9], in which case the methods based on cyclic feature detection maybe considered. Ghasemi and

Sousa [10] showed that the performance of the energy detector degrades in shadowing or fading environments, which are quite central in applications envisaged for opportunistic spectrum access.

- *Wavelet detection:* The signal spectrum over a wide frequency band can be decomposed into elementary building blocks of subbands which are well-characterized by local irregularities in frequency. Wavelet transforms maybe used to detect and estimate the local spectrum irregular structure which carries important information on the frequency locations and power spectral densities of the subband [11]. An OR has to identify the frequency locations of the nonoverlapping spectrum bands and classify them as black, grey, and white spaces, corresponding to high, medium, or low power spectral density (PSD) levels. The white spaces can be used for opportunistic purposes. The wavelet transforms are important mathematical tools that characterize the local regularities of the signals [12].

  The drawback of Fourier expansion is that although the frequencies present in the signal are known, the exact time of their presence is not known. The wavelet transform solves this problem by presenting the signal in time and frequency domain simultaneously. In wavelet analysis, a fully scalable modulated window divides the signal into separate portions, and as the window is shifted along the signal, the spectrum for every position is calculated and the process is repeated for different window sizes in each new cycle. Finally, a multiresolution analysis in wavelet terms is obtained. The continuous wavelet transform (CWT) is a two parameter expansion of a signal in terms of a particular wavelet basis function. An OR receiver that gets the signal of the shape PSD within a known wide band of spectrum with the use of CWT, the number of spectrum subbands, the boundaries of each subband, and the PSD of each spectrum subband can be deduced. This information can be used to infer the availability of spectrum holes in each spectrum subband, depending on the PSD levels of each space, categorized into high, medium, and low power [7].

An OR device may form a local view of the spectrum situation, subjected to uncertainty due to channel and receiver location. With respect to the channel uncertainty, the random nature of the wireless channel in which multipath fading and shadowing degrade the performance of the primary user detection methods also contributes to the hidden node problem. Cooperative detection or sensing schemes may be used to solve the problems due to channel uncertainty. Sensing information from multiple OR devices is merged to allow more reliable primary user detection. In absence of multiple antennas, the multiple cooperating radios will help reduce the effects of severe multipath at a single radio, since it gives multiple independent realizations of related random variables. In that case, the probability that all the users will see a deep fade is very low. Thus, cooperation will make the spectrum sensing robust to severe or poorly modeled fading environments without drastic requirements on the individual radios.

Cooperative detection may be implemented either in a centralized or in a distributed way. Ganesan and Li [13] showed that cooperation helps to reduce detection time and offers the possibility of increasing the reactivity of OR devices. But the limitation of this approach are that it assumes known location of the primary user transmitters and does not consider the mobility of the cognitive users, the presence of more than one cluster of OR devices, and the effect of shadowing.

Ghasemi and Sousa [10], however, showed that cooperative detection schemes may help cope with multipath fading and shadowing effects by improving detection probability in a heavily shadowed environment. The distributed solutions require exchanges of observations among the OR devices, as in Zhao et al. [14]. For these sensing architectures, the sensing and data communication functions are colocated within each OR device, which can lead to a suboptimal view of the spectrum situation due to conflicts between these two functions. Shankar [15] investigated a type of sensing architecture to cope with the problems related to the colocation of the data transmission and sensing functions. A sensor network for cooperative spectrum sending and an operational network for data transmission were deployed. The former is deployed in the desired target area, and a central controller processes the spectrum information collected from sensors to make a spectrum occupancy map. The operational network then uses this map to determine communication opportunities.

Though the cooperative approaches of spectrum sensing provide accurate sensing performance, they also cause adverse effects on resource-constrained networks due to additional operations and overhead traffic. Also, cooperative sensing does not overcome the problem of the second source of uncertainty mentioned before, which is due to the lack of knowledge about the location of primary receivers; it translates into the fact that the OR transmitter does not know whether its transmission based on sensing information is going to cause any interference to the primary user receivers.

Some of the issues related to sensing aspects of opportunistic radio communication are:

- *Sensing-based opportunistic channel access:* Sensing-based approaches have the advantages of simplicity and low-infrastructure requirements. Liu and Shankar [16] propose an approach that studied the secondary users who observe the available channels in a dynamic fashion and explore them opportunistically. Since the spectrum availability varies in time and is location dependent, the spatial and temporal properties are analyzed and two new metrics are defined for its characterization. Hence, a two-step approach to channel selection is proposed. The first step will determine whether the channel is idle, using algorithms to perform the accessibility check on measurements of the primary signals' strength. These algorithms offer static outage guarantee to the primary users. In the second step, the secondary user decides whether an accessible channel is a good opportunity, based on channel sensing statistics obtained in the first step. It will prefer a channel where it can finish the transmission before the primary users return. A sensing scheme is required to determine whether the particular channel is an opportunity. The occupancy of a particular channel is defined as the probability that the physical layer signatures of the current primaries are present and, using simple correlation or feature detection techniques, it is easily possible to determine the presence of the primary.

- *Information theoretic aspects concerning the cognitive radio channel:* Devroye et al. [17] proposed a new and potentially more spectrally efficient model for wireless channel employing cognitive radios, in which the main considerations are how the smart wireless devices can be used to increase the spectral efficiency and how fundamental limits on the communication possible over such a channel have been derived. Jafar and Srinivasa [18], however, used the information theoretic framework for communication with side information at the transmitter or the receiver for investigating the capacity of opportunistic communication in the presence of dynamic and distributed spectral activity. This is characterized by time-varying spectral holes sensed by the cognitive transmitter, which is correlated but not identical to those sensed by cognitive receiver. Cognitive radio communication was analyzed and found to be robust to dynamic spectral environments, even when the communication occurs in bursts of only three to five symbols. A simple model was used to facilitate the analytical reasoning results, but it can be extended to account for more realistic channels by incorporating the effects of fading, interference, and channel knowledge.

- *Trust issues and cooperation:* Mishra et al. [19] showed that collectively sensing the spectrum availability can deliver tremendous gains, even with a small to moderate number of perfectly trusted users, as long as they occupy a large enough cooperation footprint to mitigate the spatial correlative effects. But this trust is not used in real-life network deployments, in which the radios may be malicious or failing in unknown ways due to the presence of other secondary transmissions. This can also be considered as additional user uncertainty in decision making. If it is not possible to model the failure behavior, then the OR devices fail to introduce a limit on achievable sensitivity reductions. It has also been shown that cooperative gain proves to be very sensitive to failing cognitive OR devices (i.e., in a case in which one out of N users can fail in an unknown way, then the cooperation gains for the receiver are limited to what is possible for the receiver to achieve with N trusted partners).

7.3 Cooperation Protocols for Sensing

In general, OR communication systems have to perform two basic tasks: identifying communication opportunities and communicating by exploiting the identified opportunities. The former is related to context sensing, whereas the latter concerns opportunistic access to the resources. These tasks are intimately related and require reconsidering the protocol stack design in terms of protocol layer functionality, procedures, layer interactions, and protocol messaging. The sensing function design mainly involves the design of statistical signal processing algorithms or other statistical sampling algorithms to collect the needed sensing information. After this information is collected by each OR device, it is important to define how this basic functionality will be invoked and how the collected information can be circulated within the OR network in the case of centralized or distributed sensing schemes. This is basically the sensing protocol.

This section focuses on protocol aspects related to the sensing functionality, considering cooperative sensing within the network of OR devices, which necessitate the introduction of new protocol functionalities. The two main types of cooperative sensing can be classified into centralized and distributed sensing. The former considers a hierarchical network structure in which the OR base station devices centralize the collection of sensing information and the task of context identification. In the latter, the OR devices form ad hoc cooperation clusters in which sensing information is exchanged among the neighboring OR devices. The two types of protocol are described in the following subsections.

7.3.1 Centralized Sensing Approach

A typical scenario for the centralized sensing approach is shown in Figure 7.5. Here, the regulator allows a secondary system, which is an opportunistic radio system, to operate under a primary licensed UMTS cellular network, as shown in Figure 7.5 [20]. The secondary system shares the same frequency band with the UMTS cellular network, as long as it does not create harmful interference to the primary licensed system.

In this case, a centralized approach for the secondary OR network has been assumed in which all the OR mobile terminals in the system forward their local sensing measurements regarding spectrum occupancy and interference to a central entity located at the secondary OR base station serving the particular OR region or cell. The spectrum opportunities manager (SOM) is a central entity that collects all the received sensing reports from OR terminals and runs specific spectrum opportunities management algorithms to control the spectrum allocation and access procedures. It can send control information to the OR mobile terminals about the allowed transmitter power, it can order an OR terminal to switch frequency, or ask it to stop transmitting among other possible control requests. The secondary ORs have conditioned access to the shared spectrum, as opposed to primary users, and hence they need to acquire temporarily unused licensed spectrum from the primary system to be able to communicate. The central entity is responsible for assigning resources such

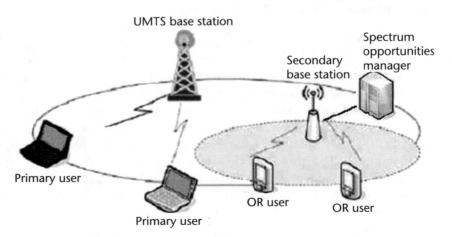

Figure 7.5 Scenario where OR network operates on licensed UMTS frequency bands.

as currently unused licensed frequencies to the OR terminals with which they are interested in communicating. This would require some form of signaling between the secondary base station and the related OR terminals. These resources have a random availability and duration, considering the primary system is a UMTS cellular network. Hence, the signaling protocol needed should be able to cope with a dynamic scenario.

Passive awareness, shown in Figure 7.6, is a form of special centralized approach for detecting opportunities in which the central station of the primary user (PU) is broadcasting information about current resource utilization [21] and has been denoted as OR-friendly behavior, since it can be utilized by receiving OR terminals to assist the local decision process [5]. But the main drawback of using this type of awareness lies in its protocol interoperability requirements. The OR terminals need to at least partially implement PHY and DLC interoperability as well as radio resource management protocols of the PU to decode messages received and to obtain the sensing information required. From the implementation point of view, it might be simpler to augment PU central stations by OR-specific protocols; however, this does not relax the interoperability requirements with respect to PHY and DLC.

Active awareness, as shown in Figure 7.7, is a type of sensing that applies to centralized and decentralized OR architectures as well. Considering the centralized case, the OR base stations perform the sensing and distribute the information toward their associated OR terminals. In the case of decentralized architecture, the network of OR terminals performs the sensing process and maintains a distributed database for current sensing parameters.

An added benefit for the PU operators is that the opportunity detection and OR decisions can be guided by some operator-controlled network management entity and can be also considered as a hybrid active/passive awareness type of sensing.

The OR mobile terminal should be able to perform spectrum sensing on its own, and it is important that the kind of information generated by the spectrum sensing operation performed by the OR terminals be exchanged between the OR mobile terminals the SOM and those at the OR base station, since the SOM uses this information to make decisions related to allocation of resources to the existing OR

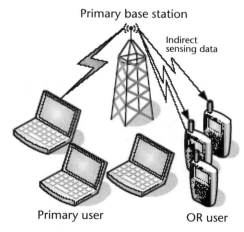

Primary base station

Indirect sensing data

Primary user   OR user

Figure 7.6   Passive awareness sensing for centralized approach.

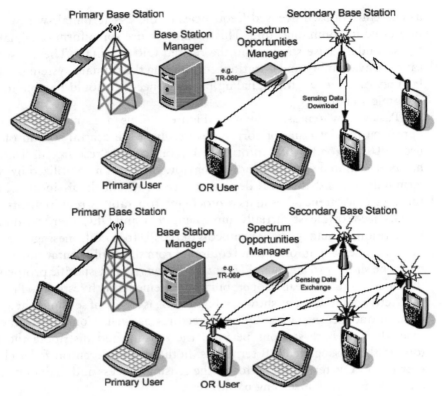

Figure 7.7   Active awareness sensing for centralized versus decentralized approach.

terminals in the opportunistic radio network. Information about the resources to be allocated to the OR terminals is important for them to communicate without causing harmful interference with the primary users which is the final aim of the OR systems. Apart from these two signals, it is important to exchange other information, such as the control signals between the OR base station and the OR terminals. More detailed description of the type of sensing information to be exchanged between the OR base stations and the OR terminals is given in the IST project ORACLE Deliverable 2.2 [20].

The framework for the development of the protocol stack of the OR system, as shown in Figure 7.8, divides the functions related to data transmission from the ones related to OR features. This separation will allow the addition of OR characteristics to already existing systems by specifying the appropriate interfaces A and B without restricting the development of new optimized PHY and/or MAC layers. The PHY layer and the link layer have certain parameters that need to be configured. Thus, the protocol stack related to the OR functionality includes, at the PHY layer, the sensing and parameter acquisition, and at the link layer it includes functionalities for link and network management. The OR PHY layer may report its measurements to the OR link layer through the interface C. At the OR link layer, some data processing may be performed, due to which there could be an adjustment of the controllable parameters of the link or PHY data transmission layers to match the environment, or the OR terminal may send control signals for negotiation purposes through the

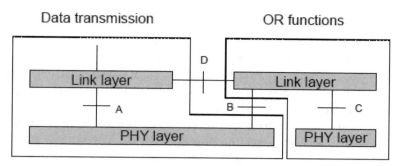

Figure 7.8   framework for protocol stack development of an OR.

data transmission plane. It is important to define channels for information exchange between the different OR terminals, since the interference computation algorithms rely on collaborative sensing and data processing. The channels may be classified hierarchically in terms of logical channels defined through the type of information they transport, and defined physically about the way that information can be effectively transported. The main challenge is faced at the physical level; also, the signaling and network broadcasting function within the OR needs to take place irrespective of the activity of the primary users. Thus, some noninterfering transmission scheme is required. One possibility could be the use of dedicated licensed bands for signaling; this does not create a significant burden, because these channels are expected to have low data rate. Alternatively, it could be considered as a noninterfering overlay method, as in UWB.

The sensing information exchange begins with the allocation of a control channel, after the channel's database lookup, to allow the OR base station to communicate with its OR mobile terminals. After the spectrum sensing is performed, the SOM at the OR base station instructs the OR mobile terminal to perform local spectrum sensing for the PUs on the channels that it found empty by broadcasting a predefined sensing request signal. The concerned OR terminal sends its local sensing measurements to the SOM by sending a sensing report that is used by the SOM to build, update, and maintain a detailed spectrum environment map (SEM) of the geographic area it covers from the gathered sensing information. This map is used as a reference for decisions to be made regarding resources management and forwards convenient decisions to the concerned OR mobile terminals. The SEM kept at the SOM of a particular OR base station can be improved further if the SOM presents an OR base station with any adjacent OR cell that could share its reference map through some interconnection between OR base stations, offering an additional enhancement to cooperative sensing.

Another possibility would be to consider secondary resources associated with the primary ones known by the OR terminals. This approach might improve the possibility of keepign continuous QoS during an unexpected need to release the primary resource used by an OR terminal. The OR terminal can immediately switch to the known secondary resource without SOM intervention, which would benefit the ongoing communication in terms of the QoS. This approach is interesting, given that the opportunities or resources behave dynamically. Some types of messages are described briefly:

- *Sensing request:* When the SOM has to build or update its SEM, it performs spectrum sensing by itself and (depending on the results) broadcasts a sensing request to OR mobile terminals for the SEM update to be made in the best cooperative way. According to the opportunities, it will verify after its spectrum-sensing procedure, and according to the geographical position of the OR terminals, the SOM will send a sensing request to one or more OR mobile terminals for them to perform sensing on one or more opportunities locally.

- *Sensing report:* After the broadcast of the sensing request, every OR mobile terminal addressed in that instruction acknowledges it and performs the requested spectrum sensing operation for the specified resources using the required sensing parameters. After the sensing procedure has been completed, the OR terminal sends the sensing report back to the OR base station. The report contains the terminal's local sensing measurements with the geographical position so that the SOM can build an accurate SEM. There should be a possibility of setting some periodicity on the sensing reports for the SOM to receive these reports from the OR terminals at a specified cadence, without the need to broadcast sensing requests too often.

- *Resources allocation:* The SOM entity maintains an available resource map (ARM) database to provide OR terminals with opportunities they can use or switch to in case of necessity or optimization, following the resource management algorithm results. After the SOM considers it necessary to allocate new resources to any OR mobile terminal because the OR terminal requested it for communications or because the currently allocated ones are no longer free to continue using them, it sends a resources allocation signal to the OR terminal.

- *Additional spatial information:* Some intelligence maybe provided to every OR mobile terminal for it to be able to decide on its own to switch to a different resource, such as frequency, after it knows it is on the border of a forbidden zone and before it can interfere harmfully with some PU communication. The SOM has to send the OR terminal some spatial information on currently forbidden zones in its neighborhood, taken from the SEM. Thus, in case of temporary access failure to SOM control information, the OR terminal may have sufficient information to decide whether a frequency switch to an alternative one is necessary when the crossing of the forbidden zone has been foreseen. This maintains the QoS of any possible ongoing communication.

- *Positioning request:* The SOM entity at the OR base station tries to prevent any possible crossing of forbidden zones by an OR mobile terminal under its control through its SEM. For this, it needs to know the exact location of all the OR terminals, and it has to send positioning request messages to them whenever it needs to verify whether they are not threatening to interfere harmfully with some PU communication. If it does, some appropriate prevention actions need to be taken.

- *Positioning report:* Every concerned OR terminal replies to the broadcasted positioning request instruction with positioning data reflecting its current geographical location in the area covered by the OR base station; the SOM can then verify whether all the OR terminals are in safe zones by crosschecking its SEM against the location of the mobile terminals and taking the necessary actions if it foresees a chance of harmful interference from any OR terminal on the licensed network.

More details about the important information exchange messages are given in IST Project ORACLE Deliverable Deliverable 2.2 [20].

7.3.2   Distributed Sensing Approach

The distributed sensing approach is described here by a specific use-case. An opportunistic radio communication system (ORCS) has been considered in which channel resources are allocated dynamically, based on information concerning the resource availability state. The system operates as a secondary user within the resource space of the primary user. Further, the system provides interference avoidance for the primary communication systems and coexistence with other secondary OR systems that may be using the same resources. The OR devices use an unlicensed short-range link for local sensing information exchange, forming local cooperation clusters. This short-range link ma ybe based on ISM band links or an UWB link.

The secondary ORCS provides data services by means of service access points at which OR user-terminal devices can connect. In this context, the radio communication equipment operates according to an ORC protocol. This protocol mainly consists of the specifications of various types of channels and the procedures for establishing and maintaining these channels. The two main types of channels are the OR pilot channels and the OR traffic channels. The OR user terminal has to connect to the OR service access point over a pilot channel to set up a traffic channel over which the communication can take place. When the OR traffic channel is lost or its quality is not according to the setup specification, then the OR user terminal reverts to the pilot channel to renegotiate the traffic channel parameters. Thus, the OR pilot channel is an opportunistically established channel on some vacant frequency band associated with a TDMA scheme. The OR service access points have to cater for the simultaneous operation of its pilot and traffic channels, which are both opportunistic. More detailed information on this, including the protocol procedures, is available in IST Project ORACLE Deliverable Deliverable 2.2 [20].

There are three different control and management protocol considered for an OR terminal as proposed by the IST project ORACLE [20]:

• *Metasignaling protocol:* This is a lightweight protocol used to detect neighboring OR terminals and to bootstrap the control channel required for the neighborhood collaboration protocol. Also, it relies on the link layer transport services and supports network identification requests, network announcements, control channel configuration announcement, and simple two way control channel configuration negotiation between an OR terminal participating in a previously established network of OR terminals and an OR terminal that wants to join the network. It is a message-oriented protocol and can be used on top of stream control transmission protocol (SCTP) [22] in association with an SCTP-based OR control protocol, which will simplify the control channel handover procedures that require intermediate metasignaling.

- *Neighborhood collaboration protocol:* This control protocol implements sensing parameter exchange procedures and negotiation of configuration parameters for opportunity detection. It is also message oriented and can use link layer transport services or network layer services. The DLC transport may be used if information must be disseminated to neighboring terminals via broadcast or multicast with best effort. Network layer transport is used if reliable end-to-end and routing communication is required.
- *Traffic flow parameter exchange protocol:* IPFIX [23] and PSAMP are used for the exchange of flow metering data sets.

7.4   Spectrum and Cooperative Sensing Algorithms

This section focuses on sensing methods and techniques and, subsequently proposed algorithms for different types of signals and for various levels of a priori information. Multiantenna systems can also be exploited for sensing. When using cyclostationarity feature detection methods, some problems may occur. The energy detector is susceptible to noise power estimations, whereas the cyclostationarity-based tests require a priori knowledge of the cyclic frequency of the signal of interest. If this frequency is unknown a priori, then an exhaustive search for the presence of cyclostationarity within a broad range of frequencies is carried out. But these exhaustive tests can be very complex and computationally extensive, thus prohibiting a commercially viable detection of free bands. This section proposes a blind and simple single-stage detection method that tests the presence of cyclostationarity within an interval of frequencies [24].

The various sensing algorithms and methods that will be discussed are the following [24]:

- Blind cyclostationarity-based detection test;
- Blind and semiblind detection algorithm for spread-spectrum signals;
- Algorithm to detect UMTS FDD signals;
- Wideband spectrum sensing for OR using wavelet-based algorithms;
- Energy detection based on multiple antenna processing.

7.4.1   Blind Cyclostationarity-Based Detection Test

The goal is to design an efficient signal-processing algorithm to identify opportunities for communicating in unused spectrum. It is important to reliably detect other signals in a band to avoid interfering with the primary owner of the band. Many detection tests are available at present; these mainly involve energy and cyclostationary detectors, but the energy detector is susceptible to noise power estimations. The IST project ORACLE [24] proposed a blind and simple single-stage detection method, which tests on the presence of the cyclostationarity within an interval of frequencies. A blind test was proposed, which was able to detect the presence of cyclostatonarity within a given frequency band. This test allows the detection of cyclostationarities by means of a single step without the need to scan for each cyclic frequency. This constitutes a significant reduction in complexity and processing time and allows the opportunistic or cognitive terminal to judge the vacancy of a given spectral band without prior knowledge of the potential primary signal. Mathematics has been used to expose various properties of the proposed test and simulation results obtained to corroborate the findings, some of which are shown in Figures 7.9, 7.10, 7.11, 7.12, and 7.13. These simulations confirm that the detection probability is strongly dependent on the size of the spectral band under test. But it is also to be noted that the probability of false alarm is virtually independent of this spectral width.

7.4.2 Blind and Semiblind Detection Algorithms for Spread Spectrum Signals

In a spread signal, each transmitted information bit is multiplied by a spreading sequence before being emitted. Several methods to detect these signals have been

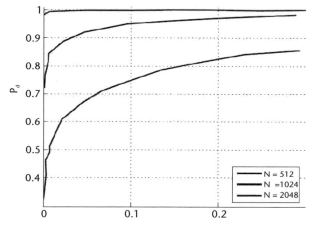

Figure 7.9 Simulated receiver operating characteristics (ROC) curves, SNR = 0 dB, M = 10.

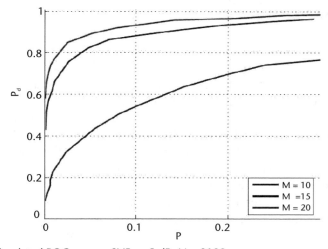

Figure 7.10 Simulated ROC curves, SNR = -5 dB, N = 8192.

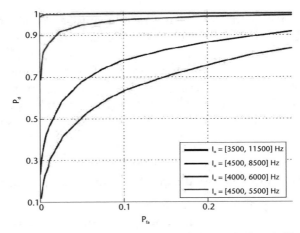

Figure 7.11 Simulated ROC curves for different interval sizes I_α, SNR = -8 dB, N = 32768 and M = 20.

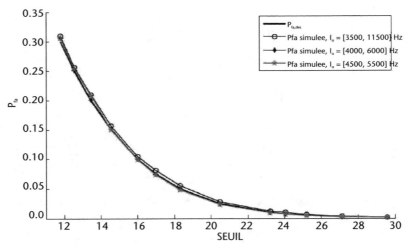

Figure 7.12 False alarm probability as a function of the detection threshold for different interval sizes I_α, N = 1024 and M = 10.

proposed. Some exploit the variations of the power detector algorithm [25], whereas some exploit subspace methods [26]. Marques et al. [27] exploit the cyclostationary property of the spread signals to build a detection criterion, and the proposed algorithm is based on a cost function of the received signal cycle-spectrums energy. In the IST project ORACLE, Deliverable 2.4 [24] describes how a new criterion function was built that exploits the cyclostationary property of the received signal; results were obtained that show that the time approach allows a better exploitation of the cycle-correlations' coefficients.

The performance of the detection algorithm for a Barker spread sequence is shown in Figure 7.14. The algorithm performs almost identically for all the cases. Based on these results, a detection algorithm for spread signal in SISO and SIMO contexts was proposed. Its performance for a Hadamard spread sequence in SISO

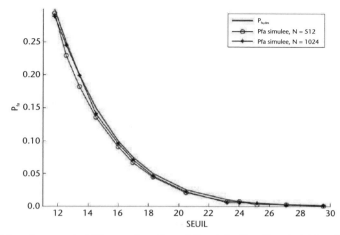

Figure 7.13   False alarm probability as a function of the detection threshold for different values of N, M = 10.

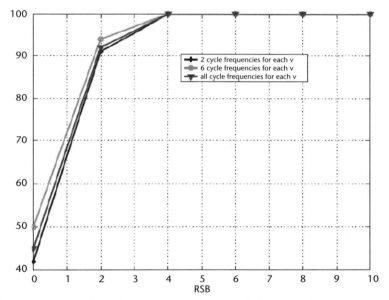

Figure 7.14   Performance of the detection algorithm for a Barker spread sequence.

and SIMO context respectively (1 × 2) is shown in Figures 7.15 and 7.16. It can be seen that the performance in the SIMO context is better.

Only some of the cycle frequencies should be taken into account to build a cost function based on the cycle-correlations' coefficients. But choosing these cycle frequencies requires some prior knowledge of the kind of spread signal searched; also, it can have an impact on the performances of the detection algorithm. For the SIMO case, a method using several terminals that perform measurements to detect spread signals has been proposed, and when the gain is compared to the SISO context with two measurements, it has been evaluated to be close to 2 dB when the SNR is high enough for detection.

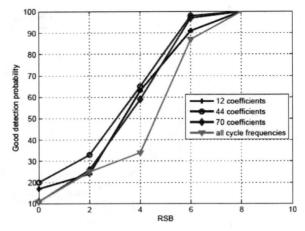

Figure 7.15 Performance of the detection algorithm for a Hadamard spread sequence in a SISO context.

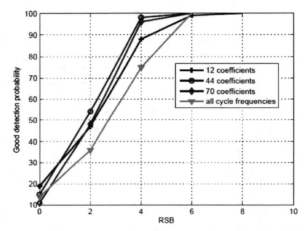

Figure 7.16 Performance of the detection algorithm for a Hadamard spread sequence in a SIMO context (1 × 2).

7.4.3 Algorithm to Detect UMTS FDD Signals

The chosen scenario for this algorithm is a cellular use-case scenario with opportunistic use of UMTS bands [24]. The significance of the algorithm proposed is that it does not need to know the spreading factor of the particular UMTS signal. This novel approach exploits the cyclostationary features from the UMTS chip rate, rather than the cyclostationary feature from the spreading operation. Since the chip rate of the UMTS signal is known, this new approach avoids spectrum search procedures over the SCD function, and as a consequence, it decreases the complexity of the detector proposed previously. Further, the impact of impairments such as the multipath channel and lack of synchronization on the cyclostationary detector performance has been analyzed.

The proposed cyclostationary frequency feature exploits the cyclic frequency common to all downlink signals in the UMTS cellular scenario. The cyclic frequency comes from the UMTS chip rate, assuming that the OR knows the UMTS carrier frequencies and bandwidths. The proposed detector also uses the periodogram approach which relies on second order statistics based on spectrum cyclic density function. The output of this detector, after all the signal processing, is a detection statistic, d, in dB which represents the ratio between the power of the cyclostationary feature measured at cyclic frequency α_c and the estimated noise floor measured at α_n. More detailed information is given in IST project ORACLE Deliverable 2.4 [24].

The simulation results show that for an SNR of -10 dB and an observation time of at least 30 ms it is possible to ensure a 99.9% probability of detection while having a negligible probability of false alarm, which is also possible for 10 ms of observation time if the SNR is at least -5 dB in the case of AWGN channel. The results for the cyclostationary detectors' receivers are shown in Figure 7.17.

When realistic impairments are considered, the proposed detector performance is moderately degraded, which points out the importance of the hardware quality and stability, namely local oscillators, for the detector to perform at its best. It is clear that the propagation channel plays an important role in the performance of this cyclostationary feature detector; thus, it should be taken into major consideration.

7.4.4   Wideband Spectrum Sensing for OR Using Wavelet-Based Algorithms

OR systems make use of the availability of spectrum holes in the primary systems. The spectrum hole availability needs to be identified through spectrum sensing procedures. Apart from spectrum sensing, the OR should be capable of detecting the primary system and moving out of the spectrum once the primary system requires the spectrum usage. Thus, spectrum sensing and detection needs to be performed efficiently for the successful deployment of an OR system. The PSD could be used for identifying the spectrum holes along with the power levels within a selected por-

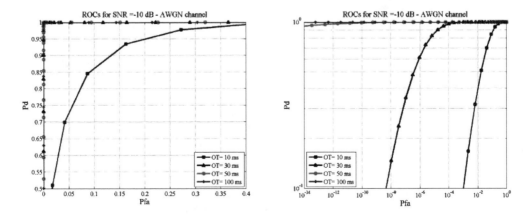

Figure 7.17   Cyclostationary detector ROCs for SNR = -10 dB (AWGN channel).

tion of the spectrum. The white spaces are considered spectrum holes that can be utilized by the OR system for opportunistic spectrum usage.

CWT techniques can be used to detect the subband edges. The approach is a two-parameter expansion of the signal in terms of a particular wavelet basis function. The wavelets have scale aspects and time aspects. The scale aspects can be presented as an idea around the notion of local regularity, whereas the time aspects can be presented as a list of domains. In ORACLE Deliverable 2.4 [24], the CWT techniques were used to detect the subband edges in a wide spectrum band. The frequency locations of the nonoverlapping spectrum subbands of a PSD signal were identified; afterwards, each PSD signal was analyzed using a large number of different wavelets to identify the best possible wavelets for subband identification. More detailed information about the wavelet transform and wideband spectrum hole detection in OR is given in ORACLE Deliverable 2.4 [24]. It also includes an evaluation study of spectrum sensing via wavelet edge detection technique. A flowchart for the spectrum subband selection algorithm using wavelet transforms is shown in Figure 7.18.

OR networks can solve the current wireless network problems, which are the result of the limited available spectrum and the inefficient use of that spectrum by exploiting the existing wireless spectrum in an opportunistic manner. These networks use the capabilities of the cognitive radio to provide an ultimate communication paradigm in wireless communications that is spectrum-aware. The cognitive spectrum identification task can be formulated as an edge-detection problem. The wavelet edge detection approach can be considered for subband identification of wideband channels. A solution was developed within the framework of the IST project ORACLE, based on the coefficients' lines of the continuous wavelet transform,

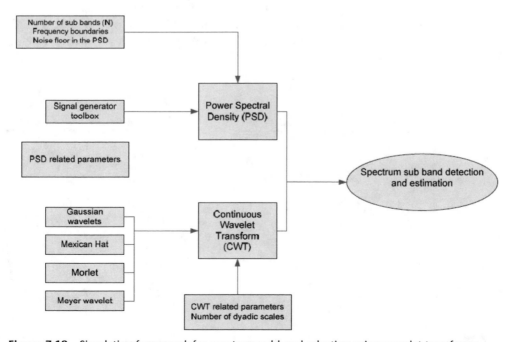

Figure 7.18 Simulation framework for spectrum subband selection using wavelet transforms.

and it was tested for different signals. The scheme proposed was able to scan a wide bandwidth to identify all the piecewise smooth subbands simultaneously within the frequency range of interest, without any prior knowledge of the number of the subbands,. It was observed that some wavelet families were more suitable in recognizing the subband edges than the other. The wavelets with compact support were found to provide more accuracy in the edge-detection approach, while others did not. Thus, wavelet families such as the Haar and Daubechies wavelet families, which have features like symmetry and compact support, were found to be more suitable for the use of subband edge-detection mechanisms. However, different wavelets should be used for edge-detection in different signals. The wavelet used will depend on the characteristics of the specific signal, and it cannot be used for all the signals. The evaluation of the PSD signal for various wavelets is given in ORACLE Deliverable 2.4 [24].

Wavelet transforms have been identified as strong candidates for the detection of nonoverlapping subband edges. It was observed that mainly the Haar, Daubechies, and Biorthogonal wavelet families were capable of detecting these frequency transitions and thus detecting edges. To obtain better results would be a challenging task, comprising the design of more specific wavelet families suitable for spectrum hole detection in the OR environment. These wavelets will need to be based on the characteristics of the signal PSDs, which need to be investigated and analyzed.

7.4.5   Energy Detection Based on Multiple-Antenna Processing

To make secondary use of a certain spectral region, a CR must first determine that there is no primary user signal present. This detection is challenging, due to the requirement of highly reliable signal detection over the wireless channels. Because wireless channels introduce signal fading, which causes low SNR conditions at the CR receiver input, this may make the CR susceptible to the hidden node problem, in which the CR fails to detect a primary user. Assuming that the spectral region is empty, the CR transmits, creating potential interference to the primary user. Thus, the primary user detection algorithm should achieve a probability of detection close to unity for a prespecified probability of false-alarm and a given SNR.

The conventional approach of using energy detection for determining whether the signal is present in a given spectrum of interest is given in Cabric et al. [28] and Van [29]. However, this approach has poor performance in low-SNR regions, where the energy detector is used in a single antenna CR, making it vulnerable to the hidden node problem. This factor motivated the investigation of multiple antennas in a CR.

ORACLE Deliverable 2.4 [24] presented a multiple-antenna processing-based energy detector. Multiple copies of the primary user signal at the antennas of the CR receiver were combined and given as input to the energy detector. Two multiple-antenna processing methods were considered, and the detection performance of each was analyzed. It was discovered that the proper combination of the multiple-antenna processing scheme with energy detection made it possible to achieve a high probability of detection, even at low to moderate SNRs. The two techniques, which were considered for processing the signals at the receiver antennas to exploit the spatial diversity offered by the wireless channel, are maximum ratio processing and selection processing. Energy detection was also used in combination. A normalized energy metric [30] corresponding to the processed signal was compared to a detection threshold. The greater the metric, the more likely it is that the signal of interest is present. Also, closed-form expressions for the probabilities of detection and false-alarms for multiple-antenna processing-based energy detection performance under maximum ratio processing and selection processing were derived in ORACLE [24].

The simulation results shown in Figure 7.19 display the efficiency of the multiple-antenna processing techniques through the detection of a prespecified probability of false alarm at a given SNR. If we compare the achieved detection probability with varying SNR for the single antenna against the two multiple-antenna processing techniques, the improvement in the detection achieved through the diversity gains offered by the multiple antennas processing in energy detection is evident. Also, more than an order of magnitude of improvement in detection performance is gained by the use of maximum-ratio antenna processing and signal processing.

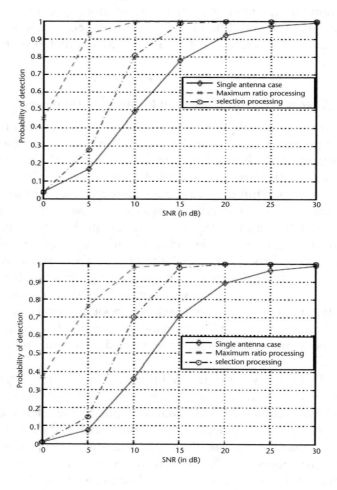

Figure 7.19 Comparison of detection performance between single- and multiple-antenna processing.

7.4.6   Cooperative Extension of the UMTS FDD Signal Detector

A network of ORs that can exchange detection decisions amongst themselves or with a centralized entity can provide good detection performance in some particular scenarios, especially in the hidden terminal problem. But when an OR experiences shadowing or fading effects, the ability to detect an active primary user transmitting is compromised; in such cases the OR is not able to differentiate between an unused band and a deep fade. It is possible that these effects can be mitigated by implementing a collaborative spectrum sensing scheme in which the sensing information from multiple ORs is gathered before a decision is made as to whether the sensed licensed spectrum band is free or being used by some primary user, according to a specific decision-making algorithm that ponders all the collected sensing metrics from the ORs belonging to the particular sensing cluster.

The cooperative sensing scenario shown in Figure 7.20 considered that the area of the OR network should be small compared to the area of the UMTS cell (i.e., $d < R$). Every OR clustered in the cooperation footprint sends its decision to a designated central sensing decision unit (CSDU) entity. This central entity can be one of the OR nodes. Also, channel correlation between the OR nodes is an important issue when cooperative schemes are analyzed. Increased correlation decreases the chance of getting an OR node with a very good channel; hence, more users need to be polled for independent readings of the same signal.

Figures 7.21, 7.22, and 7.23 show the results when evaluating the cooperative gain. Probability of detection (P_{dc}) and probability of false alarm (P_{fac}) achieved with the collaborative sensing are shown as a function of the number of consulted OR nodes (N) for different correlation shadowing levels. It is considered that every OR node spends the same observation time listening to the signal before decision.

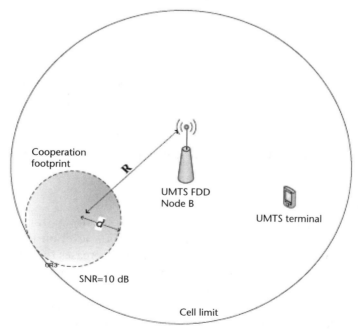

Figure 7.20   Scenario considered for cooperative sensing.

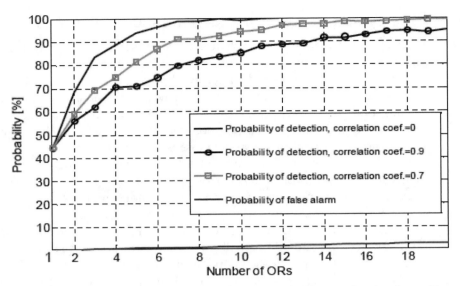

Figure 7.21 P_{dc}, P_{fac} as a function of the OR nodes consulted, SNR = 10 dB, obs. time = 10 ms.

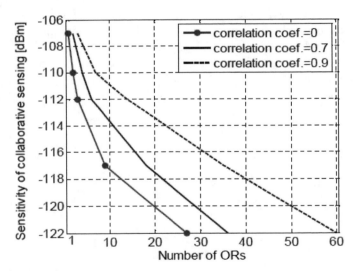

Figure 7.22   Sensitivity of cooperative sensing as function of the OR nodes consulted, obs. time = 10 ms.

The results show that P_{dc} increases monotonically as the number of OR nodes increases. The impact of shadowing correlation level on the collaborative sensing performance is clear. Shadowing correlation decreases the amount of new information that an OR node gets from the other ORs, destroying part of the collaborative gain. It was observed that for $N = 10$, the saturation for correlated shadowing occurs. P_{fac} increases slowly as the number of ORs rises, but the related increase of the P_{dc} decreases the probability of the OR network interfering with the licensed UMTS system.

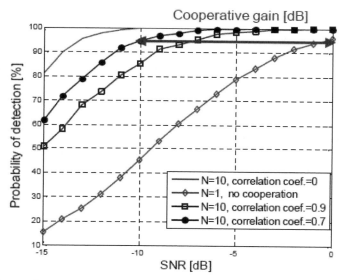

Figure 7.23 P_{dc} as a function of the nominal SNR, obs. time = 10 ms.

Figure 7.22 shows the achieved sensitivity as a function of N, in which sensitivity can be defined as the minimum signal power that allows collaborative detection to achieve P_{dc} = 99.9%.

Figure 7.23 shows the P_{dc} as a function of the nominal SNR within the cooperative footprint for $N = 1$ and $N = 10$ nodes. The comparison between these curves for the same P_{dc} target gives the cooperative gain of the sensing algorithm.

To summarize, the collaborative sensing scheme helps reduce the effect of shadowing in the decision-making process. By providing multiple independent realizations, the probability that all OR nodes are simultaneously faded is very low. Shadowing correlation decreases the amount of new information that an OR node gets from other ORs, destroying part of the theoretical collaborative gain. From the simulation results, it was observed that there is a significant decrease in the average SNR for detecting an UMTS signal when collaboration is used. It was possible to achieve a collaborative P_{dc} = 99.9% combining 10 OR nodes for an observation time of 10 ms. The collaborative spectrum sensing allowed a reduction in sensitivity requirements at individual nodes, leading to a decrease in signal processing complexity and required observation time. It also overcomes the problem of the hidden node, because in multiple ORs there is a high chance of having an OR with its SNR well above the average.

7.5   Spectrum Policies and Economic Consideration

The main objective of CR and OR technology is to enable better and more efficient use of the spectrum—licensed as well as unlicensed. The economic considerations, with respect to the CR application and OR access, are very important. They are related to the market segments over which the OR has a direct impact.

In the unlicensed spectrum bands, which are of interest to the consumer electronics and medical system segments, the capacity and communication reliability are expected to improve due to better system coexistence enabled by OR access principles. In the case of licensed spectrum bands of interest to the telecom market segment, the OR enables access to the new spectrum regions allocated to other type of services, access to spectrum resources allocated to other operations through some form of spectrum leasing/trading, and flexible access to its own allocated bands. Secondary spectrum access refers to opportunistic spectrum use of spectrum regions for which an operator does not have a statically allocated license. This enables the use of spectrum region for services other than those it is licensed to operate on this spectrum region.

In the consumer electronics domain, wireless coverage across the home is desired for delivery of multimedia content from the home and from the Internet. The deployments based on the present IEEE 802.11 systems face serious problems with providing sufficient coverage, sufficient throughput, and QoS to applications. Interference from neighbors is also a serious problem.

In the healthcare domain, sensors are used to measure continuous or bursty vital signals from a patient. In many cases, it is desired that the signals be transmitted wirelessly to give freedom of movement to patients. Also, eliminating wires reduces equipment cost, since cables can be costly. Hospitalwide wireless connectivity enables continuous monitoring during the care cycle within the hospital. This technology will also enable patient monitoring outside the hospitals, which in turn will lead to freeing-up of hospital beds and reduced medical costs.

The lighting industry domain is another example. Wireless connectivity can be used to control lights in desirable ways; additionally, wireless connectivity to various sensors can monitor changes in the environment. This promises indispensable easy deployment and reduced cost.

Therefore, the primary drivers for wireless connectivity in healthcare are reduced costs (including the installation cost), improved quality of care, workflow improvement in healthcare cycles, and error reduction. Current wireless technologies cannot live up to these expectations in many cases. Using general purpose wireless technologies based on the standards has the benefit of diversified suppliers, guaranteed interoperability, and cost reduction due to high production volumes. But these do not take into account interference from each other. Due to the nature of unlicensed technologies, which people can deploy any time and any place, as they desire, the interference can be from similar systems (in the form of self interference) and can be from dissimilar systems. Unexpected interference can cause reduced throughput and slow system response, and may even lead to system malfunctioning if it is not handled properly. The FP6 IST project ORACLE [31] contributed to research toward better interference handling and better exploitation of white space in the spectrum in unlicensed operation.

Some of the challenges in the economic considerations are related to current investments, QoS offered to customers, the impact of primary service infrastructure, and the justification of extra investments at the primary and secondary service levels. In terms of positioning of an operator with respect to OR secondary access, two conflicting interests enter into consideration. On one hand, the operator is interested in using the spectrum bands allocated to other services, while on the other hand, the operator is interested in keeping exclusive access rights in the bands allocated to it.

The ability to service higher traffic volumes translates to better service quality as well as lower prices for the customer, which also helps the operator consolidate and augment its customer base. Two of the main potential risks come from changing the rules for spectrum allocation and another from interference management.

Some evidence is required to help establish whether there is a viable business case behind OR access, under which conditions this is so, and also how to remove the operators' reluctance in reconsidering current regulation practices in favor of novel spectrum allocation schemes. This evidence may consist of the following:

- Quantification of the benefits in terms of capacity increase, spectrum utilization improvement, and so forth;
- Quantification of the interference effects, as well as proof that the interference can be controlled;
- Technical solutions that are cost effective and can be deployed in a way that justifies the additional investments.

Secondary spectrum trading refers to the transaction of spectrum usage rights between operators and/or spectrum owners to better exploit underutilized radio bands. It can potentially become a commercial beneficial concept that is mainly based on two OR features: spectrum sensing to explore underutilized bands and frequency agility to switch radio communication in these bands. Spectrum liberalization can be defined as the right of any spectrum user to change the use of any existing portion of the spectrum subject to any harmonization at EU level but also to change the technology used within such spectrum for the provision of new services or existing services. The spectrum authorities in a given geographical area monitor and control the use of frequency spectrum to ensure the secondary trading of rights to use the spectrum does not generate any harmful interference and does not lead to speculative spectrum hoarding. Also, the regulatory authorities should be able to enforce the rights and obligations of spectrum users and to make binding decisions in spectrum-related disputes. More detailed information on these issues is available in ORACLE Deliverable 1.1 [31].

Table 7.1 states the key issues and the timeline for effective radio spectrum management in the EU in the following areas [32]:

In 2002, the Spectrum Policy Task Force was formed to assist the Commission in identifying and evaluating changes in the spectrum policy that will increase the public benefits derived from the use of the radio spectrum. The task force examined the Commission's spectrum policies and rules in relation to three general models for assigning spectrum usage rights, as follows:

- *Command and control model:* This is the traditional process of spectrum management in the United States currently used for most spectrum within the Commission's jurisdiction. It allocates and assigns frequencies to limited categories of spectrum users for specific government-defined uses.
- *Exclusive use model:* This is a licensing model in which a licensee has exclusive and transferable rights to the use of specified spectrum within a defined

Table 7.1   Key Issues and Timeline for Effective Spectrum Management in EU Policy

2006	Proposals on *coherent flexibility* in spectrum bands used to provide e-communications services (WAPECS). Proposal for a common format of *usage rights* in the context of spectrum trading. Harmonization measures: Adoption of an EC decision harmonizing the use of the so-called *IMT-2000 extension band.* Adoption of one or more EC decisions harmonizing the use of *ultra wideband (UWB)* applications. Adoption of an EC cecision providing a framework for the harmonization of spectrum access for a large number of *Short-Range Devices (SRD).* Abrogation of the *ERMES Directive* and adoption of an EC decision harmonizing the use of this former paging band.
2007	Proposals regarding spectrum-related issues, including licensing, in the *Review* of the 2002 e-communications regulatory framework. Proposals for the coordinated use of part of the *broadcasting digital divided,* following the completion of the ITU RRC-06 conference. Introduction of an EU-wide approach to *license-free spectrum.* Implementing improvements to the process of *interference management.* Setting out *EU priorities* for the ITU *WRC-07* conference. Harmonization measures: Broadband wireless access and mobile satellite applications. *Assistive and medical* wireless applications. Wireless applications for the *intelligent car.*
2008	Coordinated EU introduction of relevant results of the *ITU WRC-07* conference.
2009	National implementation of *new e-communications rules.*
2010	Full establishment of a *functioning EU market* for major parts of the spectrum.
2012	Proposed date for completing *switch-off* of *analog broadcasting* in the EU.

geographic area, with flexible use rights that are governed primarily by technical rules to protect spectrum users against interference. Under this model, the exclusive rights resemble property rights in spectrum, but they do not imply or require creation of full private property rights in the spectrum.

- *Commons or open access model:* This model allows unlimited numbers of unlicensed users to share frequencies with usage rights that are governed by technical standards or etiquette but with no right to protection from interference. Spectrum is available to all users that comply with established technical etiquette or standards that set power limits and with other criteria for operation of unlicensed devices to mitigate potential interference.

More information about this topic is given in ORACLE Deliverable 1.1 [31].

7.6 Conclusions

This chapter discussed various algorithms and techniques for spectrum management and radio resource allocation performed by IST projects active in that area. The main contributions upon which this chapter was built were made by the FP6 IST Project ORACLE. Significant achievements were reported in terms of reducing complexity and processing time for an opportunistic or cognitive radio terminal, building a cost function with the help of cycle-correlation coefficients for a spread signal, hardware quality, and stability aspects of UMTS's specific cyclostationary feature detector, and so on. The use and performance of multiple-antenna processing and wavelets for detection has also been highlighted.

Since cooperative sensing is considered to be inherently static, a dynamic approach such as collaborative sensing was investigated, and the results presented are promising. This topic is open to further investigation. An overview of key issues in spectrum policies has been given. Also, the economic profitability achieved through the use of such opportunistic techniques has been discussed with special emphasis on secondary spectrum trading. Some of the issues presented here for spectrum management and resource allocation are under further research within the scope of the FP7 ICT program. The reader is referred for further reading to the European Community relevant site.

References

[1] FCC, Spectrum Policy Task Force Report, ET Docket No. 02-155, November 2002.

[2] http://www.ist-oracle.org

[3] http://www.ist-surface.org

[4] Haykin, S., "Cognitive Radio: Brain Empowered Wireless Communication," *IEEE Journal on Selected Areas in Communication*, Vol. 23, No. 2, February 2005.

[5] IST project ORACLE, "Technical Scenarios and Application Scope," WP 1 Deliverable 1.1, September 2006, http://www.ist-oracle.org

[6] Mitola, J., "Cognitive Radio for Flexible Multimedia Communications," *MoMuC*, 1999.

[7] IST project ORACLE, "Specification of Preliminary Set of Sensing Parameters," WP 2 Deliverable 2.1, April 2007, http://www.ist-oracle.org

[8] Urkowitz, H., "Energy Detection of Unknown Deterministic Signals," *Proceedings of IEEE*, Vol. 55, April 1967.

[9] Sonnenschein, A. and P. Fishman, "Radiometric Detection of Spread-Spectrum Signals in Noise of Uncertain Power," *IEEE Transactions on Aerospace and Electronic Systems*, Vol. 28, No. 3, July 1992.

[10] Ghasemi, A. and E. Sousa, "Collobrative Spectrum Sensing for Opportunistic Access in Fading Environments," *IEEE DySPAN*, 2005.

[11] Tian, Z. et al., "A Wavelet Approach to Wideband Spectrum Sensing for Cognitive Radios," *IEEE CrownCom*, 2006.

[12] Mallat, S. et al., "Singularity Detection and Processing with Wavelets," *IEEE Transactions in Information and Theory*, Vol. 38, 1992.

[13] Ganesan, G. and Y. Li., "Cooperative Spectrum Sensing in Cognitive Radio Networks," *IEEE DySPAN*, November 2005.

[14] Zhao, J. et al., "Distributed Coordination in Dynamic Spectrum Allocation Networks," *IEEE DySPAN*, November 2005.

[15] Shankar, S., "Spectrum Agile Radios: Utilization and Sensing Architectures," *IEEE DySPAN*, November 2005.

[16] Liu, X. and S. Shankar, "Spectrum Agile Radios: Utilization and Sensing Architecture," *IEEE DySPAN*, November 2005.

[17] Devroye, N. et al., *Limits on Communications in a Cognitive Radio Channel*, Harvard University, 2006.

[18] Jafar, S. and S. Srinivasa, *Capacity Limits of Cognitive Radio with Distributed and Dynamic Spectral Activity*, University of California, Irvine, September 2005.

[19] Mishra, S. et al., *Cooperative Sensing among Cognitive Radios*, University of California, Berkley ICC 2006.

[20] IST project ORACLE, "Preliminary Specification of the Basic Protocol Architecture," WP 2 Deliverable 2.2, April 2007.

[21] Fujii, T. and Y. Suzuki, "Ad-hoc Cognitive Radio Development to Frequency Sharing System by Using Multi-hop Network," *IEEE DySPAN*, November 2005.

[22] Stewart, R. et al., "Stream Control Transmission Protocol," *IETF*, October 2006.

[23] Claise, B., "Specification of the IPFIX Protocol for the Exchange of IP Traffic Flow Information," *IETF*, October 2006.

[24] IST project ORACLE, "Sensing Algorithms Part 2, Final," WP 2 Deliverable 2.4, December 2007.

[25] Cai, K. et al, "Energy Detector Performance in a Noise Fluctuating Channel," *IEEE Military Communications Conference*, Vol. 1, October 1989.

[26] Tsatsanis, M. and G. Giannakis, "Blind Estimation of Direct Sequence Spread Spectrum Signals in Multipath," *Conference Record of the 29th Asilomar*, November 1995.

[27] Marques, P. et al., "SDR for Opportunistic Use of UMTS Licensed Bands," *Software Defined Radio Technical Conference and Product Exposition*, November 2006.

[28] Cabric, D. et al., "Experimental Study of Spectrum Sensing Based on Energy Detection and Network Cooperation," *International Workshop on Technology and Policy for Accessing Spectrum*, August 2006.

[29] Van, H., *Detection, Estimation, and Modulation Theory Part 1 and 3*, Wiley-Inter Science, 2001.

[30] Urkowitz, H., "Energy Detection of Unknown Deterministic Signal," *IEEE Proc.* Vol. 55, No. 4, April 1967.

[31] IST project ORACLE, "Technical Scenarios and Application Scope," WP 1 Deliverable 1.1, September 2006.

[32] Commission of the European Communities, "A Forward Looking Radio Spectrum Policy for the European Union," Second Annual Report, June 2005.

CHAPTER 8

Reconfiguration Threats and Security Objectives

End-to-end reconfiguration capabilities have brought about flexible adaptation of reconfigurable equipment according to user and operator preferences and dynamic adaptation to changing network conditions. But these capabilities come with the need for adequate security measures to prevent misuse of the flexibility against the users, operators, and the regulatory bodies.

The major contributor in this area was the IST FP6 project E2R [1]. This chapter covers security concepts related to reconfiguration to ensure a reliable, correct operation—in particular, authorization of reconfiguration software, device-local security, and the reconfiguration process. This chapter will discuss the reconfiguration threats, security objectives, reconfiguration software authorization, authentication, and trust framework necessary for next-generation communication systems.

8.1 Introduction

A secure reconfiguration would ensure a correct, reliable operation respecting the interest of end users, operators, and regulatory bodies. This security is required to avoid the misuse of the ongoing, flexible download of software and configuration information, which could cause severe security problems. One approach to solving the security issue would be to apply restrictive security measures—for example, using a single, trusted reconfiguration manager (RM) with the capability to perform only restricted types of reconfiguration that cannot introduce any security problems. But the real challenge lies in developing security concepts for more open and decentralized approaches to reconfiguration that ensure reliable, correct operation according to the expectations of end users, operators, and concerned regulatory bodies.

Figure 8.1 shows the security issues in reconfiguration. The objective is to ensure a secure and compliant operation by accepting only authorized software. Concerning the critical functions, such as low-level radio or security-sensitive functions, the policy defining the software is to be accepted from restricted providers; concerning the less sensitive functions, an open policy is to be followed. The authorized reconfiguration managers may control reconfiguration to prevent nontrusted parties from modifying the configuration of the device. The conformity of reconfigurable terminals should be ensured, as it affects the conformance of radio emissions relevant to maintain properties asserted for type approval and also conformance with standards, operators, and user preferences.

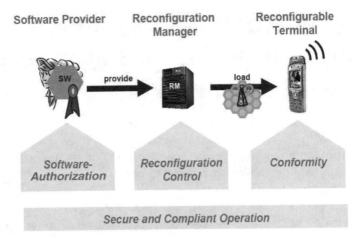

Figure 8.1   Security issues in reconfiguration [1].

8.2   Reconfiguration Threats

This section identifies the threats during reconfiguration of communication equipment. Malicious software maybe installed, but can also be configured using "good" software in an undesired way. Since reconfiguration permits changing properties of communication devices that have previously been fixed by their mere design, this flexibility may allow changes to be made to the configuration that are against the interest and expectations of the end users, network operators, service providers, equipment manufacturers, and regulatory authorities. This is a severe threat posed by the flexibility offered.

The identified security issues on architectural level are as follows:

- The various actors involved in the process should establish trust relationships that allow the download of a reconfiguration software module to take place in a secure fashion. The actors should be able to identify and authenticate themselves, the subscribers must be authenticated, and the reconfiguration support service provider and the reconfigured equipment must be mutually authenticated.
- The equipment needs to check for authorization to perform a reconfiguration that can be done implicitly or explicitly (e.g., implicit authorization would be contained in the user preferences, and the explicit case is when the user is prompted for an authorization).
- Personal user data should be protected against unauthorized access.
- Certain security requirements concerning the software should be met: the software has to be certified, the origin and the integrity of the software should be verifiable before installation, the proper installation should be guaranteed before installation, and if needed, it should be possible for the regulator to investigate the software.
- The interference control mechanism in the equipment should be protected from unauthorized access.

A few threats specific to reconfiguration would include download and execution of malicious software, invalidation of conformance requirements, undesired reconfigurations, illegitimate access to private information, no or insufficient protection of intellectual property, and easier attacks against wireless systems. These are discussed briefly as follows:

- *Download and execution of malicious software:* Malicious software downloaded by accident or by intention could cause harm. It poses a threat to reliability and availability; additionally, it can implement malicious functionality, such as dialing premium rate numbers in the background or any of the other following threats.

- *Modification of other functionality:* The functionality not intended for modification or not authorized to be reconfigured may be changed.

- *Secure and reliable operation of composed functions:* Not only are the component software modules expected to work correctly, but also the overall configuration composed of all those functions should work correctly.

- *Circumvention of security functions:* The security functions have to be trustworthy themselves but rely more on secure storage of protected access to cryptographic material and policy information. If the reconfiguration is not protected, it could weaken or circumvent the security functions and render them useless.

- *Invalidation of conformance requirements:* The regulatory bodies pose requirements on radio equipment concerning user safety, electromagnetic compatibility, and radio spectrum use and the conformance to which it is dependent on the regulatory body or an authorized testing house, or asserted by the manufacturer by stating conformance. Reconfiguration is under threat when the radio equipment is brought into the market, where the conformance requirements may be violated during the operation of the equipment. This may even endanger health and user safety. For example, when the radiated power is too high, it could result in emissions that interfere with users and radio systems. In addition to emitting in the wrong frequency bands, the access to the radio medium could be modified by using too high power or wrong modulation schemes, giving the user an unfair advantage that has a negative impact on other users.

- *Undesired reconfigurations:* The communication services used may not match the preferences and expectations of the end users concerning the available services, the provided quality of service, and the cost involved. The preferences of the service providers and the network operators may be disregarded. The information used or exchanged as part of the reconfiguration may be manipulated, thus influencing reconfiguration in illegitimate ways. If a configuration is installed and activated that does not work at all or properly, the users will be dissatisfied, and there will also be a high cost for customer care for the service provider. The reconfiguration servers, software, and configuration information required to perform a reconfiguration may not be available or function well.

- *Illegitimate access to private information:* Access to sensitive information, such as the end users' preferences, used services, or te current location and configuration, should be controlled to protect the privacy of the user. Even service providers and network operators may want to keep information about their customers or network internals with competitors confidential.

- *No or insufficient protection of intellectual property:* The reconfiguration software maybe used or copied illegally. Reconfiguration could make reverse engineering easier when reconfiguration software is developed for an open, documented platform and can be easily obtained from download servers. There are suitable solutions concerning the billing of the reconfiguration process, and the compensation received by each partner depends heavily on the business model between the involved parties [2].

- *Easier attacks against wireless systems:* Reconfiguration may make it easier to attack the wireless communication system and may bring it into the range of potential attackers. This means that the attackers will no longer need to rely on expensive equipment, such as signal generators or spectrum and protocol analyzers, or build special equipment involving reverse engineering and modification of proprietary, highly integrated devices. They will gain easy access to open interfaces and could simply reconfigure off-the-shelf equipment according to their intentions—for example, an attacker could configure his/her device as a false base station.

8.3 Security Objectives

The objectives for security identified in this section focus on the reconfiguration software download, the reconfiguration process and the compliance of radio emission.

The overall objective of security is to offer a reliable service that fulfils the expectations of the involved stakeholders.

8.3.1 Reconfiguration Software Download

This can be broken down into three main steps:

- *Protection against malicious or malfunctioning reconfiguration software:* Malicious software could invalidate properties required for type approval, but it also may be harmful, (e.g., it could circumvent other security mechanisms required for secure network access to a cellular network or a company's Intranet, send the users' private data to unauthorized parties, or simply make the device unusable). The device may be manipulated to behave against the interest of the user. The part of the software that is responsible for performing security operations has to be protected specially to prevent the erosion of security measures by replacing security functions. Hence, the downloaded software should originate from a trusted source and be executed in a controlled, restricted environment.

- *Secure storage and management of cryptographic material:* For reconfiguration, either only content from a trusted provider is accepted or the content is associated with a different protection domain, dependent on the trustworthiness of its provider. This trust relationship must be reflected by the data on the receiving client so that it can verify the providers' certificate and signature or signed piece of content and determine whether it is trusted for the respective content type. It must be defined based on the reconfiguration class and the decision, if the user, service provider, or hardware manufacturer can determine the trusted providers, will influence where the correspondent data can be stored, whether it can be modified, and by whom. When only the software provided by the hardware manufacturer is trusted, then the cryptographic material and policy information has to be stored on the reconfigurable device by the hardware manufacturer, and it needs to be assured that it cannot be changed by the end user or a service provider.

- *Prevent illegitimate use of reconfiguration software:* Digital rights management may be required to limit the illegitimate usage of software. When the software is not available free of charge, it should be ensured that the software can be used only when it has been paid for and that it cannot be copied or forwarded to other users illegitimately.

8.3.2   Reconfiguration Process

This is concerned with maintaining information about the current configuration and the environment, deciding whether and when to reconfigure and which parts, performing the actual reconfiguration, and dealing with potential failure cases. It can be broken down into the following main steps:

- *Authorization of entities triggering or performing a reconfiguration:* The restrictions on reconfiguration will ensure that only safe and working configurations are activated that are inline with the compliance and regulatory requirements and also meet end-user and network preferences and expectations. The entities triggering or performing a reconfiguration need to be authorized and trustworthy.

- *Protected reconfiguration signaling:* There is an exchange of signaling traffic between the reconfigured device and the reconfiguration support nodes. This signaling traffic needs to be protected to avoid tampering of the reconfiguration process. The communication peer entities need to be authenticated, and the authentication entity could be the subscriber, an administrative domain, or a single network element. Integrity, authenticity, and confidentiality must be ensured, which could be achieved by using well-known cryptographic means or by using closed networks that cannot be accessed by unauthorized entities.

- *Protection of reconfiguration support functions:* The additional functions will be added to the core and possibly to the access network, a public network, or a closed network to assist the reconfiguration; these support functions have to be protected from attacks to ensure their secure and reliable operation. Only authorized users or entities make use of the provided reconfiguration support services. The correct, trustworthy reconfiguration support functions must be used, or else the vicious reconfiguration support functions provided by the attacker could get access to private information and manipulate the reconfiguration. Also, the billing of the user for using the reconfiguration services should be well protected.

- *Integrity, authenticity, and confidentiality of information used in the configuration process:* The following information is needed during the reconfiguration process: user's preferences and profile (e.g., allowed and preferred access technologies, networks, and services), user's current context information (e.g., location, currently used services), device information (e.g., reconfiguration capabilities, current configuration, device status), network-based information (e.g., predefined configurations, metainformation on reconfiguration software as the most recent version or on its compatibility) and information on network (e.g., operator preferences, available modes, current network status). This information must be reliable and available, and it should not be possible to change it illegitimately; also, the private information should not be accessible to unauthorized parties.

8.3.3 Compliance of Radio Emission

Since it is possible to configure the radio parameters, it becomes necessary to control the radio emission so that no other users within the same or at another radio communication system are disturbed and that the user is safe. The limitations, from a compliance perspective, need to be observed to make sure that a device emits only radio signals permitted by its statement of conformance or its type approval. It should be clear who will be responsible for ensuring that no illegitimate radio emissions occur.

8.4 Reconfiguration Software Authorization

The authorization of reconfiguration software is one of the main concepts for secure download of reconfiguration software. Reconfigurable devices should accept only authorized reconfiguration software.

The following is the classification of the reconfiguration software authorization, according to the stakeholder:

- *Regulator* (e.g., radio software relevant for conformity of radio emissions such as transmission frequency, emission power, product responsibility);
- *Manufacturer* (e.g., bug fixes, enhancement or change of baseband algorithms, installation of optimized protocol stack components);
- *Network operator* (e.g., monitoring and selection of most suitable radio technology, handover decisions, and medium access algorithms);
- *Service provider* (e.g., branding of user interface, software needed for service-provider specific services).

The process for reconfiguration software download is shown in Figure 8.2.

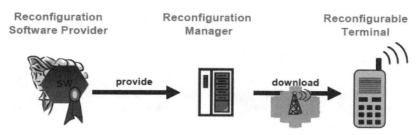

Figure 8.2   Reconfiguration software download [1].

Figure 8.2 shows a widely used security mechanism to protect software download [1]. The software module provider attaches a digital signature to the module that is verified by the target device. The signature can be verified independently from the download software. It ensures that the module is not modified and helps authenticate or certify the software provider. The software provider could either be a software developer or a separate entity as test house or operator. The signature of the received software module must be validated by the receiving device to ascertain whether it is from the trusted provider. The process of obtaining a suitable certificate depends on the restrictions enforced; authorized software providers receive a certificate that enables them to compute a valid signature for a certain type of content. The device needs to have the correct root keys of the authorized software providers so that it can verify the digital signature. These keys may be stored on the device or on the pluggable user module, depending on who is authorized to certify the reconfiguration software.

8.4.1   Relevant Security Technology

The following underlying security technologies that are relevant can be identified:

- *Digital signature:* This is based on the use of private key and public key. The private key is kept secret; only the entity that computes the digital signature possesses it. And this digital signature is verified by using a public key. In an asymmetric cryptographic system, the private key has to be kept secret, whereas the public key can be distributed freely. But authentic public keys that really correspond to the private key of the intended peer entity need to be used. This is taken care by public-key certificates issued by a Certification Authority (CA) that bind a public key to an identity or possibly to further data. The most commonly used are the certificates of X.509 standard [3] and the Public Key Infrastructures (PKI). For checking the revocation status, the Certification Revocation List (CRL) is used, or it can be checked on-line using the on-line certificate status protocol (OCSP) [4].

  RSA/PKCS#1 (Rivest-Shamir-Adleman) and Digital Signature Algorithm are some of the well-known algorithms used for digital signature [5]. Hash functions such as MD5 or SHA-1 are used to compute a digest value of the content, and then an asymmetric cryptographic digital signature algorithm is computed. A widely used format for cryptographic messages, CMS/PKCS#7 [6] supports inclusion of certificates needed to verify the signature, it also sup-

ports multiple signers; the signed content can be contained, but it is not necessary. Hence, the signature and the signed content can be encoded as single
data structure, but it is also possible that the actual software and the signature
are separate.

- *Signed content:* The digital signature is added to a piece of content, such as a
  software module, that is used to attest whether a certain software module has
  been published by the trusted provider who actually computed the digital signature. The receiving device can verify the integrity and authenticity of the digital signature. This has been used for signing MIDlet suites in MIDP2.0 [7].
  The Java archive (JAR file) is signed where the data contained in the jar file is
  protected by the digital signature including the meta information stored in the
  manifest file. The RSA/PKCS#1 signature is encoded directly in the Java application descriptor, along with the certificates needed for signature verification.
  The advantage is that the application descriptor has security information that
  can be verified even before the actual Java archive file is downloaded. The
  actual digest of the downloaded JAR file must be verified to match the reference digest as asserted by the digital signature for, as examples, Microsoft
  Authenticode [8], Symbian operating system [9], and so forth. An Internet
  standard for protecting firmware packages using cryptographic message syntax CMS is discussed in [10].

Four protection domains, namely, manufacturer, operator, trusted third party,
and the untrusted are distinguished in MIDP 2.0 in its recommended security practice for GSM/UMTS compliant devices. The restrictions enforced by the execution
environment depend on the domain into which the MIDlet has been put.

8.4.2 Software Download Authorization

A framework for software authorization was developed within the frames of the IST
project E2R [1]. Different market models have been realized for reconfiguration
software, based on the concept of signed content. These models are used to define
the legitimate authorized provider as follows:

- Vertical market model, in which radio software is certified by the device manufacturer;
- Horizontal market model with certification of each radio hardware-software
  combination by an approval authority;
- Horizontal market model with separate certification of radio hardware and
  software.

Figure 8.3 shows the concept of radio software authorization based on the
signed content by an approval authority. An approval authority authorizes the radio
software module by computing and attaching its digital signature using its private
key. The metainformation of the module includes entries to identify the module, the
authorized target device, and restrictions on the activation (e.g., it is activated only
within ITU region 1, as shown). Specific regional or time restrictions can be
encoded, too. This approach can be used to realize a vertical market model, in which

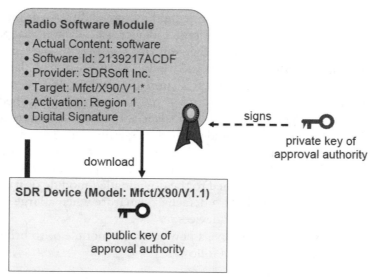

Figure 8.3   Signed radio software module [1].

only the software authorized by the device manufacturer is accepted; it can also realize a horizontal market model in which each hardware-software-combination requires authorization from a trusted approval authority.

In some cases, such as when proper operation of the certified software can no long be guaranteed, it could be necessary to revoke the authorization. The same mechanism for certificate revocation may be used for this task, or the software module may be certified for a restricted time period only.

For the vertical market model, the radio-related software will be accepted only if it is authorized by the device manufacturer, who ensures that conformance properties are met and that operations are proper and reliable. There is ongoing research on alternative approaches for secure radio software download that are suitable for horizontal market models. Some of these include combined or separate approval for radio hardware and radio software, moving the responsibility of validating a radio configuration to a network-based function, or supervising radio emissions during the operation to perform reactive measures if a malfunction is detected. The following subsections briefly discuss the two types of market models.

8.4.2.1   Vertical Market Model

In this model, the reconfigurable device accepts only software modules authorized by its device manufacturer, since the downloaded software modules are specific to a single device type. This approach has been adopted by the 3GPP 23.057 into the MExE (Mobile Execution Environment) standard [11]. Similarly, the MIDP2.0 has recommended security practices for GSM/UMTS-compliant devices to distinguish between the manufacturer, operator, third party, and untrusted domains. This security infrastructure could also be reused for other types of manufacturer-signed software. A manufacturer-signed software module is one that is authorized for a specific target device; the module is digitally signed directly, using the private key of the device manufacturer. The target device is indicated as part of the meta-information, but in general, other information may also be used to identify the target device for which the software module is authorized. A data element could be included to uniquely identify each software module, but even in the case of no explicit identification, each software module may be identified uniquely by its cryptographic diges, computed with a one-way hash function (e.g., SHA-1).

The receiving device should verify the signature and determine that the software module has been approved or authorized by its manufacturer, that it has not been manipulated, and also that it is intended for the type of target device. The digital signature could either be attached to the downloaded module or it could be a separate signature. It is better to have a detached signature when a large software module is authorized for several target devices.

The steps taken to approve a new software module or to bring a new hardware model into the market are as follows:

- *New software module:* The device manufacturer attaches his digital signature to the additional reconfiguration software module he wants to authorize. If the signature is valid, the target device will recognize this module and, hence, a new software module can be authorized after the device has already rolled out.
- *New hardware device type:* A trusted public key is stored on the device by the device manufacturer during manufacturing. The manufacturer also has to compute the digital signature for the existing reconfiguration software modules that need to be authorized for the new device model. The signatures computed for the older devices remain valid. When the new device is fully backward-compatible with some older device types, it can be programmed to accept radio software modules targeted at these older device types.

8.4.2.2 Horizontal Market Model

The horizontal market model requires approval for the steps taken by software provider in order to use the software or hardware module. These steps are listed as follows:

- *New software module:* The software provider requests authorization from approval authority to use his software on a specific hardware device. If granted, the approval authority would compute the digital signature of the software module, keeping the intended hardware model and its manufacturer as the target. The device manufacturer may also apply for the approval.
- *New hardware device type:* The hardware device manufacturer needs to store the trusted public key of the approval authority on the device during manufacturing. The approval authority has to digitally sign the already existing reconfiguration software, indicating the manufacturer and the type of the new hardware model in the metainformation. The hardware manufacturer, software manufacturer, or an independent party could apply for the hardware-software combination.

8.4.2.3   Variant for Independent Hardware-Software Authorization

This approach would involve independent approval of radio hardware and soft-ware using signed content. It would require different device models to support the same reconfiguration software execution environment, enabling the same software module to be executed on different device types. But in this case, the identifier of the intended target radio execution environment would be used, instead of indicating a single target device type that would be compared to the reference identifier by the reconfigurable device. Another option would be to use a wildcard expression matching all intended target device models. A two-step solution has been proposed, as follows:

- Authorize the software for an open radio platform.
- Perform a compatibility check of reduced complexity for each intended hard-ware model.

This combined approach is illustrated in Figure 8.4. The advantage of this approach is that the possibly complex and thorough checks of the software module against a standardized open radio platform can be combined with efficient compati-bility checks that are to be performed for each hardware device. If this approach is mapped onto a technical download solution, it would allow the steps to be executed independently.

8.4.3   Software Activation Restrictions

There could be further restrictions placed on the activation of the radio software module, and these could vary from region to region. A dynamic authorization of the radio software module by the currently used network, attesting that its functionality is supported and compatible, is also possible.

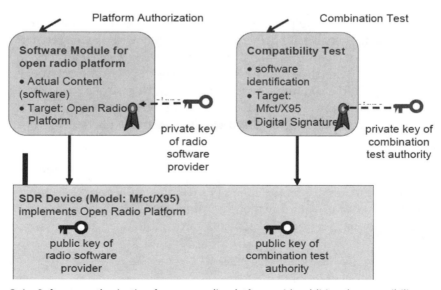

Figure 8.4   Software authorization for open radio platform with additional compatibility test [1].

It is important that when the same radio hardware model is marketed in the different market segments, it should accept only the required radio software modules.

When a single reconfigurable device is being used in different regions, the in radio certification rules will have to be distinguished. The details about the regions and access control rights could be encoded as part of its metainformation. There is a possibility that the overall policy followed for radio software authorization may vary. Thus, it is mandatory for the reconfigurable device to be aware of the location in which is it being used and accordingly switch to the corresponding radio software authorization policy, complemented by access control policies that extend the authorization policy to fine-grain levels. This would define the trusted parties who can authorize the software module and also the restrictions based on the evaluation of the software module's metainformation. Licensing-related information that limits the circumstances in which the software module will be used for licensing reasons may be included in the metainformation.

The reconfiguration manager may also authorize a software module; in this case, the device would ask the reconfiguration manager whether the specific module can be activated, following this procedure for each software module separately; alternately, a list of currently allowed software modules could be obtained.

8.4.4 Restricted Radio Execution Environment

An execution environment in which restrictions to access system functions are enforced needs to be controlled or managed. This would allow the downloaded radio software module to be isolated from other software and from user data. It is challenging to define meaningful restrictions to enforce conformance requirements, but this would have to be a compromise with the flexibility, as radio software needs to define and modify low-level radio communication behavior.

Approaches for restricted radio execution environment have been proposed by Sakaguchi [12] and by the SCOUT project [13]. Various control parameters, such as frequency, output, power, and bandwidth, can be validated to be within an authorized range. The actual radio emissions could be monitored and compared with reference data, which can be fixed or changed only with special restrictions. This needs to relate to the conformance constraints that the device enforces, independently of the currently executed radio software. The protocol behavior could be monitored by the device or the communication network. In the case of a rogue device, the reconfiguration software would be terminated, and the last correctly working software or a fixed failure mode configuration would be activated instead.

8.4.5 Reconfiguration Software Authorization Policy Framework

Digital signature for secure content download has been a well-known security mechanism used to protect download of radio software. The specifications for efficient use of known security mechanisms must be defined to implement the following policies:

- What is to be signed?
- Who is authorized to sign a download software module and authorize its use?

- Public key infrastructure required;
- Exact format.

The variation in the policies is not only according to local regulations but may also depend on the evolution of the regulatory rules, the specific market for which a reconfigurable device was intended, and the underlying business model. The policies might also change depending on the reconfiguration classes that differentiate between the properties to be modified.

8.5 Secure Execution Environment

The section has been divided into three subsections, which discuss the hardware support, software support, and equipment-related prototypes that provide security mechanisms.

8.5.1 Hardware Support

Reconfigurable equipment faces threats from attackers, who might succeed in getting unauthorized access and reconfigure the equipment according to their own needs, regardless of the restrictions.

To ensure the equipment security, the protocols, operating system, and hardware should be robust against all the threats. Persistent storage of sensitive data, mechanisms to guarantee memory integrity, random number generation, and cryptographic support are necessary. A secure environment makes it possible to load and execute operational software in a secure RAM. A secure OS mode would allow the execution of small amounts of security critical code to run as a monitored process. The protected reconfiguration software would also run in a secure OS mode.

However, a different cryptographic method would be needed to implement some of these features. In a traditional processor, there is a restricted user mode and a supervisor mode for unrestricted hardware accesses. The secure environment would provide methods for the usage of security hardware and software resources within the reconfiguration framework while protecting from security-threatening operations. The following are some key issues related to tamper-resistant and robust hardware platforms:

- A separate, secure coprocessor should be dedicated to process all sensitive information in the system.
- Sensitive information obtained from the secure coprocessor should be protected.
- Selected areas of memory subsystem should be designated as secure storage locations, protected from user space applications and certain kernel processes.
- Memory protection mechanisms and bus monitoring hardware that distinguish between legal and illegal access to the memory subsystem are needed to ensure the privacy and integrity of the data in the memory.
- Only the memory model should be executed [14].

8.5.2 Software Support

In the case of reconfigurable equipment, access control means the ability that explicitly enables or restricts the use of resources such as memory, hardware, or simply the
executable software modules. The policies define who or what has access to specific
system resources and the type of access permitted. It is challenging to manage large
distributed and networked elements where the borders of executed functionality are
diffusing.

A policy-neutral access control implementing a flexible security architecture
using the Mach microkernel was introduced earlier to support a wide range of policies in which the implemented security rules enforced by the system were defined by
a system component outside the microkernel. The Flask security architecture [15]
introduced mandatory access control within the Linux Kernel by composing two
authorization policies: "domain and type enforcement" [16] and "role-based access
control" [17].

Figure 8.5 shows the architecture of authorization policy manager. The policy
manager component forms the core of the reconfigurable authorization mechanism.
It also maintains a table for subject and object security contexts. A newly created
subject or object will be assigned a new security context according to the current
authorization model after which the permissions will be computed and assigned
depending on the authorization policy. It is then communication via several interfaces for checking, administering and reconfiguring authorization policies.

An optimized table of permissions indexed by pairs of subject and object security identifiers forms an access matrix. The administration component stores the

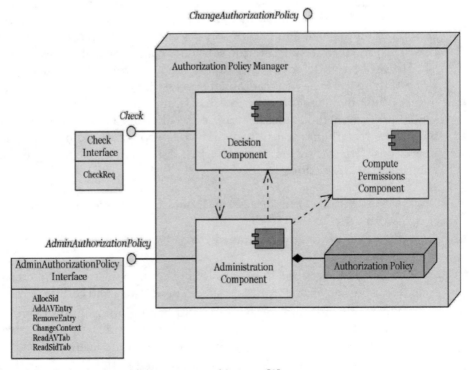

Figure 8.5 Authorization policy manager architecture [1].

access matrix and the tables of subjects and objects indexed by their security identi-fiers. This component serves as an interface to administer the authorization data-base and requires the check interface to verify whether the current subject has correct permissions to use the administration interface.

The decision component has to decide whether or not the current subject has access to the requested object. It obtains the permissions associated with the pair from the administration component. It has a check interface to check permis-sions and requires administration interface to extract permissions from the access matrix.

The authorization policy is defined by the compute permissions component. A function computes the permission, which is then included in the access matrix. If the authorization policy needs to be reconfigured, only the compute permissions com-ponent is changed, as it provides the compute interface that computes permissions, depending on the authorization policy and models. The administration and decision component are independent of the access control models and policies.

8.5.3   Equipment-Related Prototype Providing Security Mechanisms

A prototype covering the security aspects related to reconfigurable equipment has been developed to validate the concepts explained before. The reconfigurable equip-ment architecture consists of several logical and functional entities with dedicated tasks. An example for secure connection is explained in Figure 8.6.

An overview of the collaboration of the involved entities and their mapping to the prototype is given in [1]. The configuration management security module estab-lishes a secure connection through the configuration management module interface for network support services with the help of security manager. Public key infra-structure mechanisms are used, including reconfiguration manager discovery, the exchange of public keys, and the exchange of information for establishing a shared

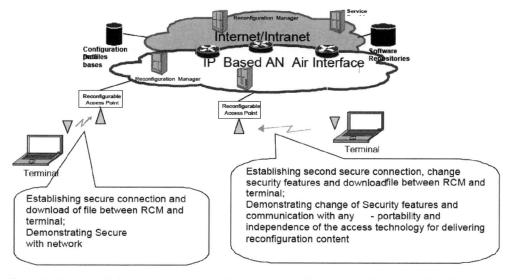

Figure 8.6   Example for secure connection between two different reconfiguration APs [1].

secret. The secure connection is available until the communication with the network is required, after which it is closed. To maintain the integrity, security, and confidentiality of the connection and data, a new connection is established when required. This approach is very good, since it has a very small overhead, making it more portable. A public key cryptographic method is used to establish a secure channel, and there is symmetric secure communication for data and signaling after a shared key is generated. Also, it has flexible security HW/SW management of the reconfiguration to support various algorithms.

8.6 Authentication and Trust Framework

Significant efforts in the areas of digital signature, authentication frameworks, and so forth are being made to assist the origin and identity of electronic data. Various aspects that are to be taken into account include:

- *Security of communications:* To ensure privacy, integrity, authentication;
- *Security of documents:* To ensure access rights, ownership, and so forth;
- *Security of program executions:* To protect against malicious programs.

The entities involved in reconfiguration need to be authenticated, whether it is done by downloading software or by modifying a set of parameters. For example, during reconfiguration, the visited network asks the home network to which the user is inherently authenticated regarding its authentication, and the reply from the home network would guarantee the authentication without disclosing any of the user's private information. Simultaneously, the home network would send to the equipment the necessary information that allows it to authenticate the visited network. The visited and home networks have a commercial agreement about the connection.

8.6.1 Security Infrastructures for Reconfiguration

Public key infrastructures are needed to ensure secure reconfiguration processes. One of the main advantages of using public key technologies is that the public key can be distributed, compared, and then used to authenticate the messages emitted by its stakeholder. However, there are some disadvantages:

- Relatively slow algorithms;
- Implicit trust relationships;
- Security of the private key;
- Architectural issues.

The following subsections discuss the authentication framework, identity framework, and access control infrastructures, which are some of the most relevant technologies.

8.6.1.1   Authentication Framework

The authentication of the reconfiguration entities uses certificates that allow asynchronous public key security mechanisms through certification paths. More details are available in E2R Deliverable D3.2 [18].

The two initial elements involved are as follows:

- *Certificate of the equipment manager:* Used to validate the reconfiguration processes involving the manufacturers' hardware, this certificate is hard-coded inside the equipment.
- *Certificate of the network operator:* Used to validate the reconfiguration processes involving network services and protocols, this certificate may be embedded inside the SIM card of the 2G/3G device.

Depending on the reconfiguration classmark and the items addressed by the reconfiguration process, one of the above mentioned certificates is used.

The execution environment providers and service providers have to authenticate the layers or underlying modules that they impact during reconfiguration. Otherwise, the authentication process can take place directly between the user application and the service configuration portal, using common technology for secure service access. When the reconfiguration process is performed through network interactions, the reconfiguration is done on-line. In that case, the process for authentication between reconfigurable device and the reconfiguration manager is as follows:

- The reconfiguration manager may verify the credentials for each of the network operator's subscribers that it can reconfigure.
- All the essential security checks are performed by the reconfiguration manager.
- Authentication of the communication between the reconfiguration manager and the limited device is done by using the certificate of the home network operator of the device's subscriber.
- Liability can be identified in the case of any failure, since all exchanges are logged.

It is a good solution, since it reduces the cost of security checks on the reconfigurable device to only two operations, which are as follows:

- Checking the validity of the certificate delivered by its home operator to the reconfiguration manager;
- Checking the correctness of the reconfiguration process by the reconfiguration manager.

In the case of off-line reconfiguration, the reconfiguration material needs to be securely packaged, along with the required certificates and authorizations, to enable a standalone authentication without requiring online access to authentication servers.

8.6.1.2 Identity Framework

The following entities are involved in a reconfiguration process:

- Device manufacturer;
- Home operator;
- Service providers;
- Users.

The manufacturer provides the cryptographic hardware components, the home operator and service providers are willing to be paid back by the owner of the reconfigurable device for the services being used, and the end users are willing to be provided with a panel of services and options that can be safely downloaded or installed on the device.

The basic authentication framework can be explained as follows: The user is registered and authenticated with only one authentication server, which recognizes the owner of the device and then allows access to services offered by third-party service providers.

This infrastructure is used for single operator, and it is challenging to extend it to domains that are managed by partner network operator for roaming services. A more global authentication framework, due to the complexity of the reconfiguration process and the diversity of the entities involved.

The concept of Identity Federation can be understood as follows:

- *WS-Federation:* The federation identity model is a high-level approach in which the low-level mechanisms for handling the authentications are not described. It covers the architecture and message exchange formats needed for managing identification and authentication in multiple domains. More details are available in a white paper, "Federation of Identities in a Web Services World," published by Microsoft and IBM in 2003 [19].
- *Liberty Identity Federation Framework (ID-FF):* The Liberty Alliance project [20] has defined an open standard designed for federated identity services for Web users and provides the following:
  - Interoperability between disparate identity systems;
  - Privacy and security for end users;
  - Scalability.

Figure 8.7 shows the four main modules that constitute the Liberty Alliance architecture, which are described briefly as follows:

- The Liberty ID-FF focuses on identity federation and management and can be used on heterogeneous platforms with all the current types of network devices as a standalone infrastructure or can be coupled with existing identity management systems.
- It is possible to adop and extend other industry standards, such as SAML, HTTP, SOAP, and so forth.

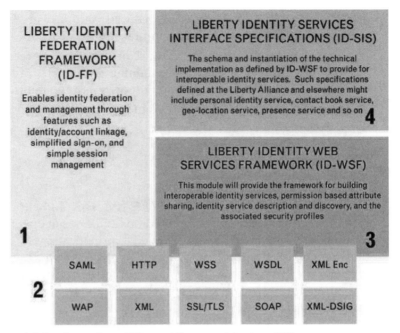

Figure 8.7 High-level overview of Liberty Alliance architecture [20].

- The Liberty Identity Web Services Framework offers a rich panel of functionalities, such as permission-based attribute sharing, identity service discovery, interaction service, security profiles, simple object access protocol binding, extended client support, and identity service templates.
- The Liberty Identity Services Interfaces Specifications enable interoperable services to benefit from identity services.

8.6.1.3 Access Control Infrastructures

The authentication, authorization, and accounting (AAA) frameworks provide suitable support for the management and control of a large number of users' subscriptions to a variety of services. With roaming and mobility on one hand, and third-party service providers on the other, new kinds of AAA protocols are needed to support the requirements.

IETF and 3GPP consortium have selected the Diameter protocol, which includes the following entities:

- *Relay agents:* These forward messages to an AAA server that is dedicated for managing the particular user.
- *Proxy agents:* These make local decisions based on policies related to resource usage and provisioning.
- *Redirect agents:* These send information on how to directly reach the AAA server in a requested domain, helping to reduce the configuration's information management.

- *Translator agents:* These translate requests and answers from legacy infrastructures into Diameter requests and responses.
- *Brokers:* These act as proxy, relay, or redirect agents between different administrative domains: AAA servers containing information about the end users and controlling authentication and authorization functions.
- *Accounting servers:* These register and collect information about the end users' service usage.

Diameter infrastructure uses some other infrastructures, which are briefly explained:

- *Extensible Authentication Protocol (EAP):* This authentication framework supports multiple authentication methods and can run directly over link layers without requiring an IP. It was designed to provide network access authentication when IP connectivity was not available. The key management framework supports the generation, transport, and usage of keying material generated by EAP authentication algorithms; the session keys could also be used for security association protocols.
- *Protocol for Carrying Authentication for Network Access (PANA):* This protocol was developed to provide a link-agnostic transport mechanism for carrying EAP-based network authentication information. This is because even though EAP was used in many different networks for authentication between clients and the network, its realization and implementation mostly depend on the type of underlying subnetwork. There is an implementation of PANA running UDP/IP: a dummy IP address is allocated to the entity who wants connection to the network, which allows connection to the authentication server; only after the authentication is a real IP address provided.

8.6.2 Trust Management and Dependencies

The technologies that enable security properties for the electronic world and the relationships between electronic entities are two complementary domains constituting the objectives of security. The following subsections discuss the mapping of the security and trust to the real world and the requirements that enforce the practical realization of technologies such as protocols, frameworks, architectures, and so forth.

8.6.2.1 Delegation Between Market Model Entities

In a market mode, the links between the entities are very important for security issues. Hence, the security model should be designed to follow the market model. This is because when an entity is liable for some interaction, the security technologies can check and ensure the liabilities. If this is not done, the entities can limit the market deployment if they are made responsible for authentication and authorization without any flexibility. For example, if the network operator has to certify each service deployed in its network, it would slow down the deployment of new value-added services, which would need to wait for the operator to certify them.

The following are bases for authentication:

- *Device manufacturer:* Responsible for the proper operations related to the hardware components;
- *Home operator:* Associated with the currently active subscription, which is responsible for the proper operation of the subscribed services;
- *Owner/user:* Decides the options and services to be used, installed, or customized on the reconfigurable device.

The following are the main entities involved in trust relationships:

- *Network operators:* They ensure the reliability and availability of their network. In the case of reconfigurable terminals, malicious software designers can create special patches to exploit protocol breaches or to use some services without paying for them. To protect the network from such misuse, the operators have to trust the communication stack of the mobile device, the device manufacturer (this means that the OS should not accept third-party software modules without any authorization record), and also the software providers that provide patches for the radio and communication layers.
- *Users:* Reliability of the device and security of their information is a must. They are interested in naturally safe reconfiguration or one that requires their explicit approval.
- *Service providers:* They need to trust the authentication mechanisms of the operator, as they have to share the authentication information with the operators. This helps to simplify the user's behavior, as they don't need to remember passwords or identification tokens for each service provider, and only one bill is produced for all the services.

8.6.2.2   Perceived Trust

The users need to have faith in the security of the reconfiguration process if they are going to use it. Reconfigurability and flexibility are key issues in security architecture. Some requirements are listed as follows:

- When a breach in the cryptographic algorithm is detected, the algorithm can either be strengthened or replaced.
- In case security material such as private or public keys has been compromised, the material should be changed securely, possibly in a coherent way.
- To deal with unexpected reconfiguration issues, a fail-safe mode and a reset operation might be used.

The reconfiguration process may be classified according to its user interactions as follows:

- *When no user interaction is requested:* This includes automatic reconfiguration, usually joint reconfiguration between the user device and the network(s). It is frequent and concerned with the reconfiguration parameters.

- *When user preferences are stored:* The stored preferences are used before and during the reconfiguration to customize the process accordingly.
- *When user interaction is requested:* This includes heavy reconfiguration processes or cases not covered by the user preferences. The request needs to be defined precisely and be user-friendly for the benefitof nontechnical users.

A fully secured OS ensures the correctness of operation and controls the interactions between the additional modules. Hence, the users will perceive the terminal as reliable. The Java technology includes mechanisms that can identify the codes by their origin. OSGi platform, a framework for installing and running distributed components, makes use of these mechanisms. The access from a code component to other code components is controlled by policies that ensure that the reliability of the terminal is maintained, irrespective of the installations of third-party modules —including those from nontrusted sources.

8.6.2.3   Comparative Analysis of Security, Usability, and Flexibility

Generally the security solutions are a compromise between the flexibility and usability of the system. Thus, depending on the level of security required for a certain operation, the security mechanisms are adapted accordingly.

Considering the terminal viewpoint, a secure OS would be the best practical basement, as all the reconfiguration requests are addressed to it and it is also aware of the level of security needed.

From the network viewpoint, the main challenge is to authenticate the user. This is because in reconfigurable terminals, it maybe possible to impersonate another user or there may be a situation in which the authorized terminals contain harmful software for the network. The following two approaches can be considered:

- *Network trusts the terminal certificates and secure OS:* The network would trust it to run only validated modules or functionalities related to network use, and also to have acquired the approval of the concerned modules or functionalities.
- *Network does not trust the terminal OS:* In this case, the terminal maybe connected through an external peripheral that can be considered to be secure. The network has no means to ensure the terminal will behave correctly if no single trust hardware module is detected. Users are identified with the authentication parameters provided by the operator, and misuses or malfunctions during an operation can be detected with the help of log traces so that appropriate action may be taken.

8.7   Challenges in the Reconfiguration Process

The network takes control of the reconfiguration process in situations where reconfiguration is too complex for the end users to understand, when the required information is not available to them, or when reconfigurations occur so often that it is inconvenient for the end users to be directly involved. If a single centralized reconfiguration manager takes control, then it should be ensured that only the single trusted reconfiguration manager can define and modify the device configuration. This demands that the communication between the reconfiguration manager and the reconfigurable device be protected. For this purpose, security protocols such as IPsec or SSL/ TLS maybe used.

The dynamic adaptation of the device configuration according to changing network conditions should correspond to the capabilities, preferences, and dynamic properties of the network currently used. This may require the currently used network to modify the current configuration of the accessing devices. A visiting network may either signal the intended changes towards the central reconfiguration manager placed in the home domain, or it maybe granted direct access so it can perform intended reconfigurations involving the home reconfiguration manager. For roaming users, it is a case of decentralized reconfiguration control in which not only can a single reconfiguration manager modify the configuration, but also the visited network can implement local adaptations by itself. This would allow the visited network to perform a coordinated reconfiguration of end-user and infrastructure equipment.

Reconfiguration Control is carried out by the reconfiguration manager, which defines and modifies the configuration of the reconfigurable device. The distributed control over the device configuration, which involves the delegation of authority to a visited domains' reconfiguration manager, is shown in Figure 8.8. The device functions are monitored, either in the network or by a watchdog, to detect operating terminals that are not behaving correctly. When a misbehaving device is detected, the network may exclude or penalize it to ensure proper operation of other devices [13]. Since the same reconfigurable device can be used in different environments, the device should be configured correctly for use in these environments. For

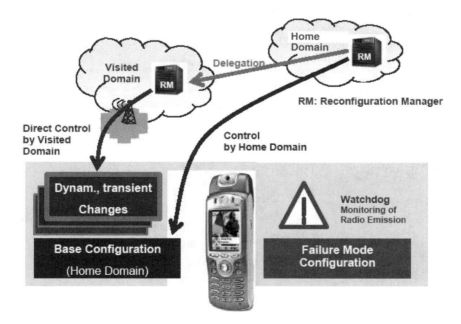

Figure 8.8   Overview of reconfiguration control.

this purpose, decentralized reconfiguration control on the device configuration of roaming users is provided [21]. Granting permissions to a visited domain reconfiguration manager is defined to realize the delegation by the home domain toward the visited domain. Several possibilities of transient configuration profiles exist for different visited network domains. The transient profiles are associated with specific networks and active only as long as that network is being used. This approach would ensure that the changes made by a specific network are in effect only as long as that network is being used and then automatically rolled back as soon as a different network is used (also useful in fault management). This way, the dynamic and transient changes to the device configuration have no effect on the configuration when a different network is used. Automatically established configuration profiles can be preserved so that when the device is used later with the same network, the proper configuration is already available and network is not required to perform parameter settings or download the software again. The use of transient configuration profiles increases the reliability and robustness. The worst effect that the specific network with a badly operating or malicious reconfiguration manager can achieve is that the affected user will not be able to use that network correctly or at all. The correct operation with the other networks is not impaired. More details on configuration profiles are given in [1].

Configuration Registration is carried out to support fault management. The registration server attests, using a digitally signed document, that the configuration is registered as such and that only duly registered configurations are activated. A configuration is registered only once, regardless of the number of devices on which it is installed. This helps to determine the responsible entity and the configuration that caused the fault when a faulty operation takes place. A network-based history function is used to determine other devices that use the same faulty configuration and also to identify the reconfiguration managers who provided the faulty configuration. Figure 8.9 shows a mobile communication system with reconfigurable terminal (R-T), reconfigurable base station (R-BS), reconfiguration manager (RM), download server (DL), registration server (Reg), history server (His). The reconfiguration

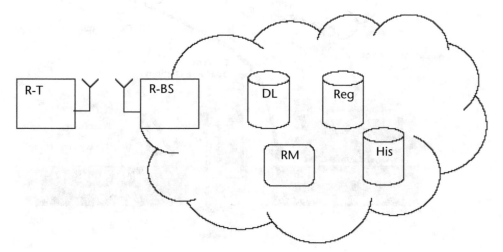

Figure 8.9 Mobile communication system for configuration registration.

manager can define a new configuration using reconfiguration software modules available on the download server. The message flow for this process is given in [1].

Configuration validation of an intended new configuration is carried out in the network to check whether it can be expected to operate correctly on the specific target device and in the current network environment. This makes it possible to detect software and configurations that would not work as expected before they are actually downloaded.

The FP6 IST project E2R has planned several validation checks that can be performed; a subset of those can be selected which are appropriate and required, depending on the type of reconfiguration, target device, and kind of software. These pragmatic checks are to detect and prevent reconfiguration attempts that can fail later; however, they do not guarantee the correct operation. The final responsibility for correct and conformant operation lies with the target device. Some examples of such validation checks include comparison of an identifier of the indicated target execution environment with the actual target environment; scanning the software calls to manufacturer-specific APIs; mimicking the signature validation checks as they will be performed by the target device; and so on. The validation functions can also perform checks that are not practical for the mobile device itself to perform. This function could be responsible for checking the revocation status of used certificates.

Due to increased reconfigurable functionality, including adaptation of lower communication layers and open reconfigurable platforms, the risk of compatibility problems of some specific software/configuration on some specific device and some specific network increases. The validation function complements the rather static security and compatibility checks, which could be performed by the target device itself with more centralized, network-based validation functions that can be updated dynamically with information about most recent software versions and device types. This function also allows "self-learning," as software configurations that do not work or are not reliable can be avoided on other devices of the same type. Flexible configuration validation [22] is useful when different kinds of reconfiguration occur, depending in the functionality to be reconfigured and the reconfiguration capabilities of the device to be reconfigured. These checks are of different complexity and affect different communication layers and functionalities; therefore, specific checks may be applicable for some reconfigurations only, or may be device specific. Besides the installation-related checks, the statistical data of the operation of certain hardware/software combinations can be exploited, for example, to prevent combinations that crash and lead to initiation of fault management procedures as activation of fallback configuration.

8.8 Conclusions

While reconfigurability enables the users, manufacturers, and network operators to modify radios after they have been manufactured, it is a technology that brings a substantial amount of risk related to security, privacy, and trust to communications and the associated applications.

Therefore, this aspect was investigated deeply within the frames of the EU-funded FP6 projects targeting advances of the state-of-the-art in the area of reconfigurability.

This chapter presented a summary of possible threats related to reconfigurability, based on a in-depth threat analysis performed by the FP6 IST project E2R, and the proposed solutions to ensuring the security, privacy, and trust of the communication and application process.

Tampering with reconfigurable software can be prevented by introducing an architecture with a single RM. However, a solution developing security concepts for more open and decentralized approaches to reconfiguration should be the next scientific objective to strengthen the foundation for next-generation wireless communication, in which radios can sense their environment and autonomously change their behavior to optimize user preferences or network efficiency.

Reconfigurable devices will need integrity measurements and a secure boot process to ensure that trust and privacy policies are enforced at all times. Additionally, in an environment in which rapid reconfiguration is possible, network operators and individual radio users will want a level of assurance that the radios with which they are communicating have appropriate configurations; therefore, radios will need what is termed *attestation* to transmit that information.

References

[1] IST project E2R, "Reconfiguration Security Concepts," Deliverable D3.5, January 2006.

[2] IST project E2R, "Business Model Elaboration and Roadmap," Deliverable D1.2, June 2005, http://e2r2.motlabs.com

[3] Housley, R. et al, "Internet X.509 Public Key Infrastructure Certificate and Certificate Revocation List (CRL) Profile," RFC3280, April 2002.

[4] Myers, M. et al, "X.509 Internet Public Key Infrastructure Online Certificate Status Protocol – OCSP," RFC2560, June 1999.

[5] Digital Signature Standard (DSS), Federal Information Processing Standards (FIPS) Publication, May 1994.

[6] Housley, R., "Cryptographic Message Syntax (CMS)," RFC3369, August 2002.

[7] Sun Microsystems, "Mobile Information Device Profile, v2.0," JSR-118, 2002.

[8] Microsoft MSDN, "Introduction to Code Signing," Microsoft Authenticode.

[9] Sander Siezen, "Symbian OS Version 9.1 Functional Description," version 1.1, February 2005.

[10] Housley, R., "Using Cryptographic Message Syntax (CMS) to Protect Firmware Packages," RFC4108, August 2005.

[11] 3GPP 23.057, "Mobile Execution Environment (MExE)," v6.2.0, 2003-9.

[12] Sakaguchi, K. et al., "ACU and RSM Based Radio Spectrum Management for Realisation of Flexible Software Defined Radio World," *IEICE Trans. Comm.*, Vol. E86-B, No. 12, pp. 3417–3424, December 2003.

[13] IST Project SCOUT, "Architecture, Functions and Security Analysis and Traffic Management Schemes for IP-Based Mobile Networks and Reconfigurable Terminals in Cellular and Ad-Hoc Networks," Deliverable D4.1.2, January 2004, http://www.ist-scout.org

[14] Lie, D. et al., "Architectural Support for Copy and Tamper Resistant Software," in Proc. ACM Architectural Support for Programming Languages and Operating Systems, (ASPLOS), pp. 168-177, 2000.

[15]   Spencer, R. et al., "The Flask Security Architecture: System Support for Diverse Security Policies," Technical Report UUCS-98-014, University of Utah, U.S.A. August 1998.

[16]   Badger, L. et al., "Practical Domain and Type Enforcement for UNIX," IEEE Symposium on Security and Privacy (S&P), 1995.

[17]   Sandhu, R. et al., "Role-Based Access Control Models," *IEEE Trans. on Computers*, Vol. 29, No. 2, 1996.

[18]   IST project E2R "Draft Reconfiguration Management Plane and Design of Network Support Functions and Signalling for Terminal and Network Element Reconfiguration," Deliverable D3.2, October 2004.

[19]   Microsoft, IBM, "Federation of Identities in a Web Services World," white paper, 2003.

[20]   Liberty Alliance Identity Architecture white paper, March 2003.

[21]   IST project E2R, "State-of-the-Art and Outlooks for Reconfiguration and Download Procedures, Network Support Functions, Protocol Architectures, Flexible Service Provision, and Enabling Platforms,"Deliverable 3.1, June 2004, http://e2r2.motlabs.com/.

[22]   IST project E2R, "Draft Reconfiguration Management Plane and Design of Network Support Functions and Signaling for Terminal and Network Element Reconfiguration," Deliverable D3.2, October 2004, http://e2r2.motlabs.com/.

CHAPTER 9

Prototyping and Requirements of the Reconfigurable Platform

This chapter describes the development of prototypes undertaken by the EU-funded IST project active in this area for proof of concept of the reconfigurability solutions presented in this book. A prototyping environment is required for demonstrating key scenarios integrating important areas such as cognitive networks, reconfigurable terminals, enhanced radio resource and spectrum efficiency, and dynamic and robust reconfigurations. Further, this chapter discusses the interfaces and specifications needed by the prototypes developed.

Various hardware and software components constitute a prototype. The hardware components range from those used in a reconfigurable mobile terminal to a multistandard base station design and a number of commercial off-the-shelf (COTS) components. The software components are used for reconfiguration, control, and service of the provisioning system, including the aspects of profile management, service discovery, and so forth, complemented by aspects of physical later emulation, radio resource management, and overall mobility support, such as proxy duplication, merging servers, and so forth.

The major contributors to this chapter are the FP6 IST projects E2R II and 4MORE. The project E2R developed a mobile terminal (MT) demonstrator, which is based on a reconfigurable RF board and a baseband (BB) board, and prototype air interfaces for software-defined radio (SDR) and cognitive radio (CR). A dynamically reconfigurable SoC was designed for performance purposes at the BB side. Advanced functions, such as flexible low density parity codes (LDPC) or complex MIMO decoding can be programmed on this platform.

The project 4MORE developed original components for a MT platform demonstrator combined with a dedicated base station emulator. The prototype has original hardware components for the radio stages, with dedicated antennas, a specific RF front-end, and a baseband modem based on an innovative and flexible architecture on SoC.

9.1 Introduction

The following scenarios were considered for the demonstration of reconfiguration concepts as described in this book:

- Flexible service deployment support (flexible service creation, authentication, billing, etc.);
- Advanced radio resource and spectrum management of a multiRAT mobile terminal in a heterogeneous network including RATs;

- Mass upgrade with partial reconfiguration of network entities up to the base station and the mobile terminal, as well as partial reconfiguration of different layers in the protocol stack, including the physical layer;
- Selection of most appropriate protocol stack responding to a given user(s) and/or multiRAT access network needs, and switching to a new mode;
- End-to-end device management to optimize the users' perceived performance.

The above scenarios would guide the selection of core technology and the direction of future development in this area. In the IST project E2R II, a multiRAT reconfigurable terminal was built with COTS components and an SDR terminal was built with a flexible radio frequency front end, flexible analog-to-digital and digital-to-analog converters (ADCs/DACs), and a reconfigurable protocol stack based on SoC technology for the base band and physical layers.

Overall, the objectives of the prototype design can be summarized as follows: to develop multimode RAT technologies, to develop the terminal hardware for flexible spectrum management, to further specialize existing RATs to incorporate into the platforms, to develop open and flexible application programming interfaces (APIs), to integrate various components into a coherent system, to develop and/or integrate the joint resource management software and integrate it with the interRAT API, and finally to implement the scenarios to demonstrate the reconfigurability concepts.

The prototype designed in the IST project E2R II integrated the multiRAT mobile terminal and provided a flexible RF design that could be used for flexible spectrum management. The goal of E2R II is to overcome these limitations. The two example scenarios used as the proof of concept were called "women on a train" and "transparent radio." The first scenario is shown in Figure 9.1. It addresses issues such as reconfigurability of the MT, advanced RRM, dynamic network planning and management (DNPM), flexible spectrum management (FSM), and mobility and QoS management in a heterogeneous network.

The second scenario covers issues such as:

- *Vertical handover subscenario:* provides seamless functionalities in a heterogeneous network at both the MT and the network level. Logical entities such as common radio resource management (CRRM) and an RRM per radio access technology are used for this purpose.

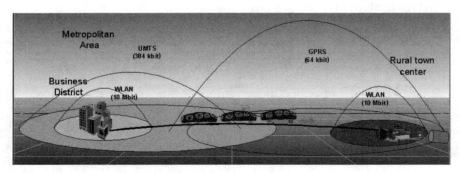

Figure 9.1 Proof-of-concept scenario for reconfigurability of applications: "Women on a Train."

- *Self-configuring protocol subscenario:* deals with a reconfiguration mechanism. It can be understood based on the following example. After a handover, the user experiences low QoS because of a wrong configuration of the protocol stack. A protocol reconfiguration process must be initialized by the decision-making mechanism. More technical details are given in E2R II Deliverable 5.1 [1]. The decision-making process uses all the static and dynamic information available as input and, considering the policies defined by various system stakeholders , it outputs a decision that is evaluated by the reconfiguration management procedure, which selects a suitable action for reconfiguration and conveys it to the remaining equipment management module.

- *Context-aware subscenario:* cognitive radio system that can adapt its behavior to its environment by analyzing the situation, making intelligent decisions per the established criteria, and using self-reconfiguration to adapt its functionality. High-level sensors would be required for sensing purposes in spectrum-related issues.

This scenario is shown in Figure 9.2.

More details of these scenarios and subscenarios including the demonstration are given in E2R II Deliverable 5.1 [1]. The MIMO aspects are not considered in the scenario, but they can be easily included.

The developed prototype platform was classified to demonstrate the reconfigurability concepts in three perspectives, namely: equipment, network, and adaptive applications.

9.2   Equipment Prototyping

For the successful development of a prototype platform, the different contributions, usually delivered by separate organizations, need to be integrated on a common platform. For the prototype developed by the FP6 IST project E2R II, a number of

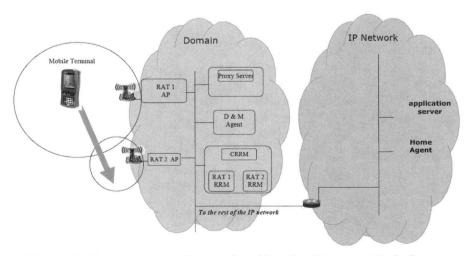

Figure 9.2   Proof-of-concept scenario for reconfigurable radio: "Transparent Radio."

organizations, such as SUPELEC, ACP, DICE, EURECOM, and CEA, provided individual contributions that were integrated on a SoC for a single demonstration of a mobile reconfigurable terminal as shown in Figure 9.3. Figure 9.4 shows the core test-bed platform. One of the fundamental limitations faced here was finding a suitable MAC.

Another challenge was interfacing between all the components. For example, additional hardware was needed to interface the DVB-H system with the CEA SoC.

To demonstrate reconfigurability in a heterogeneous environment, a platform for reconfiguration has been developed based around a core of COTS computing, and the RATs, together with the technologies and the platform, should provide sufficient interfaces to the heterogeneous RATs. APIs were developed specifically for this purpose to act as a convergence layer between discrete technologies and higher

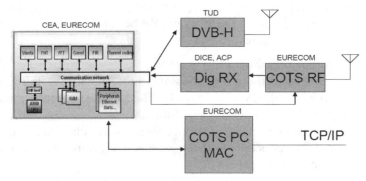

Figure 9.3   Integrating all contributions into a single mobile terminal demonstration in E2R II [1].

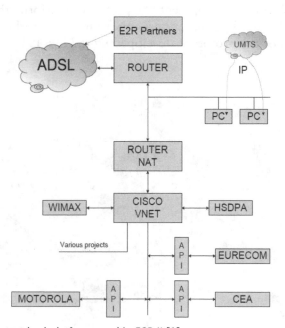

Figure 9.4   The core test-bed platform used in E2R II [1].

level software. The software aspects can be divided into reconfigurable stack, prototyping software, and emulation of missing hardware blocks.

In general, the requirements of a prototype mobile terminal can be summarized as follows:

- It should support double RAT transmission/ reception;
- It should integrate flexible RF/ ADAC/ BB/ protocol stack;
- It should cover the necessary layers required for duplex communication with the network;
- It should be as flexible/ modular as possible.

Figure 9.5 depicts the hardware architecture developed by the project E2R II. It co.nprises six modules; the RF board has four transmitters and four receiver chains that cover from 400 MHz to 6 GHz, including an analog and a digital interface with the rest of the system. The ADAC and BB board has four ADACs and two powerful FPGAs, and has interfaces with all the other components.

The following subsections describe the main components of the block diagram in Figure 9.5:

9.2.1   FAUST SoC

Faust SoC was developed by the IST project FAUST [1] as an advanced real-time prototyping platform for next-generation telecommunication standards. It provides an evolving platform with real-time features which distinguishes it from the FPGA-based platforms used for developing key concepts in SDR or cognitive radio. The features of the FAUST platform [1] are:

- Efficient performance through intensive computation units;
- Real-time support through QoS–aware communications between the computation units;

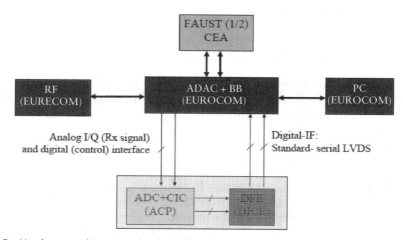

Figure 9.5   Hardware architecture developed by E2R II.

- Low power features for study of embedded features;
- Reconfigurability and flexibility of architecture through soft blocks; smart memory engines; and parameterizable, configurable, and reconfigurable IP blocks;
- Extension of communication backbone of the architecture outside the SoC for rapid prototyping of new functionalities inside FPGA or dedicated DSP;
- Support for 802.16e and 802.11n standards.

For seamless communications, the MT needs to be a multimode one—that is, it should be able to use two RATs simultaneously; or, while the stack of the second RATs is setting up communication with the base station, the first RAT must continue working to enable a seamless handover. The FAUST SoC provides the prototyping equipment with reconfigurable computing units that address the OFDM RATs. The network-on-chip (NOC) architecture allows flexible routing of data between the units and considers a communication scenario with mixed CDMA and OFDM in a MIMO scheme. The main features were implemented with high-performance computing through dedicated programmable OFDM blocks; CDMA blocks; programmable mapping and soft demapping; bit interleaving and deinterleaving; and powerful channel coding and decoding through turbocode and convolutional codes. Blocks such as the channel estimation, or DSP blocks like frame synchronization, were implanted as reconfigurable. The smart memory engines help to manage the streaming flow. It is expected that in the future, such an SoC technology will offer MIMO OFDMA computing blocks that can share resources between two completely independent computing flows at the same time [1]. The technology is based on dynamic reconfiguration support and will be well-adapted to multiple application support, including state-of-the-art low-power features to support SDR or CR scenarios with good power efficiency. It will also have highly reconfigurable blocks that allow mapping complex MIMO channel estimation, equalization, and advanced synchronization schemes. Such a development would require that around 2048 points, configurable OFDM blocks and channel coding/decoding must be added.

The FAUST platform was used also for the proof of concept within the IST project 4MORE and is shown in Figure 9.6.

9.2.2 Dual Band RF and ADDAC Board

A quasioptimal handover is possible with the help of duplication and merging approaches in which the data flow is sent from and to the MT through two radio interfaces simultaneously during the handover. For this kind of soft handover, two RF front ends and two entities of the protocol stack need to work in parallel. The following global functionalities are targeted: real-time working, a protocol stack that enables communication with the base station or the access point, the ability to communicate through two different RATs with two different bandwidths and frequency bands, the reconfigurability of each RAT, and the ability to switch seamlessly from one RAT to another.

Figure 9.7 shows the overall functional architecture for this block. The purpose of the generic radio access adaptation layer (GRAAL) module is to hide the hetero-

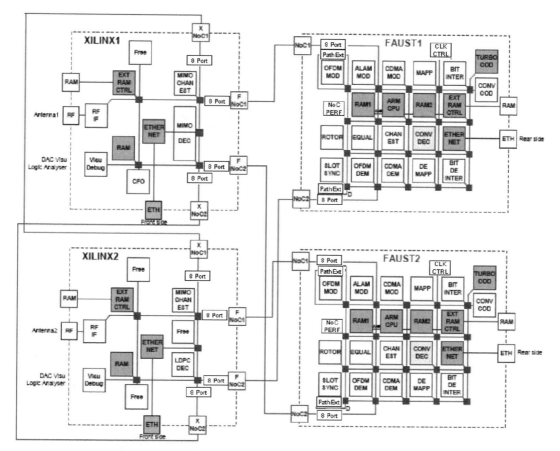

Figure 9.6 FAUST1 platform used in IST project 4MORE.

geneity of the RATs from the upper layers. The possibility of two instances of the protocol stack and RF-FE allows the MT to communicate through two different RATs during a vertical handover, or to communicate through one standard using MIMO.

Regarding the ADDAC board, it has a two-way real-time experimentation with reconfigurable broadband air interfaces and allows for experimenting with system-on-chip architectures for wireless communications. A high-level view of the ADDAC board is shown in Figure 9.8.

The RF input/output consist of two TX/RX duplexed MMCX connectors for external antennas, as shown in Figure 9.9.

The transmitter comprises a dual high-speed DAC (14-bit) with upsampling/filtering and frequency synthesis and low-speed DACs for controlling variable-gain amplifiers with a programmable configuration, a variable-gain direct upconversion chip with on-chip VCO and PLL, a low-loss 1900 MHz SAW filter, a 27-dBm power amplifier and two TX/RX switches. The receiver consists of an LNA, a SAW filter, a MAX2393 zero-IF receiver, differential unit-gain amplifiers, and a dual high-speed ADC with digital low-pass filtering and downsampling possibilities. A three-wire interface can be used to control the front-end gain settings through a software program running in the FPGA CPU. The RF chipsets derive all internal clocks

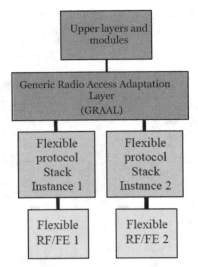

Figure 9.7   Functional architecture for double RAT transmission for seamless handovers.

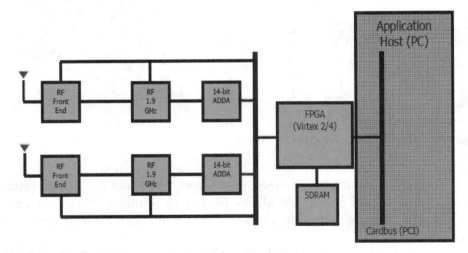

Figure 9.8   High-level view of the dual-RF CardBus/PCMCIA.

from 26 MHz and provide synthesized clocks to the dual high-speed ADC to control the sampling frequency of the base band receiver. There is an FPGA subsystem responsible for configuring the analog portion of the card; interfacing with the external ADC and DAC; and providing a high-speed interface with a laptop or embedded processing system equipped with a CardBus/PCMCIA slot, low-level signal processing, and 16-bit high-speed general-purpose I/O interface. The design makes use of Xilinx's EDK around MicroBlaze soft core and the CoreConnect OPB bus architecture. The card also supports external digital ports and external 4V 2A power supply for the baseband and RF portions of the circuit. A foreseen evolution of this RF board is shown in Figure 9.10.

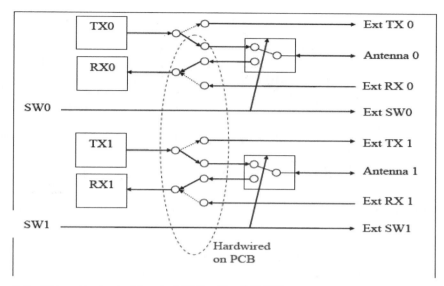

Figure 9.9 RF input/output of the dual-RF CardBus/PCMCIA.

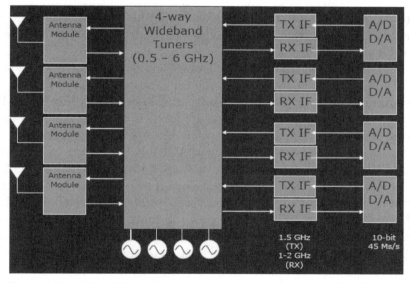

Figure 9.10 Evolution of the RF board to include four ADAC, two powerful FPGAs, and four TX and four RX RF chains with a digital interface with the rest of the system.

9.2.3 MT Local Functionalities

Some of the local functionalities targeted by the MT are: flexible or multiband antennas, a flexible RF-FE, a reconfigurable baseband, a reconfigurable proto-col-stack, and a logical entity that can handle communication through two different protocol stacks and interface the upper layers. Figure 9.11 shows possible hardware architecture for an SDR terminal that can support more than one band. The ability

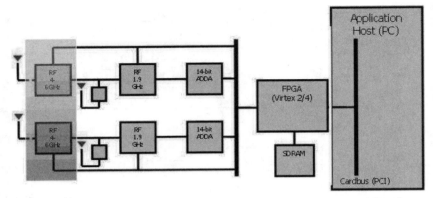

Figure 9.11   Possible HW architecture for a SDR terminal that can support multiband.

to receive and transmit data through two RATs simultaneously and dynamic reconfiguration are a must to be able to implement the proposed scenario.

9.2.4   ADC/Digital Front End

This block should be able to cope with the requirements of the targeted standards in
terms of number of bits and achievable throughput, and it should be dynamically
reconfigurable during a communication. An overview of the SDR receiver is shown
in Figure 9.12.

   If most of the dynamic range and selectivity requirements are pushed to the digital domain, it will boost the reconfiguration capabilities of the whole RFIC. The utilization of digital signal processing capabilities in an advanced CMOS technology
has advantages over a "pure analog" RFIC in terms of stability, reconfigurability,
power consumption, and costs.

   Figure 9.13 shows the proposed ACP/DICE daughter board in the E2R II project. The ADC-chip is assembled on a PCB together with an FPGA containing the
digital front end (DFE). The system clock is provided for the PLL, but there is also an
option for an external clock. There are several high-speed links to the FPGA, which
implements the digital signal processing functions and the developed interference
sensing techniques. The reconfiguration processes are triggered by the digital baseband section, and a flash memory is used to hold the FPGA configuration.

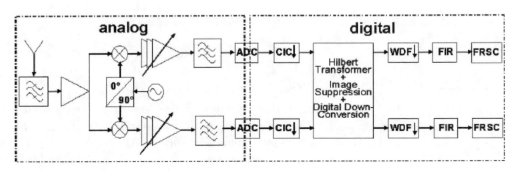

Figure 9.12   Overview of the SDR receiver.

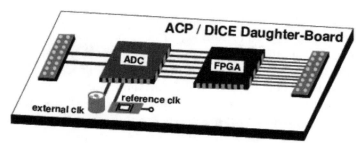

Figure 9.13 ACP/DICE daughter board as proposed in E2R II.

An oversampled $\Delta\Sigma$ technique may be used as shown in Figure 9.14; it offers a natural tradeoff between resolution and speed at low power consumption, in the range of milliwatts. This technique allows the filter specifications of the baseband filter preceding the ADC to be relaxed and offers choices for post-filtering. The discrete-time $\Delta\Sigma$ modulators consume more power than the continuous-time implementations, since they require amplifiers with high bandwidth to satisfy the settling requirements. But the former loop filters scale with the sampling frequency, unlike the latter, which is the main reason behind implementing the discrete-time $\Delta\Sigma$ ADC. The on-ship PLL generates different sampling frequencies for the different standards.

9.2.5 SAMIRA DSP

Some of the main challenges faced by digital signal processing system implementers of mobile communication systems are:

- Pace of product development;
- Increased system implementation complexity, as increasing number of standards are to be implemented;
- Flexibility issues, because the implemented standards need to be reconfigured or a complete RAT has to be changed even during runtime of the system;
- Efficiency issues;
- High performance and low battery consumption.

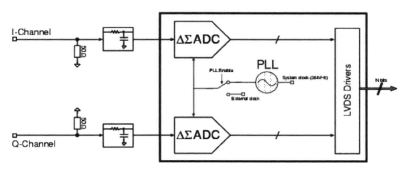

Figure 9.14 A based $\Delta\Sigma$ ADC for reconfigurable devices.

The architecture of the DSP is based on the STA, which is a high-performance, lower-power DSP architecture [2, 3]. The basic module is shown in Figure 9.15. Each module has an arbitrary number of input and output ports. The basic modules build the data processing part of the processor. Processor instructions control the functionalities of the basic modules and the input multiplexers. During each cycle, the instruction configures the multiplexing network and the functionality of the basic modules, hence forming a synchronous network. This architecture offers a high degree of data reusability, which speeds up the computations and also lowers the power consumption and the register file pressure.

The template architecture shown in Figure 9.16 supports data and instruction level parallelism; for the SAMIRA processor, the following design has been incorporated to offer a high degree of parallelism:

- Single instruction, multiple data (SIMD);
- Synchronous transfer architecture (STA);
- Specialized memory architecture;
- Compressed instruction words.

The simplicity and modularity of the STA concept enables the automatic generation of the RTL and simulation models of the processor codes, as well as the MATLAB compiler, from a machine description that enables the generation of processor cores with different characteristics, such as size of register file, memory capacity, interconnection network, functional units, data types, and amount of SIMD-vector parallelism, and enables the creation of heterogeneous and highly

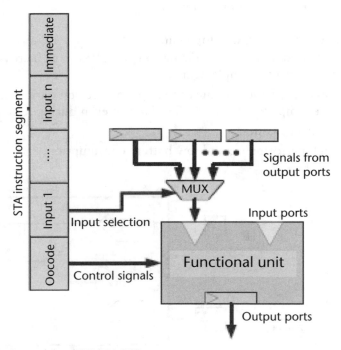

Figure 9.15   Basic module of SAMIRA DSP.

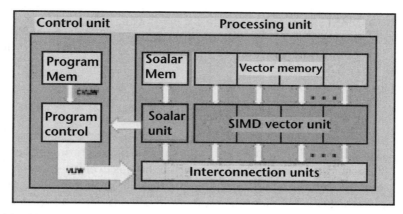

Figure 9.16   Generic architectural template used in wireless applications.

reconfigurable low-power processors capable for the mobile devices. The technical details of the implementation results are available in E2R II Deliverable 5.1 [1].

The SAMIRA DSP integration and interface overview are shown in Figure 9.17 and 9.18, respectively.

The DSP is accessed by the FAUST SoC via a wishbone to an NoC bridge placed in the FPGA, which is connected to the DSP through a wishbone bus [5]. NoC extension connects the FAUST SoC to the FPGA. In this manner, the DSP can be integrated to operate in the system as any other functional unit of the SoC.

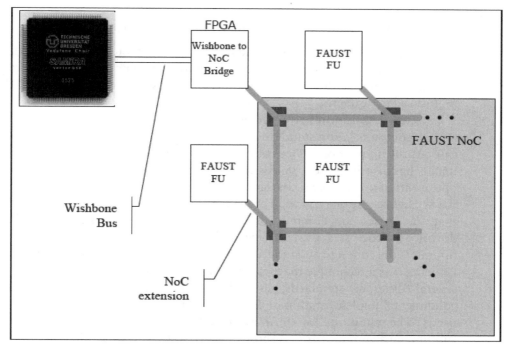

Figure 9.17   SAMIRA DSP integration.

Figure 9.18   SAMIRA interface overview.

The MASTER interfaces are cores capable of generating bus cycles, whereas the slave interfaces are cores that can receive the bus cycles.

For the demonstration purposes of the IST project E2R II, the SAMIRA DSP supplies the DVB-H functionality. It could be considered a broadcast technology for the platform that could be used as a traditional broadcast medium or as a multicast download channel for nontime-critical software downloads.

9.2.6 FPGA Dynamic Partial Reconfiguration

The partial reconfiguration of the FPGA is needed to save hardware resources and reconfiguration time. Pieces of adequate distributed hierarchical reconfiguration management that can cope with different levels of granularity can also be implemented. The FAUST NoC framework integrates this approach for few processing elements.

A few techniques and concepts that enable benefiting from partial reconfiguration of the FPGA for the SDR systems are presented in the following.

9.2.7 Parameterization and Common Operator Approach

A parameterization approach in the context of common operator use, along with hierarchical configuration management, is particularly important for heterogeneous multiprocessing platforms and reconfiguration scenarios required by the SDR multistandard systems.

Parameterization involves the identification of all the common aspects of the mobile communication standards with which the receiver is expected to communicate. As the number of services offered is increasing, the demand for multimode terminals has also increased. In this case, parameterization could be considered as an optimization process for reconfiguration. The common aspects of the different standards can become one common processing procedure and be stored on the device. In that way, when this common procedure is executed by a simple call during the download process, a gain of size and time can be achieved. Parameterization optimizes the codesign and reduces the time to market. Hence, parameterization can be defined as a technique that searches for and finds all the commonalities between several different standards, in order to optimize the resources during the equipments' implementation and/or execution process. The cost function involved needs to be minimized [6, 7]. An example for such multistandard decomposition in subfunctions and operators is shown in Figure 9.19. The detailed case of finding of the optimal path for the channelization, equalization, and OFDM demodulation for the FFT operator is available in E2R II Deliverable 5.1 [1].

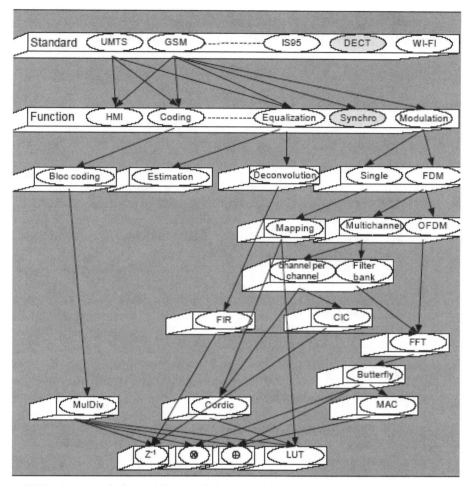

Figure 9.19   An example for multistandard decomposition in the parameterization process.

9.2.8   Hierarchical Management

The hardware in a software radio is expected to be heterogeneous to face the wide variety of processing categories and also switch from one processing mode to another. This means that the number of configuration contexts that must be managed increases as the number of standards increases, along with the number of different processing algorithms. A functional baseband analysis can classify each baseband function to reduce the overall configuration management complexity. Further, to optimize this process, the classification can be done in terms of parameters, requirements of hardware resources, and flexibility for reducing the number of configuration contexts. Then the classification of context-switching cases can be done to determine the granularity of the reconfiguration.

The multistandard functional analysis [8] initially studied the baseband functions of the uplink transmitter of three standards (GSM 900, UMTS UTRA/FDD, WLAN 802.11g mode) that involved all the conceptual challenges for defining configuration management architecture. These multistandard baseband functions, as illustrated in Figure 9.20, can be grouped into three functional classes with similar processing requirements for maximizing the reuse of hardware, as follows:

- *Coding class:* This includes functions such as cyclic redundancy check (CRC) coding, conventional clock and convolutional forward error correction (FEC) coding, and more advanced turbo-FEC coding, which have feasible software implementations but more power-efficient hardware implementations that consume less computing resources. A software implementation accelerated through power-efficient dedicated coprocessors could result in a suitable architecture.
- *Data structuring class:* This includes functions that perform all types of manipulations on data packets, such as concatenation, segmentation, and multiplexing. These impose significant requirements on memory usage. Special attention needs to be paid to the architectural design of the memory arrays through resizing, switching off unused memory blocks, and so forth to reduce the overall power consumption.
- *Modulation class:* This includes baseband functions performed just prior to the up-conversion to the carrier frequency. These functions consume a lot of computing resources and possibly could take advantage of dedicated configurable hardware accelerators.
- This would also help in configuration management, since the functions will be handled by few parameters that have to be adapted and fit into the requirements of a specific standard.

The purpose is to reduce the configuration management complexity by minimizing the resources to be reconfigured during any context switching. The reuse of the processing blocks should be enhanced as much as possible. The different types of application switches are:

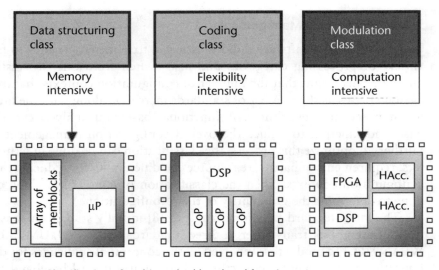

Figure 9.20   Classification of multistandard baseband functions.

- *Standard switching:* This is very demanding, since the transmitting chain between the two standards is extremely different; it would require the reconfiguration of almost all of the baseband processing. This implies the physical as well as the higher layers of the protocol stack.
- *Mode switching:* This does not involve the huge reconfiguration as in standard switch, and it can be considered as an intrastandard context switching, in which the mode parameters are usually defined by the MAC layer.
- *Service switching:* This can be performed by the parameterization and reconfiguration of some of the baseband functions. Then the partial reconfiguration of the transmitting chain is usually sufficient in this case.
- *Performance enhancement:* This type of reconfiguration needs to be carried out without disrupting the service for the user and is needed to reconfigure small parts of the processing chain to allow some performance enhancements or bug fixing.

9.2.9   Hierarchical Configuration Management Architecture

The functions of an SDR transmitting chain are mapped into several processing block units (PBUs), each of which is optimized using specific reconfigurable hardware resources. A configuration path split into several configuration manager units (CMUs) controls the reconfigurable processing path, in which each CMU is dedicated to a type of PBU and manages the configuration of a type of baseband function in the chain. This split allows the partial reconfiguration of the transmitting chain by an independent reconfiguration of each PBU. The design complexity is reduced by such distributed configuration management.

The hierarchical configuration management model shown in Figure 9.21 is needed to manage the multigranularity, if configuration is required by the different context switching.

The three levels of hierarchy are described briefly as follows:

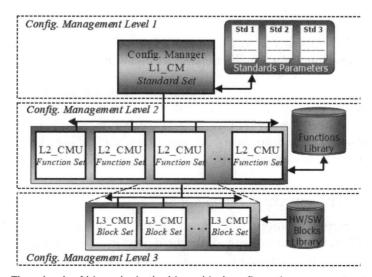

Figure 9.21   Three levels of hierarchy in the hierarchical configuration management model.

- *Hierarchical Level 1:* This level allows the control of category-specific functions to manage the parameters at the standard level. The configuration manager is responsible for choosing the functional units that constitute the entire configuration of the baseband processing chain. No hardware implementation is considered at this level, and generic functions are dealt as generic components.
- *Hierarchical Level 2:* At this level, the generic functions selected at Level 1 are parameterized in accordance with standard specifications. The configuration manager handles the set of attributes of each function to create each functional context of the entire processing chain.
- *Hierarchical Level 3:* The processing data-path architecture depends on reconfigurable computing resources. The processing path can be formed by different types of reconfigurable resources and can correspond to configurable accelerators, an array of DSP locks, or a fine-grained reconfigurable data path. The CMUs need to find the available processing resources and configure them to enable the execution of the functional context that has been created.

The hardware architecture is separated into three main hardware clusters to distinguish the configuration data path from the processing data path:

- *Coding cluster:* Though software implementations are possible, hardware will be more efficient and will consume fewer resources. Usually software implementations are underoptimized in this case, because the data width is only one bit and hardware implementation is necessary to reach the expected high-throughput performances. Software flexibility is also mandatory due to the wide variety of coding schemes. Hence, this cluster is composed of DSP aided by reconfigurable coaccelerators, accelerated through efficient dedicated coprocessors corresponding to a suitable architecture.
- *Data structuring cluster:* A power-saving architecture design is important in this case, since memory consumes a lot of power. The high flexibility offered by the GPP processors is useful to run the wide variety of handling functions for every standard.
- *Modulation cluster:* The intensive processing requirements of the modulation class functions usually necessitate a HW implementation.

A software radio usually comprises reconfigurable computing architectures, reprogrammable DSP, reconfigurable FPGA, and so forth. The heterogeneous platform constituted by GPP, DSP, and FPGA is presented here, along with the FPGA reconfiguration requirements. At present, only a few FPGA devices enable dynamic and partial reconfiguration [1].

9.2.10   FPGA Partial Reconfiguration

Partial reconfiguration of FPGA is important to be able to reconfigure a subpart of the FPGA. This feature can allow the design modules to be swapped dynamically. The module-based design flow is shown in Figure 9.22.

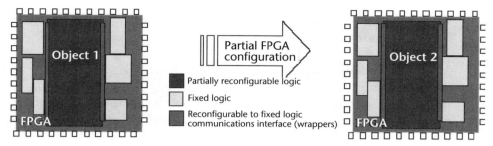

Figure 9.22 Module-based design flow.

Depending on the FPGA reconfiguration capabilities, there are two types of reconfiguration schemes: internal and external reconfiguration, which depend on the reconfiguration interface and the reconfiguration controller of the platform. The logic resources are separated into the fixed logic area and the dynamically reconfigurable. Partial reconfiguration can be managed from outside of the FPGA using an external controller such as a DSP. The internal reconfiguration enables autoreconfiguration. More details about their implementation are available in E2R II Deliverable 5.1 [1].

The partial reconfiguration feature is available on Xilinx Virtex families. Figure 9.23 shows the design flow, along with the design tools. The design description describes the HDL and synthesizes the application into modules following the partial reconfirmation design guidelines. The initial budgeting prepares the floor plan of the design and defines the time and area constraints as well as the partially reconfigurable and static areas. The implementation involves place and route procedures.

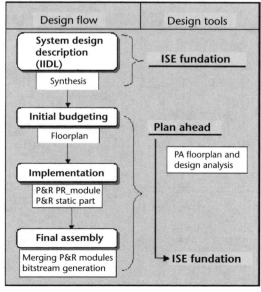

Figure 9.23 Partial reconfiguration using Xilinx tools.

The final assembly produces the top-level description and also generates the configuration files.

In the next generation of Xilinx, the partial reconfiguration flow will be based on a modular design methodology that will allow top-level description of the application to be separated into independent modules.

One of the main challenges is the design of the interface between two modules, one reconfigurable and the other fixed. The control of the routing resources of the signals that cross the two modules over the several contexts of the reconfigurable subparts is critical. The first internal connection pattern to design the interfaces between the reconfigurable module and another part of the design is shown in Figure 9.24.

9.2.11 Common Operator-Oriented Design for FPGA Partial Reconfiguration

This design separates the implementation of the operator processing part and the operator control part. The control part is located in a reconfigurable area inside the FPGA, and the operator part is static. Partial reconfiguration allows the parallel processing of operators to increase; however, it requires a nonoptimal logic resources allocation. This methodology optimizes the bitstream size for partial reconfiguration and also the delay associated with it, which is important for real-time reconfiguration needs. A local reconfiguration optimization can have a deep, beneficial impact on global handover efficiency.

9.2.12 Reconfiguration Concepts for the Physical Layer of an MSBS

Figure 9.25 shows the concept of MSBS, in which the scenario can be sketched through three main steps. On initialization, the CMM triggers the download of RAT 1 modules. Then, the GUI or external manager orders the deployment of RAT 2 through CMM. Finally, the GUI or external manager orders through the CMM, the removal of RAT 1.

A block diagram of a SDR base station is given in Figure 9.26.

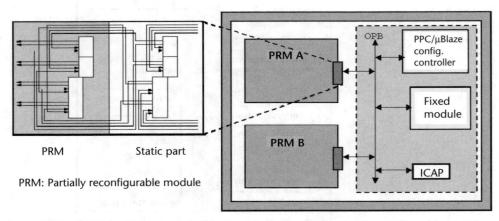

Figure 9.24 LUT-based bus macro for interfaces between the fixed and reconfigurable parts in Xilinx FPGA.

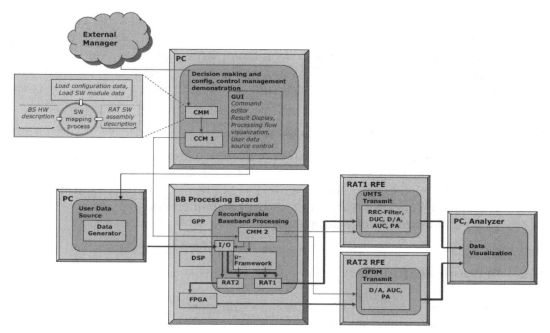

Figure 9.25 Demonstrator scenario for MSBS.

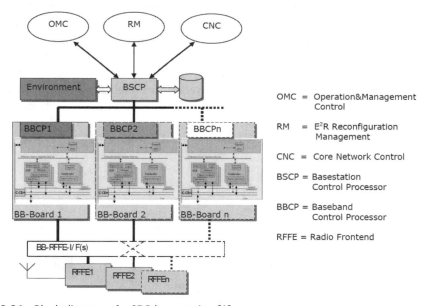

Figure 9.26 Block diagram of a SDR base station [1].

The operation and management control (OMC), reconfiguration manager (RM), and core network control (CNC) instruct the base station control processor (BSCP) which translates it accordingly to the underlying protocol stacks, telecommunication layers, and other environment controls, as well as the BBs.

The PoC prototype platform as proposed in E2R II is shown in Figure 9.27. It validates the mechanisms partly derived from the OMG platform-independent model as a reliable and economic solution for RAT deployment, and also demonstrates scenario elements, such as bug fixing and algorithm enhancement. More details are available in E2R II Deliverable 5.1 [1].

9.2.13   Detection of Vacant Radio Resources

Local scanning of the spectrum can be performed to detect spectrum use in the environment of the radio system. To accelerate the scanning, a priori information about the radio system can be used. A network entity that detects the vacant spectrum is used, and it works in collaboration with other multiple copies of the network entity to avoid the problem of hidden nodes.

9.3   Network Prototyping

Network prototyping in the context of reconfigurability was used to develop the necessary tools for demonstrating the cognitive network and joint radio resource management (JRRM) concepts. This may include network, device, and spectrum management issues as well as the identification of spectrum holes, mode monitoring, and resources brokerage between domains.

The purpose of the JRRM software is to optimize the operable spectrum capacity and end-user QoS in heterogeneous networks, as well as to manage the handoff between the technologies [1]. The issues concerning spectrum management may include a system in which the RAT adapts itself based on the knowledge of its environment obtained from centralized systems, along with the scenarios where the RAT investigates the environment and adapts itself accordingly. or maybe a combination of both such systems. Here, we mainly cover the aspects under the control of a centralized system, because the distributed management issues were dealt with in equipment prototyping, as it relies on local implementation. For a centralized cogni-

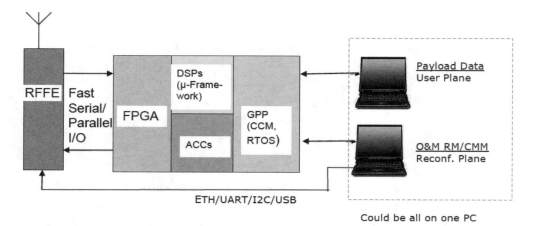

Figure 9.27   PoC prototype proposed in E2R II.

tive network and resource management system, a centralized server and the means for communication with various RATs is one of the major issues. A generic API can serve as a medium for communication for the heterogeneous systems. An open API will allow a rich set of messages to be passed between the entities on the core-network; it should be flexible and extensible to enable the inclusion of future unknown RATs into the system. It is important to address the network resources necessary for mode monitoring and brokerage between domains that allow near-optimal use of the available spectrum, which takes into account the cost of the user, guarantee of QoS, and number of users subscribed to the different systems.

Radio resource management (RRM) handles all the functionalities needed for air interface resources in mobile systems. Optimal use of RRM allows optimum coverage and maximum planned capacity, and it also guarantees the required QoS. For efficient use of spectrum, it is important to accurately estimate or predict the load and traffic demand. If the deviations from the expected are large, appropriate incentives can be applied to meet the demand. Device management allows the application of user-level policies that determine the manner in which the terminal functions and include aspects such as device security, reliability, power management, routing, and over-the-air reconfiguration. Device management software was developed for the E2R II prototype, which includes the terminal side application and policy service in addition to a network-side server. Aspects such as device security, including end-to-end security implementation, user policy management, and terminal reliability, are some of the covered issues.

The following subsections are about entities that are part of the networking prototype.

9.3.1   Reconfiguration Control and Service Provisioning Manager (RCSPM)

This entity works from the network side and acts as a service/software/content aggregator and provider. It contributes to the overall scenario as follows:

- It supports dynamic user registration.
- It provides a mechanism for advanced service discovery.
- It provides the means for dynamic reconfiguration of services, protocol stacks, and network components.
- It supports the deployment of flexible business models with novel dynamic services,
- It supports secure usage value-added services (VAS).
- It support profile management issues.

Figure 9.28 gives a technical view of the RCSPM as it was implemented in the E2R II prototype. All the modules are implemented in Java; RMI is used for intramodule communication. External actors, such as manufacturers, operators, and service providers, gain access to the platform by employing the Web interface implemented and provided by the repository management modules, which allow the actors to add, delete, or update information from the databases. The user uses the simple object access protocol [9] and invokes the interface implemented through the user interaction management module (UIMM) to access the server.

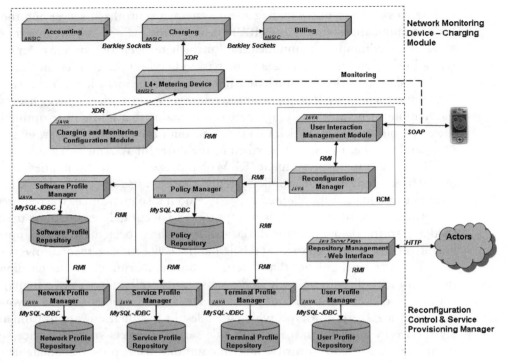

Figure 9.28 Technical overview of RCSPM implemented in E2R II.

Some of the modules are described briefly here, but more detailed information is available in E2R II Deliverable 5.1 [1].

- *Reconfiguration control manager (RCM)*: This is the central module that receives information from all the other modules; it basically encompasses and enforces policy rules for the reconfiguration and service-provisioning process.
- *EUT registration module (EUTRegM)*: The terminal manufactures or other entities may use this module to register a new type of terminal in the platform. In case of demand, the terminal capabilities derived from the registration process will be communicated to the RCM to support the service provisioning and reconfiguration processes.
- *Net registration module (NetRegM)*: This module is used for the insertion, deletion, and update of various TCP adaptation profiles.
- Software registration module (SoftRegM): The terminal manufacturers and/or other authorized entities use this module to register a new patch/bug to fix/upgrade for a specific type of terminal. The RCM may exploit the list of software for each kind of terminal for a network or perform a terminal-initiated reconfiguration.
- *User registration module (UserRegM)*: The network operators use this module to register new users who are able to alter the default user profile by themselves. RCM may exploit this information to support the service provisioning and reconfiguration processes.

- *VAS registration module (VASRegM)*: Service providers use this module to register a new service or modify the properties of an existing service. The type of service, charging scheme, hardware requirements, and so forth are registered here, and the module communicates the information to the RCM to support service provisioning and the application layer reconfiguration process.
- *Layer 4 and CAB configuration module (L4CABCM)*: This module is responsible for interacting with the Layer-4 metering system and the CAB module.

Some of the restrictions faced in this approach relate to the following issues:

- *Operating system*;
- *Server host network residence*: Either public static or a static NAT IP address must be provided to the machine that hosts the server;
- *Ports*: All the ports used for HTTP trafficking must be allowed and the ports used for FTP trafficking should not be blocked.

9.3.2   RCSPM User Agent

This entity manages the application layer of user equipment and is designed to provide support for the negotiation, download, and execution of application components and communication protocols. It contributes to the scenario as follows:

- It manages the process for service provision and adaptation.
- It manages the interaction with the other components of the RCSPM platform.
- It manages the download procedure of protocol components.
- It manages the service execution at the terminal part and supports the profile management.

The client software is a process/service that is activated at the terminal after startup and serves as the main communication channel between a terminal and the network system. The user component (UC) defines the needed interfaces, and the communication component (CC) implements these interfaces and interacts with the server via the SOAP protocol for interoperability issues and the remaining architectural entities locally. Figure 9.29 gives a technical view of the client implemented in E2R II using Java. There are two client implementations; one is done using J2SE and targets high-capability devices, whereas the second is designed for mobile devices and implemented in J2ME [10].

The interfaces are described here briefly, but the implementation was described in E2R II Deliverable 5.1 [1] in detail. The CC component acts as a SOAP client and as a SOAP listener. The UC component supports as a friendly user interface and an abstract programming interface, which enables the communication with any downloaded service and is also responsible for downloading and installing the appropriate patches. The resource manager periodically checks the CPU usage and the battery level of the mobile device. When the CMM_DMP receives the notification consisting of the patch's resource profile and a request for installation confirmation, it accesses the file system, which enables the client to retrieve the CPU and battery

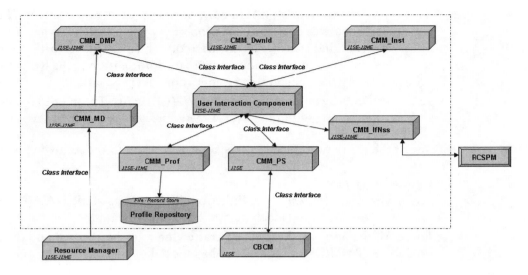

Figure 9.29   Technical overview of RCSPM client implemented in E2R II using Java.

values representing the current resource profile of the terminal. The integration scenario for the resource manager and the DMP is shown in Figure 9.30. The resource manager constantly monitors the hardware resources of the device and provides information to the decision-making module, whereas the UMTS manager triggers the subsequent events after being notified.

The CCM_PS module is responsible for the control and supervision of the protocol layer reconfiguration process and has two main functions:

- The dynamic binding of the protocol components within each protocol layer;
- The dynamic replacement of one protocol component with another.

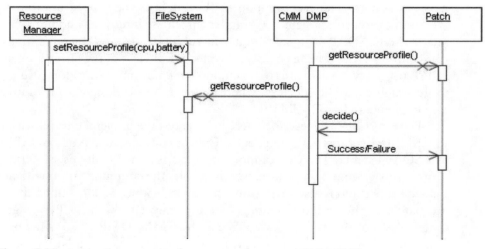

Figure 9.30   Integration scenario of resource manager and CMM_DMP.

The requirements imposed by the reconfiguration concept introduce great complexity, which should not affect the reliable operation of the protocol stack. Mechanisms that are able to seamlessly control the protocol reconfiguration process are necessary, and some proposed by E2R II are [11]:

• A component binding control module entity within each protocol. This is responsible for realizing the dynamic composition of the protocol components after verifying that composition can indeed take place. It also controls the process of dynamic replacement of protocol components and performs the integration and replacement of protocol components while simultaneously it controls and coordinates the support functions to successfully achieve the switch to the new module. This allows seamless dynamic replacement.
• A semantic layer of information for each protocol component [12].

The metadata information includes a unique identifier, the version of the component and the composition information of the components, which enables the component binding control module (CBCM) to verify the proposed binding of the components. This information helps to identify each protocol component and to prevent the binding of incompatible components. The protocol components should be defined so that when integrated, the module implements the functionality of the specified protocol. A generic reconfigurable component interface is required to represent the reconfigurable component.

The CCM_PS module controls and supervises the protocol layer reconfiguration process and communicates with the component binding control module to realize the following functions:

• Dynamic binding of protocol components;
• Dynamic replacement of one protocol component with another.

The two main restrictions imposed by this approach are to the operating system and the network residence. The device needs to be able to support the download and execution of Java applications and should be able at least to perform HTTP browsing.

9.3.3   ASM/ARRM Prototyping Demonstration Framework

The scenario assumes that several operators working in the same area or sharing part of it are using different RATs with different cell size, sectorization, and geographical distribution; hence, they operate in different coverage areas. ARRM strategies can be checked coherently, including the management of sufficient number of users with different mobility patterns to produce relevant loads in the access network. It is also important to consider the availability of different services with different QoS requirements for the existing users. Figure 9.31 shows a view of the assumed wireless heterogeneous scenario. The heterogeneous technologies are currently deployed systems, such as GPRS and UMTS.

A concrete network deployment maybe assumed in this case, in which a reference user can be selected and real-IP based applications can be run for this user. This

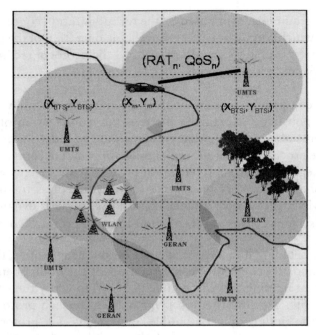

Figure 9.31 A wireless heterogeneous scenario.

will help to show the influence of ASM, DNPM, and JRRM algorithms on the QoS experienced by the reference user using a defined service and the comparison between different ARRM strategies can be performed.

The ASM/ARRM prototyping demonstration framework developed in E2R II, as shown in Figure 9.32, provides the capabilities to demonstrate some of the developed concepts and algorithms related to advanced spectrum management (ASM), dynamic network planning management (DNPM) and advanced radio resource management (ARRM) when considering multiple operators, multiple RATs, multiple cells, multiple users, and multiple services. The framework considers real users

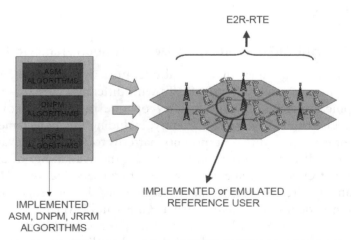

Figure 9.32 Proposed prototyping framework for ASM/DNPM/JRRM implementation in E2R II.

that generate a high load into the network, which would benefit the ASM/DNPM/ARRM strategies. The prototyping framework requires the development of real systems by using the appropriate technology to deal with real implementation aspects.

Another interesting issue is to investigate the interactions of the JRRM elements and the behavior of the radio environment with the real and reconfigurable equipment and the network mechanism to manage the reconfiguration.

In the above scenario, a reference user is selected for whom real IP-based applications are run, and the influence of ASM, DNPM, and ARRM algorithms on the user's QoS can be observed. The emulation model needs to be sufficiently realistic and should be capable of running in real time. Hence, the impact on real applications and services along with the algorithms' implementation can be validated, and the proof of concept needed to market real products can be provided. The emulation process should reproduce the system by reducing its complexity without losing its flexibility. Figure 9.33 shows the standalone demonstration framework. The functionalities required include the capacity to define and manage a specific scenario that may include:

- Multioperator, multiuser, multicell, and multiRAT situations;
- An interface from E2R-RTE emulator to ASM/DNPM/ARRM entities to provide them with appropriate measurements to control the network and also to

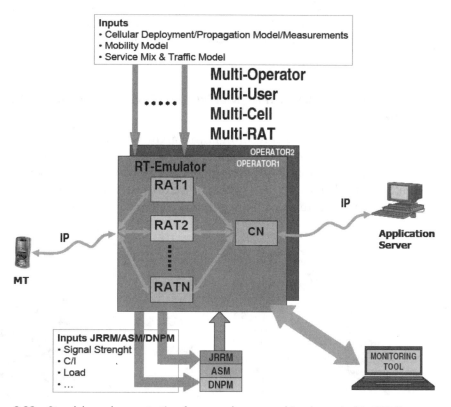

Figure 9.33 Standalone demonstration framework proposed implemented in E2R II.

understand the different ASM/DNPM/ARRM decisions and use them to modify the emulator operation accordingly;

• To monitor and configure the relevant parameters to evaluate the performance of the defined ASM/DNPM/ARRM algorithms under realistic conditions.

The considered entities inside the scenario are as follows:

• *RAT*: Every instance of a given access technology has its own set of local parameters.
• *Mobile terminal*: The terminals report the measurement depending on their current location and network configuration.

The scenario may be modified depending on the base stations belonging to a given RAT instance and the frequencies assigned to every base station within every RAT instance. The changes could be applied either all at once or step by step. More details about the core API functions and their uses are available in E2R II Deliverable 5.1 [1].

Figure 9.34 shows a coordinated demonstration, as the emulator needs to provide the suitable parameters for the adequate control of the real infrastructure used in E2R II.

Two main demonstration situations can be described, namely, intra–MS-BS RRM and inter–MS-BS RRM, as shown in Figure 9.35(a) and (b), respectively. The former addresses the situation where the ASM/JRRM/DNPM effects inside a BS cell and the latter consider a situation where the mobile under test is moving through different cells.

9.3.4 Real-Time Platform for Mobility and QoS and Reconfiguration Management

One of the main goals of a reconfigurable architecture is to allow multiRATs terminals to communicate efficiently (which means to be able to communicate at any time with the best available RAT, with the best available QoS and being able to move seamlessly in a heterogeneous system) with the rest of the system. The following are some important features of the realized reconfigurable architecture:

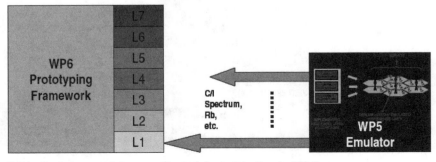

Figure 9.34 Basic approach for coordinated demonstration in E2R II.

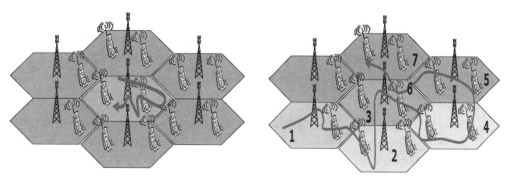

Figure 9.35 (a) Intra–MS-BS scenario and (b) Inter–MS-BS scenario.

- All-IP network infrastructure;
- Heterogeneous network with several RAT technologies;
- Joint radio resource management;
- QoS and mobility management.

The architecture proposed in E2R II is shown in Figure 9.36. The following entities are required to realize the architecture:

- Proxy server to manage the domain;
- One or more UMTS radio gateway (RG): this provides the UMTS connectivity;

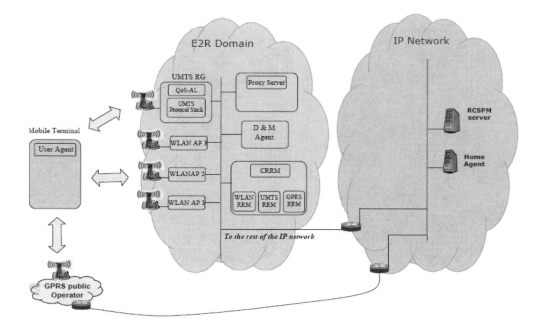

Figure 9.36 Overview of architecture proposed by E2R II.

- One or more WLAN access points;
- Duplication and merging agent for support of mobility management;
- A CRRM or JRRM component to manage the resources of several RATs in the same domain;
- An MT.

The connection to GPRS services could be made by using either a commercial public network or an experimental GPRS infrastructure. COTS components were used for the WLAN part of the prototype. The UMTS-TDD hardware used comprises a BTS with range of 500 m and power of 34 dBm as well as several mobile terminals, all of which are PC-based. The connection of the UMTS RG in a domain managed by a proxy server is shown in Figure 9.37. This architecture allows for experimentation with futuristic "all-IP" network infrastructures and has reduced the number of bearer services when compares to 3GPP protocol.

The overall protocol stack is shown in Figure 9.38. The RRC provides the interfaces and signaling procedures between the low-layer access protocols and the networking protocols and entities. Specifically it provides the following:

- Connection establishment/release;
- Radio bearer establishment/release;
- Broadcast and paging mechanisms;
- Measurement reporting.

The RG uses IP-based QoS information to route packets on the appropriate radio bearers. The UMTS drivers comprise a user plane and the control plane. The user plane acts as classifier for the radio bearer. The control plane is involved in

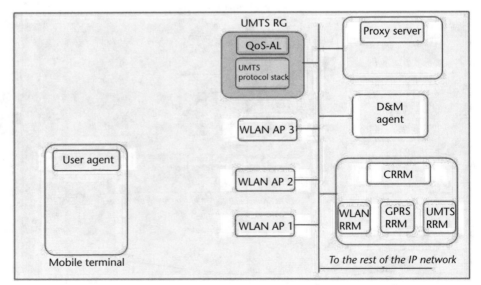

Figure 9.37 Interconnection of UMTS RG in a domain managed by a proxy server in which the RG is directly connected to the proxy server, a D&M agent, common RRM, and to the ALAN access points.

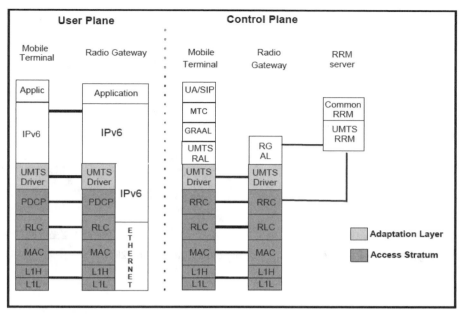

Figure 9.38 Overview of protocol stack of the UMTS part as proposed by E2R II.

establishment/release of procedure of mobile station or radio bearer, mapping QoS attributes between UMTS and differentiated services and broadcasting the neighbor discovery messages, cells list, cells quality list, and finally, the computation of the IPv6 link-local address.

9.3.5 Radio Resource Management (RRM)

The basic measurement of the performance of the radio access technologies is required as an input to the algorithms used for reconfiguration. Commercial off-the-shelf technologies, such as WiFi, HSDPA and WiMAX, have been used in the prototype setup [1].

The implementation of RRM, shown in Figure 9.39, is done in a centralized way across a group of RGs or access points that allow for joint management of radio resources in different cells and potentially across different access technologies. Some functions performed include:

- Automatic access stratum configuration based QoS-based service requests;
- Interference mitigation;
- Low-level mobility management;
- Joint resource allocation across several access technologies;
- Measurement collection.

The interaction of the RRM-related entities with the rest of the system is shown in Figure 9.40. The proxy server routes the signalization traffic. The user agent sends the communication requests to the CRRM, which chooses a RAT that can fulfill the minimum requirements for this specific service.

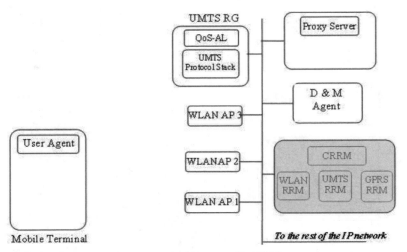

Figure 9.39   Radio resource management implemented in E2R II.

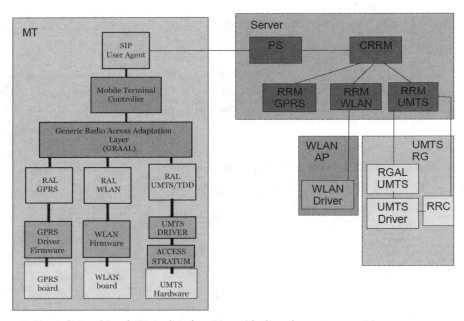

Figure 9.40   Relationship of RRM-related entities with the other system entities.

The RATs should be reconfigurable to guarantee a certain QoS as a function of traffic fluctuations, radio channel disturbances, and limitations imposed by the support network. Hence, its physical layer should adapt access techniques, channel coding, modulations, and so forth; additionally, the MAC layer should provide a scheduling mechanism in line with the QoS requirements, traffic volume, and physical layer capabilities, in terms of available bandwidth and signal quality. Hence, control layers like CRRM are developed to handle flexible spectrum allocation and smartly and jointly coordinate the pool of radio resources to ensure the continuity of service with the best possible QoS, according to the available support network and constraints introduced by the radio channel. The joint validation work in E2R II can be summarized as follows:

- Simulation of the deployment of a heterogeneous network;
- Calculation of available throughput at each position in the region of interest for each RAT;
- Creation of the best network map possible for switching zones for the joint RRM;
- Deployment of the network and validation in real time.

The approach in E2R II was developed using UMTS and WLAN. The UMTS-RRM covers the functionalities for handling the UMTS air interface resources of a radio access network. This enables supplying optimum coverage, offering maximum planned capacity, guaranteeing the required QoS, and ensuring efficient use of physical and transport resources. It involves power control, handover control, congestion control, and resource management. Since many users are operating simultaneously, power control (PC) will allow the adjustment of the transmit powers in uplink and downlink to the minimum level required to ensure the demanded QoS. Handover control ensures that a connected user is handed from one cell to another as he or she moves through the coverage area. Admission control allows the users to reconfigure a radio access bearer in such a way that it does not overload the system and only gains admission if the resources required are available. Load control (LC) prevents the system from going into an overload situation. Packet data scheduling handles non-real-time traffic, allocates optimum bit rates, and schedules transmission of the packet while maintaining the required QoS in terms of throughput and delays. The resource manager controls the physical and logical radio resources under one RNC; it also coordinates the use of available hardware resources and manages the code tree. PC, LC, and RM are available in the base station, while the user equipment has only the PC functionality included.

The strategies used in the UMTS-RRM are discussed very briefly here. UMTS TDD systems provide global mobility and a wide range of application, with different QoSs for multimedia communications that can be classified as real time (RT—guaranteed bit rate and hard requirements on delays) and non-real time (NRT—nonguaranteed bit rates and soft requirements on delays). Most of the works propose call admission control strategies to allocate resources only for real-time applications, but suitable schemes for mixed application were proposed in E2R II. The novelty lies in the joint minimum power channel allocation strategy, which determines precisely the judicious place to allocate resources for RT and NRT application among slots. In support of this mechanisms for adaptive scheduling, fairness rules were applied. Four algorithms were implemented and some interesting results were obtained. These are shown in Figure 9.41 and in Figure 9.42.

From Figure 9.41, it is seen that the Max in slot algorithm introduces a deep difference between the first slots and the end of the frame, whereas the Min in slot 1 and 2 is much more balanced. On the other hand, the power-based algorithm is the most effective in terms of power balance. Figure 9.42 shows the CDF of the powers of various algorithms, and it is seen that power distribution is not centered on the

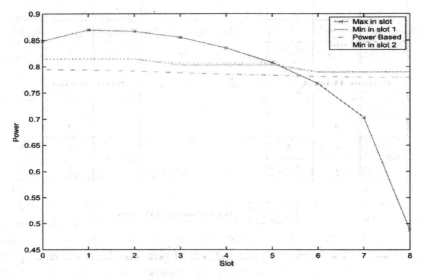

Figure 9.41　Average power per slot and high load.

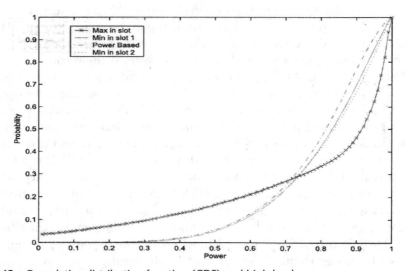

Figure 9.42　Cumulative distribution function (CDF) and high load.

middle of the frame. With the exception of the Max in slot algorithm, the other algorithms have a better balance and the probability that they will have unused slots is very low. Further technical information and implementation details are available in E2R II Deliverable 5.1 [1].

The WLAN-RRM is done mainly by using an estimation of the available throughput for each AP in the network. More information about interfaces, technical details about the implementation, and the integration of the WLAN-RRM into an access point is available in E2R II Deliverable 5.1 [1].

9.4   Adaptive Applications

The successive RATs on which the reconfigurable terminal is connected can have very diverse capabilities. The services provided should be able to cope with variable QoS conditions. This capability could be demonstrated using a multiparty video conference call that goes on seamlessly during reconfiguration operations, regardless of the reconfiguration effects in terms of bit-rate capability.

One of the goals of E2R II was to build a prototype to validate the effectiveness and benefits of end-to-end reconfigurability. The demonstration architecture used for this purpose is shown in Figure 9.43. Another important feature is the definition and software implementation of specific APIs, so that the RPE can be remotely accessed by the functional components developed by individual partners. The API has several functions defined to obtain the parameters needed from the client functionalities or to set the new reconfigurations in the RPE platform. A high-level description of all the interactions involved in demonstrating the DNMP component is available in E2R II Deliverable 5.1 [1].

Portions of the analog transceiver functionality are slowly being moved to the digital domain to fully exploit the advantages of using deep submicron CMOS for RF-FE [1]. Digital correction functionality was used to cope with the impairments of the RF-FE. One of the factors driving the implementation of digital functions locally in the RF-FE is the ever-increasing demand for functionality at, simultaneously, lower power consumption and higher integration level.

For demonstrating the provision of optimum video quality for a given channel QoS level, the H264/AVC video coder would be a good choice. It has the ability to exploit redundant slices and thus resists bit error and packet losses.

The use of multicast transmission in multimedia applications can optimize resource consumption if either broadcast programs or large multiparty conferences are involved. The choice between unicast and multicast transmissions depends on which one can be used to achieve video scaling. Unicast transmissions use fine-grained video scaling, whereas in multicast transmissions, the scalability consists in splitting the multimedia transmission into several streams; the receivers may join as many multicast streams as they can, depending on their receiving capability.

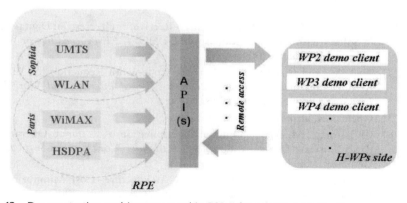

Figure 9.43   Demonstration architecture used in E2R II for adaptive applications.

Simulcast scalability makes nonoptimal use of resources, since a unique video content is sent through several redundant streams, which is not so in multilayer scalability. But multilayer scalability is not efficient when a large difference magnitude exists between minimum and maximum bit-rate capabilities.

More details on the scalability mechanisms are available in E2R II Deliverable 5.1 [1].

9.5 Conclusions

This chapter gathered demonstration scenarios and physical setup for proofs of concept related to reconfigurability. It showed the progressive building of the prototype equipment by first deriving scenarios and an architecture for the prototype and then integrating the various solutions. The described prototype platforms are able to show innovative reconfiguration processes and correctly demonstrate the concepts described in this book.

In conclusion, with the ceaseless advance of wireless communications and the gradual introduction of adaptive networks, more flexible network architecture and advanced management mechanisms can be achieved and programmable network management can be carried out. In the future, network management functions should not only consider the features and capabilities of the actual network segments, but should also include traffic demand, resource, and traffic scalability, as well as enlarged cooperation between coexisting network providers to efficiently allocate the overall available resources. It is anticipated that by using such an approach, system performance can be significantly improved—which in turn will help to reduce the deployment and operational cost of networks.

References

[1] IST Project E2R II, "Requirements, Interfaces and Specification of the Proof of Concept," WP 5 Deliverable 5.1, August 2006, http://e2r2.motlabs.com

[2] Matus, E. et al., "Software Reconfigurable Baseband ASSP for Dual Mode UMTS/WLAN 802.11b Receiver," *IST Mobile and Wireless Communications Summit*, Lyons, France, June 2004.

[3] Hosemann, M. et al., "Implementing a Receiver for Terrestrial Digital Video Broadcasting in Software on an Application-Specific DSP," *IEEE Proceedings in Workshop on Signal Processing Systems*, 2004.

[4] Weiâ, M. and G. Fettweis, "Dynamic Code-width Reduction for VLIW Instruction Set Architectures in Digital Signal Processors," *3rd International Workshop in Signal and Image Processing*, January 1996.

[5] Opencores, http://www.opencores.org

[6] Christrophe, M. et al,. "Optimal Determination of Common Operators for Multi-Standards Software-Defined Radio," *4th Karlsruhe Workshop on Software-Defined Radios*, Germany, March 2006.

[7] Rodriguez, V. et al., "Optimal SDR Architecture: Cost Minimizing Operators under Latency Constraints," *IST Mobile Summit*, Greece, June 2006.

[8] Jean-Philippe, D. et al., "Software Radio and Dynamic Reconfiguration on a DSP/ FPGA Platform," *SDR Forum Technical Conference*, U.S.A., November 2005.

[9]   http://www.w3.org/TR/soap/.

[10]   http://www.w3.org/TR/soap12-part0/.

[11]   Patouni, E. et al., "A Framework for Protocol Reconfiguration," *7th International Conference on Mobile and Wireless Communication Networks*, Morocco, September 2005.

[12]   Gazis, V. et al., "Metadata Design for Introspection-Capable Reconfigurable System," *3rd IFIP-TC6 Networking Conference*, Athens, May 2004.

CHAPTER 10

Concluding Remarks

This book summarizes the innovative research results of a number of IST projects related to reconfigurability funded under the EU FP6 program in the period 2004–2008.

Reconfigurability is a key enabler for efficient interworking of multiple distinct standards, technologies, and equipment for a seamless user experience. It is also a vast area that requires the integration of emerging technologies, such as cognitive and opportunistic radio; cooperative communications; flexible radio and spectrum resource management; multimode terminals; context and service provisioning; and discovery. The largest contributor within the FP6 to the general concept of reconfigurability was the project E2R. Therefore, the majority of the presented concepts and results are based on the work performed within this project.

Enablers for radio resource management (RRM) and dynamic spectrum access (DSA) procedures across heterogeneous technologies were analyzed, with a focus on specific technologies and mechanisms such as spectrum sensing and intra- and interdomain information provision mechanisms, including service and context provisioning.

In particular, this book defines the state of the art and its advancements in the coexistence of heterogeneous standards, including the possible solutions of SDR and CR as background information and the software download–related aspects as an important technique supporting heterogeneous network convergence. Typical scenarios for service applications and resource utilization in heterogeneous networks were also presented, taking into account the service application and spectrum optimization by cognitive radios in a heterogeneous wireless networks context.

Physical layer technologies, which play an important role for supporting reconfigurability, include OFDM, MC-CDMA, MIMO, smart antennas, multiple access space time coding, and so forth.

Reconfigurability is an asset for future mobile devices and associated network elements, enabling dynamic adaptation of utilized radio interfaces dependent on the propagation conditions while also facilitating the optimization of spectrum usage and many other benefits. Cognitive radio promises many more advantages through reconfigurability, which can further be assisted by artificial intelligence and awareness mechanisms achieved by dynamic reconfiguration of software-based components. A system with autonomic capabilities will be capable of making decisions without human intervention, thus reducing the complexity involved. Autonomic computing has revolutionized the IT research community through autonomic capabilities that aim to keep up with the ever-increasing complexity of the systems. Self-configuration, self-management, and autonomic decision making are the main steps to achieving autonomic communications. Further, various profiles, ontology, and context models have been successfully developed and implemented to assist device management and operation support systems and lead to a unified scenario on autonomic communications systems for a seamless experience.

Future work involves a more advanced definition of the reconfigurable network architecture as well as methodology on how to refine control and management aspects while considering the functional and performance requirements.

System capabilities must be successfully incorporated into reconfigurable terminals and network elements to alleviate the need for strict intransigent policies; this incorporation can boost the performance stability and scalability of the next-generation networks dependent on locally or temporally varying requirements and observations. The next-generation mobile network architectures are expected to be integrated with emerging wireless local, metropolitan, and regional access networks. The advances in core networks are paving the way for full exploitation of the IP transport and mobility mechanisms, on top of heterogeneous wireless access networks. These system capabilities also serve as a basis for the definition of the functional architecture. The various system capabilities that have been addressed in this book are policy management; cognitive service provision and discovery; emergency services; context interpretation; self-configuring protocols; mass upgrade of mobile terminals; handover; formation of network compartments and base station reconfiguration; traffic load prediction and balancing; network resource management; and RAT discovery and selection. Architectural capabilities that offer seamless communications despite the resources and connectivity fluctuations should be further investigated, along with protocol extensions and developments to govern and perform the reconfiguration process, which requires advanced device management solutions.

In the basics of the reconfigurable architectures are various physical layer reconfigurable elements, both analog and digital. An overview of the partitioning and the architectural concept of the physical layer, which is applicable to the mobile terminal as well as the base station, has been provided in this book. In the high-level view, the physical layer can be partitioned into main modules, such as configuration control, operational software environment, and hardware resources. A connection element of the configurable executable elements to the configuration control module and an appropriate control of the processing data flow has been proposed. The configuration and management plane connect to the physical control interfaces of the configurable hardware resources and can be tied together by heterogeneous communication elements or possibly by a network-on-chip connection plane, according to the required processing functionality and data flow. The radio frequency front ends of the reconfigurable equipment will most likely include multiple transmit and receive paths, which can be operated simultaneously to accommodate the corresponding user applications, such as voice or data transmissions and location finding or tracking. The configuration possibilities in the analog front end can be minimized to reduce the silicon area and, thus, the cost, whereas the digital front end provides additional flexibility to adapt the filtering and sampling functions to the desired standard and correct analog imperfections simultaneously. A concrete receiver architecture based on these principles has been developed and, since the channelization is done in the digital domain and a resampling capability is foreseen, any standard with modulation bandwidth within a given range can be processed; and a suitable data rate has been realized at the interface. The architecture supports the implementation of multiple standards and allows flexible spectrum management to be realized; however, the current RF solutions do not support this sort of flexibility and need to be developed further. The digital baseband architecture highlights the mapping of the reconfigurable processing elements to functional modules by means of appropriate communication elements and reconfiguration strategies based on different handling of the mapping procedure; this has been investigated corresponding to various levels of reconfiguration complexity.

Communication elements play an important role in the flexible hardware architecture, and in the provided overview of the technology and architectures, a special emphasis was given to network-on-chip (NOC) technology. The logical partitioning between the RF front end and the baseband processing part, as well as the interface, has been examined, and technical solutions have been presented.

Areas related to the functional architecture of the reconfigurable equipment consist of the reconfiguration management, reconfiguration control, and reconfigurable elements. Implementation issues have been dealt with extensively within the FP6 IST research. The prototype and the framework developed within the project E2R and their implementation have been described. The implementation requires both a network perspective and an equipment perspective. The network perspective comprises an adaptive security framework, policy-supported device management, secure OTA download, and network support for negotiation functionality, whereas the equipment perspective comprises context-aware security mechanisms, policy-based device management, reconfiguration negotiation and selection, reconfigurable terminal QoS architecture and reconfiguration management function for multistandard base stations. The section on reconfiguration control presents the CCM simulations and verifications, the configuration control for multistandard base stations, a functional description language interpreter, and spatial scheduling. One of the key challenges for reconfiguration control is to propose a unified, platform-independent model vision clarifying respective assumptions. The section on reconfiguration elements covers the CEM-HAL implementation, the CEM implementation of SAMIRA DSP, the adaptive execution environment, and software architecture for embedded real-time processors. Important precisions have been raised related to this. More research for an efficient proof-of-concept demonstration that supports the coherent architecture is required. An exhaustive platform-independent model that depicts all the important functional requirements ruling the interactions between the various modules, including the interfaces, operations, arguments, and virtual attributes of the main modules of the architecture, needs to be defined.

Issues concerning spectrum management and radio resource allocation are very important for the spectrum-sensing technology. The candidate sensing parameters can be classified into direct and indirect parameters. The direct parameters, such as spectrum cyclic density, interference temperature, interference power vector, and terminal location, are those obtained directly from the OR sensors; the indirect parameters, such as bit error rate, packet loss rate, access delay, end-to-end delay, and throughput are computed based on the direct parameters. Various algorithms have been presented with the simulation results. To summarize few key results, a blind test has been proposed that significantly reduces the complexity and processing time and allows the opportunistic or cognitive terminal to judge the vacancy of a given spectral band without prior knowledge of the potential primary signal. It was found that detection probability is strongly dependent on the size of the spectral band under test. A detection algorithm for spread signal was proposed; it highlighted that only some of the cycle frequencies should be taken into account for building a cost function, based on the cycle-correlation coefficients. A UMTS-specific cyclostationary feature detector was elaborated, but it was observed that when realistic impairments are considered, the performance of the detector degrades moderately, which points out the importance of the hardware quality and stability. With respect to multiple-antenna processing techniques, it was shown that there is more than an order of magnitude improvement in detection performance with the use of maximum-ratio antenna processing and selection processing. It was also shown that different wavelets should be used for edge detection in different signals. A significant challenge is posed when cooperative extensions to local sensing are considered.

Cooperative sensing is inherently static in the sense that the same state of the radio environment will lead to identical decisions for local OR users. Collaborative sensing, however, can be considered as dynamic, since the decision of the local OR users depends on the behaviors and decisions of OR users in their vicinity. The goals and strategies of local OR users adapt to the changing radio environment and also to the specific requirements of local applications or human users. This topic is open to further investigation. The collaborative sensing scheme reduces the effect of shadowing in the decision-making process, and it can also be used to overcome the hidden node problem. Collaborative sensing enables sensitivity requirements to be reduced at individual nodes, which leads to a decrease in signal processing complexity and required observation time. The opportunistic radio algorithms can help reduce the intersystem interference in the unlicensed frequency bands. Widely available ISM bands are also potential candidates for OR applications, and there should be more research on this subject. The economic profitability achieved by introducing OR has been highlighted in terms of upcoming technologies to the competitive market using OR and economical benefits from the secondary spectrum trading concept. With respect to cellular communication scenario, the operators could benefit from OR techniques due to the potential for exploitation of their spectrum or extra revenues generated from spectrum trading. OR techniques enable secondary spectrum trading that inherently offers the opportunity for spectrum owners to realize extra incomes in situations when their own spectrum is highly underutilized; also, temporary rights maybe sold to operators who wish to use parts of their spectrum.

Security is required to prevent the abuse of flexibility provided by reconfigurability. Fault management refers to the use of preventive means complemented by reactive means as monitoring and fallback. Solutions for the relevant security issues relate to reconfiguration software download, reconfiguration processes, and compliance of radio emission. This book also explains the authorization for reconfiguration software for configuration validation and for decentralized control on the reconfiguration. Other issues, such as the secure local execution environment and authentication and trust framework, are also important. The existing security technology, implementing security services such as encryption, authentication, and nonrepudiation, is available, but further secure security mechanisms have to be developed. Some of the interesting and challenging areas under investigation and requiring further work are reconfiguration security framework, reconfiguration software authorization framework, flexible validation concept, and distributed reconfiguration control with transient configuration profiles.

Finally, prototyping provides additional requirements to a reconfigurable platform. The interfaces and specifications for the relevant prototyping activities are important, as they serve as a means for communicating with the outside world and define a responsible partner on each side of the interface. The proposed prototype platform described in this book consists of both hardware and software components. A realistic reconfigurable mobile terminal has been created, and other hardware aspects related to multistandard base station design and COTS components have been addressed. The software components consist of the reconfiguration, control, and serving provisioning system, and include aspects of profile management, service discovery, and so forth. These components are complemented by aspects related to physical layer emulation and radio resource management. Overall mobility support is also an important aspect and must be provided.

In summary, future research should focus on integrating cognitive wireless systems in the Beyond 3G (B3G) world and on evolving current heterogeneous wireless system infrastructures into an integrated, scalable, and efficiently managed B3G cognitive system framework. Work on this concept has already begun within the European Union–funded Information and Communication Technologies FP7 research and development program. Several major challenges to be overcome in this respect include identifying the roles of traditional and new stakeholders as well as the respective business models in the framework of reconfigurable and cognitive heterogeneous wireless technologies. This will require an enlargement of the portfolio of reference use-cases, which should refer to a multioperator scenario.

In the context of a heavily heterogeneous, cognitive wireless systems landscape, the convergence of independently managed radio access systems towards an efficient and integrated framework is a key challenge and requires as a first step the identification and definition of an enabler to help the terminal discover the spectrum allocation and develop cognition enabler schemes for reconfigurable/cognitive systems focusing on spectrum-sensing and information provision mechanisms.

As a long-term approach, effort must be put into standardization toward the convergence of future systems.

About the Authors

Ramjee Prasad was born in Babhnaur (Gaya), India, on July 1, 1946. He is now a Dutch citizen. He received his B.Sc. in engineering from the Bihar Institute of Technology, Sindri, India, in 1968, and his M.Sc. and Ph.D. from Birla Institute of Technology (BIT), Ranchi, India, in 1970 and 1979.

Dr. Prasad has a long path of achievement and rich experience in the academic, managerial, research, and business spheres of the mobile communication areas.

He joined BIT as a senior research fellow in 1970 and became an associate professor in 1980. While with BIT, he supervised a number of research projects in the areas of microwave and plasma engineering. From 1983 to 1988, he was with the University of Dar es Salaam (UDSM), Tanzania, where he became a professor of telecommunications in the Department of Electrical Engineering in 1986. At UDSM, he was responsible for the collaborative project Satellite Communications for Rural Zones with Eindhoven University of Technology, the Netherlands. From February 1988 through May 1999, he was with the Telecommunications and Traffic Control Systems Group at Delft University of Technology (DUT), where he was actively involved in the area of wireless personal and multimedia communications (WPMC). He was the founding head and program director of the Center for Wireless and Personal Communications (CWPC) of International Research Center for Telecommunications—Transmission and Radar (IRCTR).

Since June 1999, Dr. Prasad has held the chair of Wireless Information and Multimedia Communications at Aalborg University, Denmark (AAU). He was also the codirector of AAU's Center for Person Kommunikation until January 2004, when he became the founding director of the Center for TeleInfrastruktur (CTIF), established as a large multiarea research center on the premises of Aalborg University.

Dr. Prasad is a worldwide established scientist, which is evident from his many international academic, industrial, and governmental awards and distinctions, his over 25 published books, his numerous journal and conference publications, a sizeable amount of graduated Ph.D. students and even larger amount of graduated M.Sc. students. Under his initiative, international M.Sc. programs were started with the Birla Institute of Technology in India, the Insititute of Technology Bandung in Indonesia. Recently, cooperation was established with the Athens Information Technology (AIT) in Greece.

Under Dr. Prasad's successful leadership and extraordinary vision, CTIF currently has more than 150 scientists from different parts of the world and three CTIF branches in other countries: CTIF-Italy (inaugurated in 2006 in Rome), CTIF-India (inaugurated on December 7, 2007 in Kolkata), and CTIF-Japan (inaugurated on October 3, 2008).

Dr. Prasad was a business delegate in the Official Business Delegation led by Her Majesty The Queen of Denmark Margarethe II to South Korea in October

2007. He is a Fellow of the IEE, a Fellow of IETE, a senior member of the IEEE, and a member of NERG. He was the recipient of the Telenor Nordic Research Award (2005), the Samsung Electronics Advisor Award (2005), the Yearly Aalborg-European Achievements Award (2004), and the IEEE Communication Society Award for Achievements in the area of Personal, Wireless, and Mobile Systems and Networks (2003). Ramjee Prasad is a member of the steering, advisory, and program committees of many IEEE international conferences.

Dr. Prasad is the founding chairman of the European Centre of Excellence in Telecommunications, known as HERMES, and now is an honorary chair of HERMES. HERMES currently has ten member organizations from Europe. He is the founding cochair of The International Symposium on Wireless Personal Multimedia Communications (WPMC), which has taken place annually since 1999.

Dr. Prasad has been strongly involved in European research programs. He was involved in the FP4-ACTS project FRAMES (Future Radio Wideband Multiple Access Systems), which set up the UMTS standard, as a DUT project leader. He was a project coordinator of EU projects during FP5 (CELLO, PRODEMIS), and FP6 (MAGNET and MAGNET Beyond), and is currently involved in FP7.

He was the project leader for several international industrially funded projects with NOKIA, SAMSUNG, Ericsson Telebit, and SIEMENS, to name a few. Dr. Prasad is a technical advisor to many industrial international companies.

Dr. Prasad is the founder of the IEEE Symposium on Communications and Vehicular Technoliógy (SCVT) in Benelux. He was the chairman of SCVT in 1993.

He is the founding editor-in-chief of the *Springer International Journal on Wireless Personal Communications*. He is a member of the editorial board of other international journals and is the series editor of the Artech House Universal Personal Communications Series.

Albena Mihovska was born in Sofia, Bulgaria. She completed her B.Sc. in engineering at the Technical University of Sofia, Bulgaria, in 1990, followed by her M.Sc. in engineering at the Technical University of Delft, the Netherlands (1999). Since 1999, Ms. Mihovska has been with Aalborg University, Denmark, where she is currently an associate professor at the Center for TeleInfrastruktur (CTIF).

During her years of employment at Aalborg University, Ms. Mihovska has gained extensive experience in the administrative and technical management of EU-funded research projects. Further, she gained experience initializing industrial research cooperation as well as research cooperation funded by the EU.

She joined Aalborg University as a research engineer in July 1999 and was appointed to the European Union–funded technical management team within the FP4 ACTS project ASAP until its successful completion in 2001.

From September 2001 until April 2005, she was the project coordinator of the European Union—funded FP5 IST project PRODEMIS as a special support action instrument until its successful completion. The project was a main supporting project of the EU IST projects within the mobile and satellite area. The outcome of the project was published as two books by Artech House in 2005, as well as in a number of technical research publications in peer-reviewed journals and conferences, an e-conference on mobile communications, a joint workshop, and a technology roadmap for the future development of mobile communications.

From January 2004 until December 2005, Ms. Mihovska was the research coordinator of the research team within the EU-funded IST FP6 project WINNER, which continued from January 2006 to December 2007 as WINNER II. The main objective of the project was the design of a new air interface that could be a competitive candidate for next-generation systems, in the scope of standardization activities within the IMT-Advanced ITU group. The project was a part of the WWI initiative and, as such, had close and required cross-issue collaboration with the rest of the WWI projects. Ms. Mihovska was part of the research teams working toward the identification of the system requirements and the design of interworking mechanisms between the newly designed system and other systems. Within the project, she proposed a concept for cooperation between different systems based on an autonomous decision framework. Based on this research idea, the theoretical approach was put forward as a development activity in the second stage of the project and was successfully demonstrated at a number of international events, including the Wireless Radio Communication (WRC) '07 Conference held in Geneva from October through November 2007. The experimental setup is now being considered for use in other projects, such as the CELTIC project WINNER+ and the FP7 project FUTON, working toward an architecture design for converging heterogeneous systems and service provisioning in which AAU is a consortium member. Ms. Mihovska was part of the research group within the project consortium who developed the final system concept requirements for the air interface.

From September 2006 to March 2008, Ms. Mihovska was the deputy technical manager of MAGNET Beyond. Therein, she contributed to the overall technical work progress and to the finalization of the MAGNET Beyond Platform system requirements. Further, she was involved in AAU-related research activities in the area of security for personal networks (PNs).

Since April 2008, Ms. Mihovska has been involved in research activities within the CELTIC project WINNER+, working toward advanced radio system technologies for IMT-Advanced systems. She is conducting research activities within the area of advanced radio resource management, cross-layer optimization, and spectrum aggregation.

The work proposed in pursuit of her Ph.D. degree from Aalborg University is a novel concept for interworking between radio resource management entities in the context of next-generation mobile communication systems. It is based on research activities commenced prior to and continued within the frames of the WINNER project. The concepts proposed within her Ph.D. thesis have been successfully implemented in the overall WINNER concept and have resulted in a number of peer-reviewed journal and conference publications, including the demonstration activities mentioned above. Further, she has a number of project-related publications, presentations, and various international and EU events.

Ms. Mihovska is a reviewer for *IEEE Communication Letters* and *The Springer Journal of Telecommunication Systems*. She has been part of the organizing and TPC committees of a number of international conferences, such as WPMC 2002, WCNC 2007, VTC 2008 Spring, the IST Mobile Summits 2002-2007, ATSMA-NAEC 2009, IEEE Mobile WiMax 2009 Symposium, and several workshops.

Index

The Artech House Universal Personal Communications Series

Ramjee Prasad, Series Editor

Towards the Wireless Information Society: Heterogeneous Networks,
  Ramjee Prasad, editor

Towards the Wireless Information Society: Systems, Services, and Applications,
  Ramjee Prasad, editor

Universal Wireless Personal Communications, Ramjee Prasad

WCDMA: Towards IP Mobility and Mobile Internet, Tero Ojanperä and Ramjee Prasad,
  editors

Wideband CDMA for Third Generation Mobile Communications,
  Tero Ojanperä and Ramjee Prasad, editors

Wireless Communications Security, Hideki Imai, Mohammad Ghulam Rahman and
  Kazukuni Kobara

Wireless IP and Building the Mobile Internet, Sudhir Dixit and Ramjee Prasad, editors

WLAN Systems and Wireless IP for Next Generation Communications, Neeli Prasad and
  Anand Prasad, editors

WLANs and WPANs towards 4G Wireless, Ramjee Prasad and Luis Muñoz

For further information on these and other Artech House titles, including previously
considered out-of-print books now available through our In-Print-Forever® (IPF®)
program, contact:

Artech House	Artech House
685 Canton Street	46 Gillingham Street
Norwood, MA 02062	London SW1V 1AH UK
Phone: 781-769-9750	Phone: +44 (0)20 7596-8750
Fax: 781-769-6334	Fax: +44 (0)20 7630-0166
e-mail: artech@artechhouse.com	e-mail: artech-uk@artechhouse.com

Find us on the World Wide Web at: www.artechhouse.com